D1257923

Ökonometrie und Unternehmensforschung

Econometrics and Operations Research

VII

VNR

Boolean Methods in Operations Research

and Related Areas

Peter L. Hammer (Ivănescu)
Sergiu Rudeanu

With a preface by
Professor Richard Bellman

With 25 Figures

Springer-Verlag New York Inc. 1968

Professor Dr. Peter L. Hammer (Ivănescu)
TECHNION-Israel Institute of Technology
Department of Industrial and Management Engineering
Haifa

Professor Dr. Sergiu Rudeanu
Academy of S. R. Roumania
Institute of Mathematics
Bucharest

Professor Dr. Richard Bellman
University of Southern California,
Los Angeles (USA)

Library of Congress Catalog Card Number 67-21932

Titel-Nr. 6482

To our Professor

GRIGORE C. MOISIL

with respect and affection

Preface

In classical analysis, there is a vast difference between the class of problems that may be handled by means of the methods of calculus and the class of problems requiring combinatorial techniques. With the advent of the digital computer, the distinction begins to blur, and with the increasing emphasis on problems involving optimization over structures, the distinction vanishes.

What is necessary for the analytic and computational treatment of significant questions arising in modern control theory, mathematical economics, scheduling theory, operations research, bioengineering, and so forth is a new and more flexible mathematical theory which subsumes both the classical continuous and discrete algorithms. The work by Hammer (Ivănescu) and Rudeanu on Boolean methods represents an important step in this direction, and it is thus a great pleasure to welcome it into print. It will certainly stimulate a great deal of additional research in both theory and application.

Richard Bellman
University of Southern California

Foreword

It was in the late forties that the theory of Boolean algebras was given the first practical application to the study of switching circuits. This approach to a problem of techniques became possible due to the fact that, as switching circuits are generally made up of two-state devices, they may be studied by means of a mathematical apparatus using bivalent variables. This apparatus was found to be the two-element Boolean algebra.

It is a natural idea to use bivalent variables whenever we are faced with problems involving situations with only two possible outcomes. Such problems of "binary decisions" are frequently found in operations research, theory of graphs, combinatorial mathematics, etc. The importance and width of this class of problems was first pointed out by G. B. DANTZIG in 1957.

Meanwhile many papers have been published on these topics and the research is in full progress. Some of the works in the field apply *Boolean techniques*, that is techniques actually using properties of Boolean algebra, while other research is mainly combinatorial.

The present book deals almost exclusively* with applications of Boolean techniques (in the above sense) to a field which, in lack of a better term, was called "operations research and related areas".

The main tools which will be described and employed in the book are: Boolean matrix calculus, Boolean equations and pseudo-Boolean programming.

The core of the book is pseudo-Boolean programming, a method for solving bivalent (0, 1) programs which was developed by I. ROSENBERG and the authors in 1963, using an idea of R. FORTET. The method, which, in fact, is a combination of R. BELLMAN's principle of dynamic programming with Boolean procedures, is presented here in an improved form, due to the present authors, along with various applications of it. Such applications were made, in addition to the authors, by I. ROSENBERG, Y. INAGAKI and K. SUGINO, U. S. R. MURTY, G.-B. IHDE, J. KRÁL et al.; these contributions are based on the first (unimproved) version of pseudo-Boolean programming.

As there exists a wide literature on Boolean methods in switching theory, this topic was not included in our book, except for an application of pseudo-Boolean programming to certain minimization problems of switching theory. We have also omitted non-Boolean approaches

---

* With the exception of §§ 1—4 in Ch. III and of §§ 1—2 in Ch. V.

to problems of mathematical programming with bivalent variables. However, we have included a bibliography of books on switching theory.

Our book is intended for people interested in operations research, on the one hand, and for those interested in applications of Boolean algebras, on the other hand. Therefore the content will be as follows.

In the first chapters of the book we present the necessary mathematical background: Boolean functions and equations, pseudo-Boolean functions (i.e. real-valued functions with bivalent (0, 1) variables), pseudo-Boolean equations and inequalities, pseudo-Boolean programming.

The method of pseudo-Boolean programming enables the solution of linear and nonlinear bivalent programs as well as of various generalizations including integer polynomial programming.

The last chapters are devoted to applications to the theory of graphs, networks, partially ordered sets, sequencing problems, assignment and transportation problems, time-table scheduling, optimal plant location, minimization problems in switching circuits, etc.

The book includes also an Appendix written by I. Rosenberg on a generalization of pseudo-Boolean programming.

Besides the main bibliography, two supplementary bibliographies (A and B) on Boolean equations and switching theory are also included. Since the number of papers on switching algebra is tremendous, the latter bibliography contains only books pertinent to the field.

Any omission in these bibliographies is due to the ignorance of the authors and is not to be considered as the result of a valuation.

We have tried to make this book as self-contained as possible. The reader is supposed to be familiar with the usual symbols of set theory; we mention here that notations like $A \cup B$ or $\bigcup_{i=1}^{n} A_i$ mean set-theoretical joins, while $a \cup b$ or $\bigcup_{i=1}^{n} a_i$ denote Boolean disjunction of elements.

Formulas, theorems, lemmas, etc., are numbered separately in each chapter. References to formulas, theorems, lemmas, etc. in the same chapters indicate only their numbers, while references to other chapters indicate the number of the corresponding chapter (for instance, formula (II.13), Theorem IV.1, etc.).

The authors would much appreciate receiving observations from the readers.

Bucharest/Haifa, August 1967

Peter L. Hammer (Ivănescu) · Sergiu Rudeanu

Acknowledgements

Our interest in the application of Boolean algebra originates through participation in the research seminar led by Professor GRIGORE C. MOISIL, who has steadily stimulated the work on pseudo-Boolean programming.

The publication of this book is due to Professor HANS PAUL KÜNZI, who suggested its writing and included it into the series of monographs on operations research of the Springer-Verlag.

We were much encouraged in our research by the positive appreciations of Professor RICHARD BELLMAN, who had the kindness of prefacing this book.

The ellaboration of pseudo-Boolean programming and much of the research in this field was carried out in collaboration with our friend Doctor IVO ROSENBERG (Polytechnical Institute of Brno, Czechoslovakia), who has also written two sections and the Appendix of the present book.

Computer codes for pseudo-Boolean programming are being ellaborated by R. SOWELL and R. VAUGHAN in the team of Professor ANDREW WHINSTON (University of Virginia, USA), by Doctor A. K. DIXIT (Massachusetts Institute of Technology, USA), by TOMA GASPAR (Polytechnical Institute of Timișoara, Romania), and by PETRE DRĂGAN, CSABA FÁBIÁN and GHEORGHE WEISZ (Computing Center of Hunedoara, Romania).

The criticisms of CSABA FÁBIÁN and GHEORGHE WEISZ helped us in improving our methods.

Dr. MANFRED KOEGST (Jena University, Germany) has kindly read the manuscript and suggested several corrections.

The editorial assistance of Springer-Verlag was most helpful for the publication of this monograph.

The manuscript was typed with much caie and competence by Mrs. ANNIE BĂLESCU.

To all of them we express our deep gratitude.

August 1967.

PETER L. HAMMER (IVĂNESCU) · SERGIU RUDEANU

Table of Contents

Introduction

We are going to present some typical problems leading either to the minimization of certain pseudo-Boolean functions, or to the solution of certain Boolean equations, in order to justify the abstract study of Boolean algebras, Boolean equations, pseudo-Boolean functions, etc. given in the first chapters of the book.

Example 1 (L. R. FORD JR., D. R. FULKERSON [4]). One of the basic problems of the theory of flows in networks is that of determining their minimal cuts. A (directed) network (N, A) consists of a collection $N = \{n_0, n_1, \ldots, n_p, n_{p+1}\}$ of elements (called its nodes) together with a set A of arcs joining certain couples of nodes; the arcs are directed, i.e. each arc has an original node and a terminal node. Two nodes are distinguished: a source n_0 having the property that no arc is coming towards it, and a sink n_{p+1} with the property that no arc is departing from it. To each arc a non-negative integer value is assigned, called its capacity. Assume that the set N is divided into two disjoint subsets N' and N'', so that* $n_0 \in N'$ and $n_{p+1} \in N''$. We say that N' and N'' determine a cut of the network, and we define the capacity of the (N', N'')-cut as being the sum of the capacities of all arcs joining nodes of N' to nodes of N''. A cut with smallest capacity is called a minimal cut.

In order to determine the minimal cuts of a network, let us characterize a cut by means of a vector $(x_0, x_1, \ldots, x_p, x_{p+1})$ of zeros and ones so that

$$x_i = \begin{cases} 1 & \text{if } n_i \text{ belongs to } N' \\ 0 & \text{if } n_i \text{ belongs to } N''; \end{cases}$$

obviously $x_0 = 1$ and $x_{p+1} = 0$. Let us denote the capacity of the arc joining n_i to n_j (if any) by c_{ij}, setting $c_{ij} = 0$ if there is no arc from n_i to n_j. Then the capacity of the cut is simply equal to

$$\sum_{i=0}^{p+1} \sum_{j=0}^{p+1} c_{ij} x_i (1 - x_j), \tag{1}$$

because $x_i(1 - x_j)$ is 1 if and only if $n_i \in N'$ and $n_j \in N''$. Thus the original problem has been reduced to that of minimizing the pseudo-Boolean function (1), where $x_0 = 1$ and $x_{p+1} = 0$.

Example 2 (P. L. HAMMER, S. RUDEANU [1], [2]). In the Hungarian method for solving the transportation problem (J. EGERVÁRY [1]; see also H. W. KUHN [1, 2]), at a certain step the following problem arises. An $m \times n$ matrix and a set of distinguished elements of it (in fact, its zeros) are given. To each row i, and to each column j, a positive integer number a_i and b_j, respectively, is assigned. The problem is to find a system

$$S = \{i_1, \ldots, i_p; j_1, \ldots, j_q\},$$

* $n \in N$ means "the element n belongs to the set N".

of rows and columns, covering the distinghuished elements, and such that the sum

$$f(S) = \sum_{h=1}^{p} a_{i_h} + \sum_{k=1}^{q} b_{j_k}$$

is minimized.

We assign to each row i (respectively, to each column j), a bivalent variable x_i (respectively, y_j), equal to 1 or to 0 according to the presence or absence of row i (respectively, of column j) in the system S to be found. Then S will be a covering if and only if, for each distinguished element (i, j), either row i, or column j, or both, belong to S, i.e. the relation

$$x_i + y_j \geqq 1$$

holds for every distinguished element (i, j): The above relation may also be written in the form

$$x_i \cup y_j = 1, \tag{2}$$

where $\cup$ denotes the Boolean "disjunction", defined by $0 \cup 0 = 0$, $0 \cup 1 = 1 \cup \cup 0 = 1 \cup 1 = 1$.

On the other hand, $f(S)$ becomes

$$f(x_1, \ldots, x_m, y_1, \ldots, y_n) = \sum_{i=1}^{m} a_i x_i + \sum_{j=1}^{n} b_j y_j, \tag{3}$$

so that the original problem has been transformed into the equivalent problem of minimizing the pseudo-Boolean function (3) under the Boolean conditions (2).

Example 3. J. von Neumann and O. Morgenstern (cf. C. Berge [1] Ch. V, § 32), defined the solution of an n-person coalition game with a domination relation $\prec$ between the strategies, as being a subset K of the set $S = \{s_1, \ldots, s_t\}$ of all strategies, with the following properties:

(4) relation $a \prec b$ does not hold for any $a, b \in S$;

(5) for every $a \notin S$, there exists an element $b \in S$ such that $a \prec b$.

In order to determine a subset K with the above two properties (if any), let us introduce a $t \times t$ matrix (a_{ij}) defined by

$$a_{ij} = \begin{cases} 1, & \text{if } s_i \prec s_j, \\ 0, & \text{if } s_i \nprec s_j, \end{cases}$$

and let us assign to the set K to be found, a vector $(x_1, \ldots, x_t)$ of zeros and ones, where

$$x_i = \begin{cases} 1, & \text{if } s_i \in K, \\ 0, & \text{if } s_i \notin K, \end{cases}$$

for $i = 1, \ldots, t$. We see that condition (4) becomes "if $x_i = x_j = 1$, then $a_{ij} = 0$", while condition (5) says that "if $x_i = 0$, then there exists an index j such that $a_{ij} = x_j = 1$". It is proved (see Chap. X, § 1) that the above conditions may be translated into the following system of Boolean equations:

$$a_{ij} x_i x_j = 0 \quad (i, j = 1, 2, \ldots, t),$$

$$x_i \cup \bigcup_{j=1}^{t} a_{ij} x_j = 1 \quad (i = 1, \ldots, t),$$

where $\bigcup$ plays with respect to $\cup$ the same role as $\sum$ with respect to $+$. Thus the original problem has been reduced to that of solving a system of Boolean equations.

Part I

Chapter I

Boolean Algebra

In this chapter the basic definitions and properties of the Boolean algebra, of Boolean functions, of Boolean matrices and of the Boolean rings will be presented, in order to facilitate the understanding of Chapters II—VI, where the basic techniques used in this book will be developed.

§ 1. The Two-Element Boolean Algebra

We have seen in the Introduction (Example 2), how practical reasons lead to the consideration of the Boolean disjunction $\cup$, defined for couples of elements $0, 1$:

$$0 \cup 0 = 0, \quad 0 \cup 1 = 1 \cup 0 = 1 \cup 1 = 1. \tag{1}$$

We have also noticed in the first example of the Introduction that, x being a variable which takes the values 0 and 1, the function $1 - x$ plays a distinguished rôle. Since it frequently appears, it deserves a special notation, $\bar{x} = 1 - x$:

$$\bar{0} = 1, \quad \bar{1} = 0; \tag{2}$$

$\bar{x}$ will be called the "negation" of x.

As shown by the same examples, a third important operation in the two-element set $\{0, 1\}$ is the usual multiplication:

$$0 \cdot 0 = 0 \cdot 1 = 1 \cdot 0 = 0, \quad 0 \cdot 0 = 0. \tag{3}$$

Now, the algebraic structure introduced by these three operations in the two-element set $\{0, 1\}$ is called the "two-element Boolean algebra" B_2, i.e.:

Definition 1. By the *two-element Boolean algebra* we mean the set

$$B_2 = \{0, 1\},$$

consisting of the numbers 0 and 1, together with the operations of *disjunction* ($\cup$), *conjunction* ($\cdot$) (also denoted by juxtaposition), and *negation* ($^-$), defined by formulas (1), (3) and (2), respectively.

Remark 1. As a matter of fact, in the usual definition of the two-element Boolean algebra, 0 and 1 are not required to be numbers, but they are interpreted as abstract elements. However, as remarked by T. Gaspar, we need the above assumption in the study of pseudo-Boolean functions.

It is not difficult to prove that the following properties hold:

$$\left.\begin{array}{ll}(4') & x \cup y = y \cup x, \\ (4'') & x\, y = y\, x,\end{array}\right\} \text{(commutativity)}$$

$$\left.\begin{array}{ll}(5') & (x \cup y) \cup z = x \cup (y \cup z), \\ (5'') & (x\, y)\, z = x\,(y\, z),\end{array}\right\} \text{(associativity)}$$

$$\left.\begin{array}{ll}(6') & x \cup x = x, \\ (6'') & x \cdot x = x,\end{array}\right\} \text{(idempotency)}$$

$$\left.\begin{array}{ll}(7') & x \cup x\, y = x, \\ (7'') & x\,(x \cup y) = x,\end{array}\right\} \text{(absorption)}$$

$$\left.\begin{array}{ll}(8') & x \cup y\, z = (x \cup y)\,(x \cup z), \\ (8'') & x\,(y \cup z) = x\, y \cup x\, z,\end{array}\right\} \text{(distributivity)}$$

$$(9') \quad x \cup 1 = 1,$$

$$(9'') \quad x \cdot 0 = 0,$$

$$(10') \quad x \cup 0 = x,$$

$$(10'') \quad x \cdot 1 = x,$$

$$(11') \quad x \cup y = 0 \quad \text{if and only if} \quad x = y = 0,$$

$$(11'') \quad x\, y = 1 \quad \text{if and only if} \quad x = y = 1,$$

$$(12') \quad x \cup y = 1 \quad \text{if and only if} \quad x = 1 \quad \text{or} \quad y = 1,$$

$$(12'') \quad x\, y = 0 \quad \text{if and only if} \quad x = 0 \quad \text{or} \quad y = 0,$$

$$(13') \quad x \cup \bar{x} = 1,$$

$$(13'') \quad x\, \bar{x} = 0,$$

$$\left.\begin{array}{ll}(14') & \overline{x \cup y} = \bar{x}\, \bar{y}, \\ (14'') & \overline{x\, y} = \bar{x} \cup \bar{y},\end{array}\right\} \text{(the de Morgan laws)}$$

$$(15) \quad \bar{\bar{x}} = x, \quad \text{(the law of double negation)}$$

$$\left.\begin{array}{ll}(16') & x \cup \bar{x}\, y = x \cup y, \\ (16'') & x\,(\bar{x} \cup y) = x\, y,\end{array}\right\} \text{(Boolean absorption)}$$

where x, y, z denote arbitrary elements of B_2.

The above properties are proved by direct verification for all possible values of x, y, z. For instance, relation (7′) is checked as follows: if $x = y = 0$, then $x \cup x\,y = 0 \cup 0.0 = 0 \cup 0 = 0 = x$; if $x = 0$, $y = 1$, then $x \cup x\,y = 0 \cup 0.1 = 0 \cup 0 = 0 = x$; if $x = 1$, $y = 0$, then $1 \cup 1.0 = 1 \cup 0 = 1 = x$; if $x = y = 1$, then $x \cup x\,y = 1 \cup 1.1 = 1 \cup 1 = 1 = x$. The proofs can be facilitated by a reasonable organization of the computations. For instance, if relations (9′) and (10′) have been previously verified, then the proof of (8′) can be carried out as follows. If $x = 0$, then $x \cup y\,z = 0 \cup y\,z = y\,z$ and $(x \cup y)(x \cup z) = (0 \cup y)(0 \cup z) = y\,z$; if $x = 1$, then $x \cup y\,z = 1 \cup y\,z = 1$ and $(x \cup y)(x \cup z) = (1 \cup y)(1 \cup z) = 1.1 = 1$.

We notice that the usual ordering $\leqq$ of the numbers $0, 1$ verifies also the following properties:

(17′) $\quad x \leqq y$ if and only if $x \cup y = y$,

(17″) $\quad x \leqq y$ if and only if $x\,y = x$,

(18′) $\quad x \leqq x \cup y$ and $y \leqq x \cup y$,

(18″) $\quad x\,y \leqq x$ and $x\,y \leqq y$,

(19′) $\quad$ if $x \leqq z$ and $y \leqq z$, then $x \cup y \leqq z$,

(19″) $\quad$ if $t \leqq x$ and $t \leqq y$, then $t \leqq x\,y$,

and, in fact,

(20′) $\quad x \leqq z$ and $y \leqq z$ if and only if $x \cup y \leqq z$,

(20″) $\quad t \leqq x$ and $t \leqq y$ if and only if $t \leqq x\,y$;

further,

(21′) $\quad x \leqq y$ implies $x\,z \leqq y\,z$,

(21″) $\quad x \leqq y$ implies $x \cup z \leqq y \cup z$,

(22′) $\quad x \leqq 1$,

(22″) $\quad 0 \leqq x$,

(23′) $\quad x \leqq y$ if and only if $\bar{x} \cup y = 1$,

(23″) $\quad x \leqq y$ if and only if $x\,\bar{y} = 0$,

(24′) $\quad x = y$ if and only if $x\,\bar{y} \cup \bar{x}\,y = 0$,

(24″) $\quad x = y$ if and only if $(\bar{x} \cup y)(x \cup \bar{y}) = 1$,

for every $x, y, z, t \in B_2$.

Some of the above properties are easily extended to n elements, $x_1, \ldots, x_n$, instead of two; we leave this task to the reader.

It is important to notice the

Principle of duality. *If a property of B_2, expressed in terms of the operations $\cup$, $\cdot$, $^-$, of relations $\leqq$, $\geqq$, and of the constants* 0, 1, *holds then the "dual property", obtained by interchanging $\cup$ with $\cdot$, $\leqq$ with $\geqq$ and* 0 *with* 1, *also holds.*

For instance, most of the properties listed above are grouped into dual couples; in other words, the dual property of (n') is (n''), and conversely.

The principle of duality is one of the central theorems of Boolean algebra. Its proof requires an axiomatic treatment; the reader is referred, for instance, to B. H. ARNOLD [B 1], M. DENIS-PAPIN, R. FAURE, A. KAUFMANN [B 1], R. DUBISCH [B 1], J. E. WHITESITT [B 1], et al.; see also § 4 of this Chapter.

Remark 2. We call the reader's attention to the fact that the *two-element Boolean algebra,* as defined above, is not to be confused with the general notion of *Boolean algebra.* The latter term denotes any set B (finite or not), in which three operations: disjunction ($\cup$), conjunction ($\cdot$) and negation ($^-$) are defined, satisfying the above properties (4)—(11) and (13)—(24), whatever the concrete definition of the three operations may be (see B. H. ARNOLD [B 1], M. CARVALLO [B 1], M. DENIS-PAPIN, R. FAURE, A. KAUFMANN [B 1], R. DUBISCH [B 1], H. G. FLEGG [B 1], D. E. RUTHERFORD [B 1], S. RUDEANU [B 1], J. E. WHITESITT [B 1] and § 4 of this Chapter).

A special feature of B_2 is the fact that its Boolean operations may be expressed in an *arithmetical form*:

$$x \cup y = x + y - x\,y, \tag{25}$$

$$\bar{x} = 1 - x, \tag{26}$$

where $+$ and $-$ have their usual arithmetical meaning. These formulas, together with their consequences

$$\overline{x\,y} = 1 - x\,y, \tag{27}$$

$$\overline{x \cup y} = \bar{x}\,\bar{y} = (1 - x)(1 - y). \tag{28}$$

etc., enable us to express Boolean functions (i.e. two-valued functions with bivalent variables; see § 2) in terms of the arithmetical operations of addition, subtraction and multiplication.

§ 2. Boolean Functions

Definition 2. By a *Boolean function* f, we mean a mapping

$$f : B_2^n = \underbrace{B_2 \times \cdots \times B_2}_{n \text{ times}} \to B_2, \tag{29}$$

i.e. a function whose arguments and values are in B_2.

Boolean functions are, for instance, $x \cup y\,\bar{z}$, $y[\bar{x} \cup \overline{y\,z(t \cup \overline{x\,y})}]$, etc. It is not by accident that in our examples the functions are built up by superpositions of the Boolean operations $\cup$, $\cdot$, and $^-$: each Boolean function has this property. For example, let us consider the function $f(x, y, z)$ defined by the following table:

Table 1

x	y	z	$f(x, y, z)$
0	0	0	1
0	0	1	0
0	1	0	1
0	1	1	0
1	0	0	1
1	0	1	1
1	1	0	1
1	1	1	0

In order to express this function by means of Boolean operations, we proceed as follows. We consider the five 3-tuples of values of x, y, z, for which $f(x, y, z) = 1$, namely $(0, 0, 0)$, $(0, 1, 0)$, $(1, 0, 0)$, $(1, 0, 1)$, $(1, 1, 0)$, and we associate to them the following Boolean functions: $x^0\,y^0\,z^0 = \bar{x}\,\bar{y}\,\bar{z}$, $x^0\,y^1\,z^0 = \bar{x}\,y\,\bar{z}$, $x^1\,y^0\,z^0 = x\,\bar{y}\,\bar{z}$, $x^1\,y^0\,z^1 = x\,\bar{y}\,z$, $x^1\,y^1\,z^0 = x\,y\,\bar{z}$. Here and in the sequel, we will use the following notation:

$$x^0 = \bar{x}, \qquad x^1 = x. \tag{30}$$

Now notice that $x^1\,y^1\,z^0$ is equal to 1 for $x = 1$, $y = 1$, $z = 0$, and is equal to 0 for the other values of x, y, z; similar results hold for the other above listed products. We infer that

$$f(x, y, z) = \bar{x}\,\bar{y}\,\bar{z} \cup \bar{x}\,y\,\bar{z} \cup x\,\bar{y}\,\bar{z} \cup x\,\bar{y}\,z \cup x\,y\,\bar{z}. \tag{31}$$

Indeed, consider a 3-tuple for which $f(x, y, z)$ must be 1, in accordance with the definition given by the table; say, $x = y = z = 0$. We see that for this 3-tuple, the right-hand side of (31) is also 1; the same equality holds for all 3-tuples (x, y, z) for which f is 1. Now, if (x, y, z) is a 3-tuple for which f must be 0 according to its tabular definition, i.e. if $x = 0$, $y = 0$, $z = 1$, or $x = 0$, $y = 1$, $z = 1$, or $x = 1$, $y = 1$, $z = 1$, then all the products in the right-hand side of (31) are equal to 0, so that relation (31) does again hold.

Now we shall prove that every Boolean function has a "Boolean expression" like (31). To do this, we need some specifications.

Roughly speaking, a Boolean expression is made up of a finite number of Boolean variables linked by Boolean operations. More precisely (see, for instance, Gr. C. Moisil [2]):

Definition 3.

1) 0 and 1 are Boolean expressions;

2) the indeterminates $x_1^1, x_1^0, x_2^1, x_2^0, \ldots, x_n^1, x_n^0$ are Boolean expressions;

3) if E_1 and E_2 are Boolean expressions, then $E_1 \cup E_2$, $E_1 E_2$ and $\bar{E}_1$ are Boolean expressions;

4) any Boolean expression is formed by a repeated application of the rules 1), 2), 3).

Definition 4′. Boolean expressions which do not contain disjunctions are called *elementary conjunctions*.

Elementary conjunctions are, for instance, $x^1 y^0 z^0$, $x^1 y^1 z^0$, x^1, y^0, 0, 1, $x_1^1 \ldots x_n^1$, etc.

Definition 4″. Boolean expressions which do not contain conjunctions (juxtapositions) are called *elementary disjunctions*.

Elementary disjunctions are, for instance, $x^1 \cup y^0 \cup z^0$, $x^1 \cup y^1 \cup z^0$, x^1, y^0, 0, 1, $x_1^1 \cup \ldots \cup x_n^1$, etc.

Definition 5′. A disjunction of elementary conjunctions, that is an expression of the type

$$c_1 \cup \ldots \cup c_r, \tag{32}$$

where $c_1, \ldots, c_r$ are elementary conjunctions, is called a *disjunctive form*.

Definition 5″. A conjunction of elementary disjunctions, that is an expression of the type

$$d_1 \ldots d_r, \tag{33}$$

where $d_1, \ldots, d_r$ are elementary disjunctions, is called a *conjunctive form*.

Definition 6. The function f_E *generated* by a Boolean expression E is the Boolean function obtained from E by interpreting the letters $x_1, \ldots, x_n$, occuring in the indeterminates $x_1^1, x_1^0, \ldots, x_n^1, x_n^0$, as variables in B_2, the exponents $0, 1$ as the functions defined by (30), the connectives $\cup$, . (or concatenation) and $\bar{}$ as the functions defined by (1), (3) and (2), respectively, and the letters 0 and 1 as the constant functions $0(x) = 0$, $1(x) = 1$, respectively.

We have emphasized the distinction between Boolean functions and Boolean expressions. The reason for doing so was that, on the one hand, a Boolean expression generates a single Boolean function, while, on the other hand, a Boolean function is generated by several Boolean expressions. For instance, the expressions $x \cup y$ and $x \cup \bar{x} y$ are different, but generate the same function $f(x, y) : f(0, 0) = 0$, $f(0, 1) = f(1, 0) = f(1, 1) = 1$.

However, we shall adopt the usual convention of using the same notation for a Boolean expression as well as for the Boolean function generated by it. For instance, we shall write $x \cup \bar{x} y = x \cup y$ instead of $f_{x \cup \bar{x}y} = f_{x \cup y}$, etc.

We are now in a position to carry out our promise and prove first that every Boolean function is generated by (at least) a Boolean expression. More precisely:

Theorem 1. *Every Boolean function f can be written in the form*

$$f(x_1, \ldots, x_n) = \bigcup_{\alpha_1, \ldots, \alpha_n} f(\alpha_1, \ldots, \alpha_n) \, x_1^{\alpha_1} \ldots x_n^{\alpha_n}, \tag{34}$$

where $\bigcup_{\alpha_1, \ldots, \alpha_n}$ means that the disjunction is extended over all 2^n possible values of the vector $(\alpha_1, \ldots, \alpha_n) \in B_2^n$.

Proof. Using formulas

$$x^0 = \bar{x}, \qquad x^1 = x, \tag{30}$$

and

$$\bar{0} = 1, \qquad \bar{1} = 0, \tag{2}$$

we conclude that, for any $\alpha, \beta \in B_2$,

$$\alpha^\beta = \begin{cases} 1, & \text{if } \alpha = \beta, \\ 0, & \text{if } \alpha \neq \beta. \end{cases} \tag{35}$$

Hence, for $\alpha_1, \ldots, \alpha_n \in B_2$, we have

$$x_1^{\alpha_1} \ldots x_n^{\alpha_n} = \begin{cases} 1, & \text{if } x_1 = \alpha_1, \ldots, x_n = \alpha_n, \\ 0, & \text{otherwise.} \end{cases} \tag{36}$$

It follows, that for any system of values $x_1 = \beta_1, \ldots, x_n = \beta_n$ given to the variables $x_1, \ldots, x_n$, the right-hand side of (34) reduces to $f(\beta_1, \ldots, \beta_n) \beta_1^{\beta_1} \ldots \beta_n^{\beta_n} = f(\beta_1, \ldots, \beta_n)$, thus completing the proof.

Remark 3′. Formula (34) may be written in the form

$$f(x_1, \ldots, x_n) = \bigcup'_{\alpha_1, \ldots, \alpha_n} x_1^{\alpha_1} \ldots x_n^{\alpha_n}, \tag{37}$$

where $\bigcup'_{\alpha_1, \ldots, \alpha_n}$ means that the disjunction is extended over those values $(\alpha_1, \ldots, \alpha_n) \in B_2^n$ for which $f(\alpha_1, \ldots, \alpha_n) = 1$.

For instance, by applying formula (37) to the function given in Table 1, we obtain relation (31).

Notice that (34) is a disjunctive form with the special property that in each of its conjunctions, all the variables $x_1, \ldots, x_n$ do appear.

Definition 7′. The right-hand side of relation (37) is said to be the *disjunctive canonical form* of the function $f(x_1, \ldots, x_n)$; each conjunction of the form $x_1^{\alpha_1} \ldots x_n^{\alpha_n}$ (i.e., containing all the variables $x_1, \ldots, x_n$) is called a *complete elementary conjunction* in $x_1, \ldots, x_n$.

Theorem 2. *Every Boolean function f may be written in the form*

$$f(x_1, \ldots, x_n) = \prod_{\alpha_1, \ldots, \alpha_n} [f(\bar{\alpha}_1, \ldots, \bar{\alpha}_n) \cup x_1^{\alpha_1} \cup \cdots \cup x_n^{\alpha_n}], \tag{38}$$

where $(\alpha_1, \ldots, \alpha_n)$ runs over all 2^n possible values of B_2^n.

Proof. Similar to that of Theorem 1.

Another proof is obtained by applying formula (34) to the function $\bar{f}$ and then taking the negation in both sides.

Remark 3''. Formula (38) may be written in the form

$$f(x_1, \ldots, x_n) = \prod''_{\alpha_1, \ldots, \alpha_n} (x_1^{\alpha_1} \cup \cdots \cup x_n^{\alpha_n}), \tag{39}$$

where $\prod''_{\alpha_1, \ldots, \alpha_n}$ means that the conjunction is extended over those values $(\alpha_1, \ldots, \alpha_n) \in B_2^n$ for which $f(\bar{\alpha}_1, \ldots, \bar{\alpha}_n) = 0$.

For instance, by applying formula (39) to the function f given in Table 1, we obtain

$$f(x, y, z) = (x \cup y \cup \bar{z})\,(x \cup \bar{y} \cup \bar{z})\,(\bar{x} \cup \bar{y} \cup \bar{z}). \tag{40}$$

Definition 7''. The right-hand side of relation (39) is said to be the *conjunctive canonical form* of the function $f(x_1, \ldots, x_n)$; each disjunction of the form $x_1^{\beta_1} \cup \cdots \cup x_n^{\beta_n}$ (i.e., containing all the variables $x_1, \ldots, x_n$) is called a *complete elementary disjunction* in $x_1, \ldots, x_n$.

Formulas (34) and (38) are usually called *interpolation formulas.*

§ 3. Boolean Matrices and Determinants

Definition 8. A *Boolean matrix* is a matrix of zeros and ones.

Definition 9. Given two $m \times n$ Boolean matrices (a_{ij}) and (b_{ij}), their *disjunction* is the $m \times n$ Boolean matrix

$$(a_{ij}) \cup (b_{ij}) = (a_{ij} \cup b_{ij}). \tag{41}$$

Definition 10. Given two $m \times n$ Boolean matrices (a_{ij}) and (b_{ij}), their *conjunction* is the $m \times n$ Boolean matrix

$$(a_{ij})\,(b_{ij}) = (a_{ij}\, b_{ij}). \tag{42}$$

Definition 11. Given an $m \times n$ Boolean matrix (a_{ij}), its *negation* is the $m \times n$ Boolean matrix

$$\overline{(a_{ij})} = (\overline{a_{ij}}). \tag{43}$$

Definition 12. Given an $m \times n$ Boolean matrix (a_{ij}) and an $n \times p$ Boolean matrix (b_{jk}), their *product* is the $m \times p$ Boolean matrix

$$(a_{ij}) \times (b_{jk}) = \left(\bigcup_{j=1}^{n} a_{ij}\, b_{jk}\right). \tag{44}$$

Definition 13. Given two $m \times n$ Boolean matrices (a_{ij}) and (b_{ij}), we say that (a_{ij}) is *less than or equal* to (b_{ij}):

$$(a_{ij}) \leqq (b_{ij}) \quad \text{if and only if} \quad a_{ij} \leqq b_{ij} \quad \text{for all } i, j; \tag{45}$$

Example 1. The disjunction of the matrices

$$A = \begin{pmatrix} 0 & 1 & 0 & 1 \\ 1 & 1 & 1 & 0 \\ 0 & 0 & 1 & 1 \end{pmatrix}, \quad B = \begin{pmatrix} 0 & 0 & 0 & 1 \\ 1 & 0 & 0 & 1 \\ 1 & 1 & 1 & 1 \end{pmatrix} \tag{46}$$

is the matrix

$$A \cup B = \begin{pmatrix} 0 & 1 & 0 & 1 \\ 1 & 1 & 1 & 1 \\ 1 & 1 & 1 & 1 \end{pmatrix},$$

their conjunction is

$$A\,B = \begin{pmatrix} 0 & 0 & 0 & 1 \\ 1 & 0 & 0 & 0 \\ 0 & 0 & 1 & 1 \end{pmatrix},$$

while their negations are

$$\bar{A} = \begin{pmatrix} 1 & 0 & 1 & 0 \\ 0 & 0 & 0 & 1 \\ 1 & 1 & 0 & 0 \end{pmatrix}, \quad \bar{B} = \begin{pmatrix} 1 & 1 & 1 & 0 \\ 0 & 1 & 1 & 0 \\ 0 & 0 & 0 & 0 \end{pmatrix}.$$

The product of the matrix A by the matrix

$$C = \begin{pmatrix} 0 & 1 \\ 0 & 0 \\ 1 & 1 \\ 0 & 1 \end{pmatrix}$$

is the 3×2 matrix

$$A \times C = \begin{pmatrix} 0\cdot 0 \cup 1\cdot 0 \cup 0\cdot 1 \cup 1\cdot 0 & 0\cdot 1 \cup 1\cdot 0 \cup 0\cdot 1 \cup 1\cdot 1 \\ 1\cdot 0 \cup 1\cdot 0 \cup 1\cdot 1 \cup 0\cdot 0 & 1\cdot 1 \cup 1\cdot 0 \cup 1\cdot 1 \cup 0\cdot 1 \\ 0\cdot 0 \cup 0\cdot 0 \cup 1\cdot 1 \cup 1\cdot 0 & 0\cdot 1 \cup 0\cdot 0 \cup 1\cdot 1 \cup 1\cdot 1 \end{pmatrix} = \begin{pmatrix} 0 & 1 \\ 1 & 1 \\ 1 & 1 \end{pmatrix}.$$

It is easy to see that the set of all 2^{mn} Boolean matrices with m rows and n columns has the properties (4)—(11) and (13)—(24), if $\cup$, $\cdot$, $^{-}$ and $\leqq$ are given the meaning described in Definitions 9, 10, 11, and 13, respectively, while the rôles of 0 and 1 are played by the $m \times n$ Boolean matrices

$$0 = \begin{pmatrix} 0 & \cdot & \cdot & \cdot & 0 \\ \cdot & \cdot & \cdot & \cdot & \cdot \\ 0 & \cdot & \cdot & \cdot & 0 \end{pmatrix} \tag{47}$$

and

$$I = \begin{pmatrix} 1 & \cdot & \cdot & \cdot & 1 \\ \cdot & \cdot & \cdot & \cdot & \cdot \\ 1 & \cdot & \cdot & \cdot & 1 \end{pmatrix} \tag{48}$$

respectively.

Thus we have an example of a Boolean algebra different from B_2 (see Remark 2 and § 4).

We notice also the following properties:

$$A \leqq A, \tag{49}$$

$$\text{if } A \leqq B \text{ and } B \leqq A, \text{ then } A = B, \tag{50}$$

$$\text{if } A \leqq B \text{ and } B \leqq C, \text{ then } A \leqq C, \tag{51}$$

which were trivial in case of B_2.

We call to the attention of the reader the fact that two $m \times n$ Boolean matrices A and B are not necessarily comparable. E.g., for the matrices of (46), neither $A \leqq B$, nor $B \leqq A$ holds.

The product is usually employed within the set $\mathcal{M}_n$ of all square Boolean matrices of a fixed order n. It has the following main properties:

$$(A \times B) \times C = A \times (B \times C), \tag{52}$$

$$A \times (B \cup C) = (A \times B) \cup (A \times C), \tag{53}$$

$$(A \cup B) \times C = (A \times C) \cup (B \times C), \tag{54}$$

$$\text{if } A \leqq B, \text{ then } A \times C \leqq B \times C, \tag{55}$$

and

$$A \times E = E \times A = A, \tag{56}$$

where

$$E = \begin{pmatrix} 1 & & 0 \\ & \ddots & \\ 0 & & 1 \end{pmatrix}. \tag{57}$$

The product is not commutative; for instance, taking

$$A = \begin{pmatrix} 1 & 1 \\ 0 & 1 \end{pmatrix}, \quad B = \begin{pmatrix} 0 & 0 \\ 0 & 1 \end{pmatrix},$$

we have

$$A \times B = \begin{pmatrix} 0 & 1 \\ 0 & 1 \end{pmatrix},$$

but

$$B \times A = \begin{pmatrix} 0 & 0 \\ 0 & 1 \end{pmatrix}.$$

Definition 14. The *powers* of a square matrix A are

$$A^0 = E, \quad A^p = A^{p-1} \times A \quad (p = 1, 2, \ldots). \tag{58}$$

It is important to note that, if $A = (a_{ij}) \in \mathcal{M}_n$, then its powers $A^p = (a_{ij}^p)$ $(p = 1, 2, \ldots)$ are given by

$$a_{ij}^p = \bigcup_{i_1, \ldots, i_{p-1}} a_{i i_1} a_{i_1 i_2} \ldots a_{i_{p-1} j}, \tag{59}$$

where $\bigcup\limits_{i_1, \ldots, i_{p-1}}$ means that the disjunction is extended over the n^{p-1} possible values of the indices. The recursive proof is at hand.

In the case of a matrix having only ones as diagonal elements, at most $n-1$ powers of it are distinct, as is shown by the

Theorem 3* (A. G. LUNC [1, 2]). *Let* $A = (a_{ij})$ *be a Boolean* $n \times n$ *matrix with* $a_{ii} = 1$ $(i = 1, \ldots, n)$. *Then*

$$E \leqq A \leqq A^2 \leqq \cdots \leqq A^{n-1} = A^n = A^{n+1} = \cdots \tag{60}$$

Proof. Relations $a_{ii} = 1$ show that $E \leqq A$. Therefore, in view of the isotony (55), we deduce $E \leqq A \leqq A^2 \leqq \cdots \leqq A^{n-1} \leqq A^n \leqq \leqq A^{n+1} \leqq \cdots$.

Thus it suffices to prove that $A^n \leqq A^{n-1}$. In fact, consider a term

$$a_{i i_1} \ldots a_{i_{n-1} j} \tag{61}$$

belonging to the right-hand side of the expansion (59) of a_{ij}^n. Since there cannot be $n+1$ distinct indices $i, i_1, \ldots, i_{n-1}, j$, there is an $h < k$ such that $i_h = i_k$. Then the term (61) is

$$a_{i i_1} \ldots a_{i_{k-1} i_k} a_{i_k i_{k+1}} \ldots a_{i_{h-1} i_h} a_{i_h i_{h+1}} \ldots a_{i_{n-1} j} \leqq \\ \leqq a_{i i_1} \ldots a_{i_{k-1} i_k} a_{i_h i_{h+1}} \ldots a_{i_{n-1} j}. \tag{62}$$

In the right-hand side of (62) there are at most $n-1$ factors; if there are less, we can complete the product by factors of the form $a_{i_k i_h} = a_{i_k i_k} = a_{i_h i_h} = 1$, so that

$$a_{i i_1} \ldots a_{i_{n-1} j} \leqq \underbrace{a_{i i_1} \ldots a_{i_{k-1} i_k} a_{i_k i_k} \ldots a_{i_k i_h} a_{i_h i_{h+1}} \ldots a_{i_{n-1} j}}_{n-1 \text{ factors}}. \tag{63}$$

But the right-hand side of (63) is a term belonging to the expansion (59) of a_{ij}^{n-1}. Since the relation (63) holds for all possible values of the indices $i_1, \ldots, i_{n-1}$, we conclude that $a_{ij}^n \leqq a_{ij}^{n-1}$. Since this is true for all $i, j = 1, \ldots, n$, we deduce that $A^n \leqq A^{n-1}$, thus completing the proof.

Definition 15. The first exponent e for which a square Boolean matrix A of order n satisfies $A^e = A^{e+1}$ will be called the *characteristic exponent* of A.

Remark 4. According to Theorem 3, the characteristic exponent of a square Boolean matrix A satisfies

$$e \leqq n - 1; \tag{64}$$

---

* *Note added in proofs.* An interesting generalization of Theorem 3 to the case of arbitrary Boolean matrices was given by D. ROSENBLATT [1].

furthermore, e is the characteristic exponent if and only if

$$A < A^2 < \cdots < A^{e-1} < A^e = A^{e+1} = \cdots = A^n = \cdots, \tag{65}$$

where $A < B$ means $A \leqq B$ but $A \neq B$.

Notice also that the matrix A^e satisfies $(A^e)^2 = A^e$; in other words, the characteristic exponent of the matrix A^e is 1.

More generally, we introduce the following

Definition 16. A square Boolean matrix A will be called *idempotent*, if $A = A^2$ $(= A \times A)$.

Remark 5. The matrix A is idempotent if and only if its characteristic exponent is $e = 1$.

A characterization of idempotent Boolean matrices can be given in terms of "Boolean determinants".

Definition 17. The (*Boolean*) *determinant* of a Boolean $n \times n$ matrix $A = (a_{ij})$ is

$$|A| = |(a_{ij})| = \bigcup_{h_1, \ldots, h_n} a_{1h_1} a_{2h_2} \ldots a_{nh_n}, \tag{66}$$

where $\bigcup_{h_1, \ldots, h_n}$ means that the disjunction is extended over all permutations $(h_1, h_2, \ldots, h_n)$ of the indices $1, 2, \ldots, n$.

The following properties are immediate:

1) the determinant is not altered by a permutation of the rows or of the columns;

2) the determinant can be expanded by a given row:

$$|A| = \bigcup_{j=1}^{n} a_{ij} |A_{ij}| \qquad (i = 1, \ldots, n), \tag{67}$$

(where $|A_{ij}|$ means, in analogy to the usual case of determinants, the determinant of the matrix A_{ij} obtained from A by deleting row i and colum j), or by a given column, or according to a Laplace-type formula;

3) the determinant is a linear function of each of its rows:

$$|(\alpha_i a_{ij} \cup \beta_i b_{ij})| = \alpha_i |(a_{ij})| \cup \beta_i |(b_{ij})| \qquad (j = 1, \ldots, n); \tag{68}$$

a similar relation holds for each column.

However, the determinant of a matrix with two identical rows (or columns) is not necessarily equal to zero.

Definition 18. The *adjoint* of a square Boolean matrix A is the Boolean matrix $adj\, A = (\alpha_{ij})$ defined by

$$\alpha_{ij} = |A_{ji}| \qquad \text{for all } i, j, \tag{69}$$

where $|A_{ji}|$ is the determinant of the minor A_{ji} of a_{ji} in A.

Theorem 4 (A. G. Lunc [1, 2]). *Let $A = (a_{ij})$ be a Boolean $n \times n$ matrix with $a_{ii} = 1$ $(i = 1, \ldots, n)$. Let e be the characteristic exponent of A and $adj\, A$ the adjoint of A (Definitions 15, 18). Then:*

$$adj\, A = A^e. \tag{70}$$

Proof. Since $A^e = A^{n-1}$, it suffices to prove that $a\,d\,j\,A = A^{n-1}$ or, equivalently, that

(71) $$|A_{ji}| = a_{ij}^{n-1}.$$

Now, it follows from the expansion formula

(72) $$|A| = \bigcup_{k=1}^{n} a_{jk}|A_{jk}|,$$

that $|A_{ji}|$ is obtained from $|A|$ by replacing a_{ji} by 1 and every other a_{jk} in the j-th row by 0. We can perform these transformations on formula (66), obtaining thus the following rule for calculating $|A_{ji}|$: write down each term of $|A|$ containing a_{ji}, delete a_{ji} and take the disjunction of the modified terms. But a term $a_{1h_1} a_{2h_2} \ldots a_{nh_n}$ of $|A|$ contains a_{ji} if and only if $h_j = i$, that is, if and only if the permutation

$$\begin{pmatrix} 1 & 2 & \ldots & n \\ h_1 & h_2 & \ldots & h_n \end{pmatrix}$$

includes a cycle of the form $(i, k_1, \ldots, k_r, j)$. Therefore, each term of $|A_{ji}|$ is included in a term of the form

$$a_{ik_1} a_{k_1k_2} \ldots a_{k_rj} \leqq a_{ij}^{r+1} \leqq a_{ij}^{n-1},$$

so that

(73) $$|A_{ji}| \leqq a_{ij}^{n-1}.$$

On the other hand, any term of a_{ij}^{n-1} is of the form

(74) $$a_{i i_1} \ldots a_{i_{n-2} j}.$$

If the indices i, $i_1, \ldots, i_{n-2}, j$ are distinct, then, according to the above discussion, (74) is a term of A_{ji}. Now assume that two indices, say i_r and i_s with $r < s$, coincide; then

$$a_{i i_1} \ldots a_{i_{n-2} j} \leqq a_{i i_1} \ldots a_{i_{r-1} i_r} a_{i_s i_{s+1}} \ldots a_{i_{n-2} j}$$
$$= a_{i i_1} \ldots a_{i_{r-1} i_r} a_{i_r i_{s+1}} \ldots a_{i_{n-2} j}.$$

Therefore, after applying several times this type of evaluation, we obtain an inequality of the form

(75) $$a_{i i_1} \ldots a_{i_{n-2} j} \leqq a_{i l_1} a_{l_1 l_2} \ldots a_{l_t j},$$

where the indices i, l_1, $l_2, \ldots, l_t$, j are distinct. Relation (75) may be written in the form

(76) $$a_{i i_1} \ldots a_{i_{n-2} j} \leqq a_{i l_1} a_{l_1 l_2} \ldots a_{l_t j}\, a_{m_1 m_1} \ldots a_{m_{n-t-2},\, m_{n-t-2}},$$

where $(i, l_1, \ldots, l_t, j, m_1, \ldots, m_{n-t-2})$ is a permutation of $(1, 2, \ldots, n)$. But, as we have seen in the first part of the proof, the right-hand side of (76) is a term of $|A_{ji}|$, so that (76) implies in turn $a_{i i_1} \ldots a_{i_{n-2} j} \leqq$

$\leqq |A_{ji}|$ and

(77) $$a_{ij}^{n-1} \leqq |A_{ji}|.$$

This, together with (73), completes the proof of (71).

Corollary 1. A Boolean $n \times n$ matrix $A = (a_{ij})$ with $a_{ii} = 1$ $(i = 1, \ldots, n)$ is idempotent if and only if

(78) $$A = adj\, A$$

(see Definition 16 and Remark 5).

§ 4. The General Concept of Boolean Algebra*

As we have already pointed out in Remark 2, the two-element Boolean algebra studied in § 1 is just a particular case of the general concept of Boolean algebra. The latter can be defined in various ways (see, for instance, S. RUDEANU [B 1]); we choose here the following

Definition 19. By a *Boolean algebra* we mean a set** B in which two elements, 0 and 1, are distinguished, and three operations: $\cup$ (disjunction), $\cdot$ (conjunction) and $^{-}$ (negation) are defined, satisfying the following properties:

(4′) $$x \cup y = y \cup x,$$

(4″) $$x\, y = y\, x,$$

(5′) $$(x \cup y) \cup z = x \cup (y \cup z),$$

(5″) $$(x\, y)\, z = x (y\, z),$$

(7′) $$x \cup x\, y = x,$$

(7″) $$x (x \cup y) = x,$$

(8′) $$x \cup y\, z = (x \cup y)\,(x \cup z),$$

(8″) $$x (y \cup z) = x\, y \cup x\, z,$$

(9′) $$x \cup 1 = 1,$$

(9″) $$x.0 = 0,$$

(13′) $$x \cup \bar{x} = 1,$$

(13″) $$x\, \bar{x} = 0.$$

Example 2.

1) In the particular case where $B = \{0, 1\}$ and the operations are defined by relations (1), (3), and (2), we obtain the two-element Boolean algebra B_2 studied in § 1.

2) We have already noticed in § 3 that the set of Boolean matrices of given order n is itself a Boolean algebra with respect to the operations $\cup$, $\cdot$ and $^{-}$ defined by (41), (42) and (43), the elements 0 and 1 being the matrices (47) and (48), respectively.

* This section is needed only in Chapter II, §§ 1—4, and Chapter IX, § 8.

** Finite or not.

3) We leave to the reader the task of verifying that the set of all Boolean functions of n variables becomes also a Boolean algebra, if 0 and 1 are the Boolean functions of n variables identically equal to 0 and to 1, respectively, while the disjunction, the conjunction and the negation are constructed as follows; if f and g are Boolean functions of n variables, their disjunction will be the Boolean function $f \cup g$ of n variables defined by

$$(f \cup g)(x_1, \ldots, x_n) = f(x_1, \ldots, x_n) \cup g(x_1, \ldots, x_n), \tag{79}$$

their conjunction will be the function $f g$ defined by

$$(f g)(x_1, \ldots, x_n) = f(x_1, \ldots, x_n) \cdot g(x_1, \ldots, x_n), \tag{80}$$

while the negation of f will be the function $\bar{f}$ defined by

$$\bar{f}(x_1, \ldots, x_n) = \overline{f(x_1, \ldots, x_n)}. \tag{81}$$

For instance, in the Boolean algebra of all Boolean functions of two variables, consider the function f and g given in Table 2:

Table 2

x	y	$f(x, y)$	$g(x, y)$
0	0	0	1
0	1	0	1
1	0	1	0
1	1	1	1

The disjunction $f \cup g$, the conjunction $f g$ and the negations $\bar{f}$, $\bar{g}$ are the functions given in Table 3:

Table 3

x	y	$(f \cup g)(x, y)$	$(f g)(x, y)$	$\bar{f}(x, y)$	$\bar{g}(x, y)$
0	0	$0 \cup 1 = 1$	$0 \cdot 1 = 0$	$\bar{0} = 1$	$\bar{1} = 0$
0	1	$0 \cup 1 = 1$	$0 \cdot 1 = 0$	$\bar{0} = 1$	$\bar{1} = 0$
1	0	$1 \cup 0 = 1$	$1 \cdot 0 = 0$	$\bar{1} = 0$	$\bar{0} = 1$
1	1	$1 \cup 1 = 1$	$1 \cdot 1 = 1$	$\bar{1} = 0$	$\bar{1} = 0$

Using formulas instead of tables, we have $f(x, y) = x$, $g(x, y) = \bar{x} \cup y$, hence $(f \cup g)(x, y) = f(x, y) \cup g(x, y) = x \cup (\bar{x} \cup y) = 1$, $(f g)(x, y) = f(x, y)\, g(x, y)$ $= x(\bar{x} \cup y) = x\, y$, $\bar{f}(x, y) = \overline{f(x, y)} = \bar{x}$, $\bar{g}(x, y) = \overline{g(x, y)} = \overline{\bar{x} \cup y} = x\, \bar{y}$.

Now compare Definition 19 with § 1. It was proved there that the two-element Boolean algebra satisfies certain properties (4)—(24); some of these properties, namely (4), (5), (7), (8), (9) and (13) are taken in Definition 19 as *axioms*. As a matter of fact, *each Boolean algebra satisfies* (4)—(11), (13)—(24), i.e. not only the properties which were taken as axioms. We give the proof of this assertion below.

We have to deduce from (4), (5), (7), (8), (9) and (13), the remaining properties (4)—(11), (13)—(24). In order to facilitate the proofs, we notice first that we can apply the *principle of duality*, mentioned in § 1. In fact, assuming that we have already proved a certain property, say (24′), we obtain the proof of the dual property (24″) by replacing each axiom used in the proof of (24′) by its dual.

We prove first the idempotency. It follows from (7″) that $x \cup x = x \cup x(x \cup y)$; setting $x \cup y = z$, we have $x \cup x = x \cup x z = x$, by (7′), thus proving (6′); property (6″) follows by duality.

Further $x \cup 0 = x \cup x.0 = x$, by (9″) and (7′); dually, $x.1 = x$. It follows that $0 \cup 0 = 0$; conversely, if $x \cup y = 0$, then $x = x(x \cup y) = x.0 = 0$, by (7″) and (9″), so that $y = y \cup 0 = y \cup x = x \cup y = 0$, by (10′) and (4′). Property (11″) holds by duality.

Properties (12) *hold only in the two-element Boolean algebra*; for instance, as we have already noticed, in the Boolean algebra of all Boolean functions of two variables, the elements f and g defined by $f(x, y) = x$, $g(x, y) = \bar{x} \cup y$, satisfy $f \cup g = 1$, although $f \neq 1$ and $g \neq 1$.

In order to prove the de Morgan laws and the law of double negation, we need the following lemma: *the system of equations*

$$(82) \qquad a \cup z = 1, \quad a z = 0$$

has the single solution $z = \bar{a}$. In fact, axioms (13) show that the system (82) has the solution $z = \bar{a}$; we shall prove that if z is any solution, then $z = \bar{a}$. We have $z = z.1 = z(a \cup \bar{a}) = z a \cup z \bar{a} = a z \cup \bar{a} z = 0 \cup \bar{a} z = \bar{a} a \cup \bar{a} z = \bar{a}(a \cup z) = \bar{a}.1 = \bar{a}$.

Therefore, in order to prove (14′), it suffices to prove that $z = \bar{x}\,\bar{y}$ satisfies the system of equations $x \cup y \cup z = 1$, $(x \cup y) z = 0$. But $x \cup y \cup \bar{x}\,\bar{y} = (x \cup y \cup \bar{x})(x \cup y \cup \bar{y}) = (y \cup x \cup \bar{x})(x \cup y \cup \bar{y}) = (y \cup 1)(x \cup 1) = 1.1. = 1$ (by (8′), (4′), (13′), (9′) and (10″)), while $(x \cup y)\,\bar{x}\,\bar{y} = x \cdot \bar{x}\,\bar{y} \cup y \cdot \bar{x}\,\bar{y} = x\,\bar{x}\,\bar{y} \cup \bar{x} \cdot y\,\bar{y} = 0 \cdot \bar{y} \cup \bar{x} \cdot 0 = 0 \cup 0 = 0$ (by (8), (4), (5), (13) and (10)). Relation (14″) is proved by duality.

Now apply axioms (13) to $\bar{x}$: it follows that $\bar{x} \cup \bar{\bar{x}} = 1$ and $\bar{x}\,\bar{\bar{x}} = 0$; in other words, the system of equations $\bar{x} \cup z = 1$, $\bar{x} z = 0$ has the solution $z = \bar{\bar{x}}$. However, this system has also the solution $z = x$, by (13) and (4); it follows from tne above lemma that $\bar{\bar{x}} = x$.

The identities (16) are easy consequences of (8), (13), (4) and (10).

We notice that

$$(83) \qquad x y = x \quad \text{if and only if} \quad x \cup y = y,$$

because $x y = x$ implies $x \cup y = x y \cup y = y \cup y x = y$ and, conversely, $x \cup y = y$ implies $x y = x(x \cup y) = x$.

Therefore the following definition is meaningful:

Definition 20. In a Boolean algebra, we define

$$(17') \qquad x \leqq y \quad \text{if and only if} \quad x \cup y = y,$$

or equivalently,

$$(17'') \qquad x \leqq y \quad \text{if and only if} \quad x y = x.$$

The reader may notice that in the case of the two-element Boolean algebra, relations (17) were proved, while here they are taken as a definition. As a matter of fact, the situation is the following. As we shall prove below, the relation defined above is a *partial order*, i.e. it satisfies

(84) $$x \leqq x,$$

(85) $$x \leqq y \quad \text{and} \quad y \leqq x \quad \text{imply} \quad x = y,$$

(86) $$x \leqq y \quad \text{and} \quad y \leqq z \quad \text{imply} \quad x \leqq z.$$

Thus in the two-element set $\{0, 1\}$ we have a priori two partial orders: relation $\leqq$ defined above and the usual partial order, say $\leqslant$: $(0 \leqslant 0,\ 0 \leqslant 1,\ 1 \leqslant 1)$. It was proved in § 1 that $\leqslant$ satisfies also (17); in other words, it was proved that *in the case of the two-element Boolean algebra, relation* $\leqq$ *defined above coincides with the usual ordering of the set* $\{0, 1\}$.

Now, let us prove (84), (85) and (86). Since $x \cup x = x$, it follows that $x \leqq x$. Further, if $x \leqq y$ and $y \leqq x$, then $x \cup y = y$ and $x \cup y = y \cup x = x$, so that $x = y$. Finally, if $x \leqq y$ and $y \leqq z$, then $x \cup z = x \cup (y \cup z) = (x \cup y) \cup z = y \cup z = z$, so that $x \leqq z$.

Further, $x \cup (x \cup y) = (x \cup x) \cup y = x \cup y$, showing that $x \leqq x \cup y$; similarly, $y \leqq x \cup y$. If $x \leqq z$ and $y \leqq z$, then $(x \cup y) \cup z = x \cup (y \cup z) = x \cup z = z$, so that $x \cup y \leqq z$. Properties (18″) and (19″) follow by duality; further, (20) are easy consequences of (19), (18) and (86). If $x \leqq y$, then $x\,z \cup y\,z = (x \cup y)\,z = y\,z$, hence $x\,z \leqq y\,z$ and similarly, $x \cup z \leqq y \cup z$.

Relations (22) are merely a translation of (9).

Now, $x \leqq y$ implies $\bar{x} \cup y = \bar{x} \cup (x \cup y) = (\bar{x} \cup x) \cup y = 1 \cup y = 1$, while $\bar{x} \cup y = 1$ implies, in turn, $x = x.1 = x(\bar{x} \cup y) = x\,y$, hence $x \leqq y$; dually, we have (23″). Finally, $x = y$ if and only if $x \leqq y$ and $y \leqq x$ (by (84) and (85)), hence $x = y$ is equivalent to $x\,\bar{y} = 0$ and $y\,\bar{x} = 0$, i.e. to $x\,\bar{y} \cup y\,\bar{x} = 0$, by (11′); a dual proof holds for (24″).

Thus properties (4)—(11), (13)—(24) hold not only for the two-element Boolean algebra, but also for arbitrary Boolean algebras. Moreoveor, if a *Boolean matrix* over a Boolean algebra B is defined as an $m \times n$ tableau $A = (a_{ij})$ with elements $a_{ij} \in B$, then, as the reader may easily verify, the whole discussion of § 3 remains valid (because the proof used only general properties of all Boolean algebras). We have thus demonstrated:

Theorem 5. *Let* B *be an arbitrary Boolean algebra. Then properties* (4)—(11), (13)—(24) *hold. Further, the Boolean matrices over* B *satisfy all the properties studied in* § 3.

We have noticed before (Example 2.2) that the set of all $n \times n$ matrices with elements $0, 1 \in B_2$ is a Boolean algebra with respect to the operations (41), (42) and (43). According to the above Theorem 5, the same result is valid for matrices with elements in an arbitrary Boolean algebra.

The reader may remark a certain lack of symmetry: while the results of §§ 1 and 3 were extended to arbitrary Boolean algebras, § 2 remains valid only for functions with variables and values in the two-element Boolean algebra. Indeed, the main results in § 2, namely Theorems 1 and 2, are no longer valid for functions with arguments and values in an arbitrary Boolean algebra (see, for instance, S. RUDEANU [2, 4]).

§ 5. Boolean Algebras and Boolean Rings

Given an arbitrary Boolean algebra B (as defined in § 4), it is often convenient to consider, in addition to the Boolean operations of disjunction, conjunction and negation, also the operation

$$x \oplus y = x \bar{y} \cup \bar{x} y. \tag{87}$$

Definition 21. The operation $\oplus$, defined by (87), is called the *symmetric difference*, or *sum modulo* 2.

It follows from (87) that

$$0 \oplus 0 = 0, \quad 0 \oplus 1 = 1 \oplus 0 = 1, \quad 1 \oplus 1 = 0. \tag{88}$$

The following properties of $\oplus$ are easily proved:

$$x \oplus y = y \oplus x, \tag{89}$$

$$(x \oplus y) \oplus z = x \oplus (y \oplus z), \tag{90}$$

$$x \oplus 0 = x, \tag{91}$$

$$x \oplus 1 = \bar{x}, \tag{92}$$

$$x \oplus x = 0, \tag{93}$$

$$x(y \oplus z) = x y \oplus x z, \tag{94}$$

$$x \leqq y \quad \text{if and only if} \quad x \oplus x y = 0. \tag{95}$$

Properties (89), (90), (91), (93), (4″), (5″), (10″) and (94) show that the operations $\oplus$ and $\cdot$ fashion the Boolean algebra B into a "Boolean ring" (see below).

The transition from Boolean operations to ring operations is expressed by the formulas (87) and $x y = x y$; the converse transition is given by formulas $x y = x y$, (92) and

$$x \cup y = x \oplus y \oplus x y. \tag{96}$$

More generally, it can be proved that there exists a one-to-one correspondence between Boolean algebras and *Boolean rings* [i.e., commutative rings with unity, satisfying (93) and (6″)].

In the case of the two-element Boolean algebra B_2, the Boolean ring associated with it is even a *field*, usually termed the *Galois field GF* (2).

In this case, the operation $\oplus$ is the so-called *sum modulo* 2; it can be also expressed in terms of arithmetical operations:

$$x \oplus y = x + y - 2xy. \tag{97}$$

§ 6. Pseudo-Boolean Functions

Definition 22. Let R be the field of the real numbers; by a *pseudo-Boolean function*, we shall mean a function

$$f: B_2^n \to R, \tag{98}$$

i.e. a real-valued function of bivalent variables.

As a matter of fact, in most of our previous papers, we used another definition of pseudo-Boolean functions, we defined them namely as the mappings of B_2^n into the ring Z of integers. However, most of the results in this book being valid when the mapping of B_2^n into R is considered, we have adopted here Definition 22.

From a practical point of view, however, this distinction has no special importance, since the data found in applications are usually not real, but rational numbers, and the latter can be transformed into integers by multiplying all of them by a suitable integer. For this reason, in the examples, we can always assume the data to be integers.

R. Fortet calls these functions "integer algebraic functions". Our term is justified by the following remark: if the elements 0 and 1 of B_2 are identified with the numbers 0 and 1 — and this will be tacitly assumed in the sequel — then every Boolean function

$$\varphi: B_2^n \to B_2 \tag{99}$$

is also a pseudo-Boolean function.

With regard to the properties of pseudo-Boolean functions, let us notice first that such a function is always linear in each of its variables. Indeed, if we set

$$\begin{aligned} &g(x_1, \ldots, x_{i-1}, x_{i+1}, \ldots, x_n) \\ &\quad = f(x_1, \ldots, x_{i-1}, 1, x_{i+1}, \ldots, x_n) - f(x_1, \ldots, x_{i-1}, 0, x_{i+1}, \ldots, x_n), \end{aligned} \tag{100}$$

$$h(x_1, \ldots, x_{i-1}, x_{i+1}, \ldots, x_n) = f(x_1, \ldots, x_{i-1}, 0, x_{i+1}, \ldots, x_n), \tag{101}$$

then

$$\begin{aligned} &f(x_1, \ldots, x_n) \\ &\quad = x_i\, g(x_1, \ldots, x_{i-1}, x_{i+1}, \ldots, x_n) + h(x_1, \ldots, x_{i-1}, x_{i+1}, \ldots, x_n); \end{aligned} \tag{102}$$

conversely, relation (102) implies (101) and (100).

More generally, we have the following result, due to T. Gaspar:

Theorem 6. *Every pseudo-Boolean function may be written as a polynomial, which is linear in each variable, and which, after the reduction* *of the similar terms, is uniquely determined up to the order of the sums and products.*

The proof by induction is immediate.

On the other hand, every pseudo-Boolean function has also an expansion, analogous to the canonical disjunctive form of a Boolean function.

Setting

(30) $$x^1 = x, \quad x^0 = \bar{x},$$

we obtain an *interpolation formula* for pseudo-Boolean functions, according to the following:

Theorem 7*. *Every pseudo-Boolean function may be written in the form*

(103) $$f(x_1, \ldots, x_n) = \sum_{\alpha_1, \ldots, \alpha_n} c_{\alpha_1 \ldots \alpha_n} x_1^{\alpha_1} \ldots x_n^{\alpha_n},$$

where the sum $\sum_{\alpha_1, \ldots, \alpha_n}$ *is extended over the* 2^n *values of the vector* $(\alpha_1, \ldots, \alpha_n) \in \in B_2^n$, *and the coefficients* $c_{\alpha_1 \ldots \alpha_n}$ *are uniquely determined by the relations*

(104) $$c_{\alpha_1 \ldots \alpha_n} = f(\alpha_1, \ldots, \alpha_n).$$

The proof is analogous to that of the corresponding theorem for Boolean functions.

Example 3. The pseudo-Boolean function

(105) $$f(x_1, x_2, x_3) = 2 x_1 \bar{x}_2 + 6 x_1 x_3 - 5 \bar{x}_2 \bar{x}_3,$$

which can also be defined by the following

Table 4

x_1	x_2	x_3	$f(x_1, x_2, x_3)$
0	0	0	−5
0	0	1	0
0	1	0	0
0	1	1	0
1	0	0	−3
1	0	1	8
1	1	0	0
1	1	1	6

may also be written in the form

(105′) $$f(x_1, x_2, x_3) = -5 + 2 x_1 + 5 x_2 + 5 x_3 - 2 x_1 x_2 + 6 x_1 x_3 - 5 x_2 x_3,$$

which is obtained from (105) via formula $\bar{x} = 1 - x$; the function f can also be written in the form

(105″) $$f(x_1, x_2, x_3) = -5 \bar{x}_1 \bar{x}_2 \bar{x}_3 - 3 x_1 \bar{x}_2 \bar{x}_3 + 8 x_1 \bar{x}_2 x_3 + 6 x_1 x_2 x_3,$$

which is obtained from the above table by applying Theorem 7.

* Also stated by M. Carvallo ([2], p. 102) in a matrix formulation.

Chapter II

Boolean Equations

Since the solutions of Boolean equations will be used as an important tool in our book, we devote this chapter to their study.

In §§ 1—4 we will deal with the general case when the considered Boolean equations are to be solved in an arbitrary Boolean algebra (not necessarily the two-element one). We show first that any system of Boolean equations and (or) inequations is equivalent to a single equation of the form $f = 0$ (or, if desired, $g = 1$). We shall then prove the theorem on the consistency of such an equation and give several procedures for solving it: the method of successive eliminations with its variants (parametric form and recurrent-inequalities form of the general solution, and listing of all the solutions), as well as LÖWENHEIM-type formulas for the general solution.

In the next two sections we shall consider equations in the two-element Boolean algebra $B_2 = \{0, 1\}$.

In § 5 we will describe the method of bifurcations which is very suitable for hand computation, but also adequate for an electronic computer.

When the equation is of the form $f(x_1, \ldots, x_n) = 1$ the solutions are immediately obtained, grouped into "families of solutions", as it will be shown in § 6.

The reader interested only in solving equations in the two-element Boolean algebra can omit §§ 2—4; also, § 6 can be understood without the knowledge of § 5.

The chapter terminates with a section containing some indications concerning the complete bibliography on Boolean equations given at the end of the book (Bibliography A).

A. Equations in an Arbitrary Boolean Algebra

§ 1. The Single-Equation Form of a System of Boolean Equations and/or Inequalities

For the sake of generality, we shall be concerned in the first four sections with equations in an arbitrary Boolean algebra B (not necessarily the two-element one; see § I.4), whose basic operations of disjunction, conjunction and negation will be denoted by $\cup$, by $\cdot$ or by juxtaposition, and by $^-$, respectively.

We have seen in § I.2 that any function with arguments and values in the two-element Boolean algebra B_2 is generated by a Boolean expression (Definitions I.3 and I.6), or, equivalently is characterized by the property that it can be expressed both in the disjunctive and conjunctive canonical forms (Theorems I.1, I.2 and Definitions I.7′, I.7″).

This property is no longer valid for a Boolean algebra B containing more than two elements. For instance, if $B = B_4 = \{0, a, \bar{a}, 1\}$ is the four-element Boolean algebra, then the function $f: B_4 \to B_4$, defined by

$$f(x) = \begin{cases} a, & \text{if } x = a, \\ 0, & \text{if } x \neq a, \end{cases} \tag{1}$$

cannot be expressed in the disjunctive canonical form $f(x) = f(1)\, x \cup \cup f(0)\, \bar{x}$, because $f(1)\, a \cup f(0)\, \bar{a} = 0a \cup 0\bar{a} = 0 \neq f(a) = a$.

Therefore, the following definition is not trivial:

Definition 1. Given an arbitrary Boolean algebra B, by a *Boolean function* defined on B, we mean a function

$$f: B^n = \underbrace{B \times \ldots \times B}_{n \text{ times}} \to B \tag{2}$$

which can be expressed in the disjunctive canonical form

$$f(x_1, \ldots, x_n) = \bigcup_{\alpha_1, \ldots, \alpha_n} f(\alpha_1, \ldots, \alpha_n)\, x_1^{\alpha_1} \ldots x_n^{\alpha_n}, \tag{3}$$

or, equivalently, in the conjunctive canonical form

$$f(x_1, \ldots, x_n) = \prod_{\alpha_1, \ldots, \alpha_n} [f(\bar{\alpha}_1, \ldots, \bar{\alpha}_n) \cup x_1^{\alpha_1} \cup \cdots \cup x_n^{\alpha_n}], \tag{4}$$

where the vector $(\alpha_1, \ldots, \alpha_n)$ runs over all 2^n possible systems of $0 - 1$ values and x^α is defined by

$$x^0 = \bar{x}, \quad x^1 = x. \tag{5}$$

It is easy to prove — and, in fact, it is well-known — that the above definition is equivalent to the following one:

Definition 1′. By a *Boolean function* defined on a Boolean algebra B, we mean a function of the type (2) which is *generated* by a Bolean expression E (Definition I. 3), in the following sense: f is the function abtained from E by interpreting the letters $x_1, \ldots, x_n$ occuring in the indeterminates $x_1^1, x_1^0, \ldots, x_n^1, x_n^0$, as variables in B, the exponents $0, 1$ as the functions defined by (5), the connectives $\cup$, $\cdot$ (or concatenation) and $^-$ as the basic functions (disjunction, conjunction and negation, respectively) of B, and the letters 0 and 1 as the constant functions $0(x) = 0$, $1(x) = 1$.

In other words, the disjunction, conjunction and negation are Boolean functions, and any Boolean function is obtained by *superpositions* of these three basic operations. For the equivalence of Definitions 1 and 1′, as well as for further properties of Boolean functions, see, for instance, S. RUDEANU [2, 4].

Definition 2. By a *Boolean equation* (*inequality*), we mean an equation (inequality) of the form

$$f(x_1, \ldots, x_n) = g(x_1, \ldots, x_n) \tag{6}$$

[respectively, of the form

$$f(x_1, \ldots, x_n) \leqq g(x_1, \ldots, x_n)], \tag{7}$$

where f and g are Boolean functions.

Definition 3. Two (systems of) Boolean equations (inequalities) are called *equivalent* if they have the same solutions.

Lemma 1. *A Boolean equation $f = g$ is equivalent to the Boolean equation $f\bar{g} \cup \bar{f} g = 0$, (and also to $f g \cup \bar{f}\bar{g} = 1$) while a Boolean inequality $f \leqq g$ is equivalent to the Boolean equation $f\bar{g} = 0$ (and also to $\bar{f} \cup g = 1$).*

Proof. Immediate from (I.24) and (I.23).

Lemma 2. *A system of Boolean equations of the form $h_j = 0$ $(j = 1, \ldots, m)$, is equivalent to the Boolean equation $\bigcup_{j=1}^{m} h_j = 0$, while a system of Boolean equations of the form $k_j = 1$ $(j = 1, \ldots, m)$, is equivalent to the Boolean equation $\prod_{j=1}^{m} k_j = 1$.*

Proof. Immediate from an obvious generalization of relations (I.11).

Theorem 1. *Any system of Boolean equations and (or) inequalities is equivalent to a single Boolean equation of the form $h = 0$ (and also to an equation of the form $k = 1$).*

Proof. Immediate from Lemmas 1 and 2.

In view of this theorem, we can restrict ourselves, without loss of generality, to the study of the Boolean equation of the form $h(x_1, \ldots, x_n) = 0$, where h is an arbitrary Boolean function.

Example 1. Consider the following system of Boolean equations and inequations with three unknowns x, y, z:

$$y \cup x z = b \cup c a,$$
$$z \cup x y = c \cup a b,$$
$$a \cup b c \leqq x \cup y z,$$
$$x \leqq a \cup b c.$$

By applying Lemma 1, we obtain the following system of Boolean equations:

$$(y \cup z x)\, \bar{b}(\bar{c} \cup \bar{a}) \cup \bar{y}(\bar{x} \cup \bar{z})\,(b \cup c a) = 0,$$
$$(z \cup x y)\, \bar{c}(\bar{a} \cup \bar{b}) \cup \bar{z}(\bar{x} \cup \bar{y})\,(c \cup a b) = 0,$$
$$(a \cup b c)\, \bar{x}(\bar{y} \cup \bar{z}) = 0,$$
$$x \bar{a}(\bar{b} \cup \bar{c}) = 0,$$

By applying Lemma 2 and the law of distributivity, we obtain a single equation, namely

$$\bar{b}(\bar{c} \cup \bar{a})\, y \cup \bar{b}(\bar{c} \cup \bar{a})\, x\, z \cup b\, \bar{x}\, \bar{y} \cup b\, \bar{y}\, \bar{z} \cup c\, a\, \bar{x}\, \bar{y} \cup$$
$$\cup\, c\, a\, \bar{y}\, \bar{z} \cup \bar{c}(\bar{a} \cup \bar{b})\, z \cup \bar{c}(\bar{a} \cup \bar{b})\, x\, y \cup c\, \bar{x}\, \bar{z} \cup c\, \bar{y}\, \bar{z} \cup a\, b\, \bar{x}\, \bar{z} \cup a\, b\, \bar{y}\, \bar{z} \cup$$
$$\cup\, a\, \bar{x}\, \bar{y} \cup a\, \bar{x}\, \bar{z} \cup b\, c\, \bar{x}\, \bar{y} \cup b\, c\, \bar{x}\, \bar{z} \cup \bar{a}(\bar{b} \cup \bar{c})\, x = 0.$$

After performing all the absorptions, we see that the initial system is equivalent to the following equation:

$$\bar{a}(\bar{b} \cup \bar{c})\, x \cup \bar{b}(\bar{c} \cup \bar{a})\, y \cup \bar{c}(\bar{a} \cup \bar{b})\, z \cup (b \cup c)\, \bar{y}\, \bar{z} \cup (a \cup c)\, \bar{x}\, \bar{z} \cup (a \cup b)\, \bar{x}\, \bar{y} = 0.$$

§ 2. The Boolean Equation in one Unknown: Consistency and Solutions

It is convenient to begin our study with the simple case of a Boolean equation in one unknown. In view of Theorem 1, such an equation may be written in the form

(8) $$f(x) = 0,$$

or, equivalently,

(9) $$a\, x \cup b\, \bar{x} = 0,$$

where

(10) $$a = f(1), \quad b = f(0).$$

The results in this section are due to pioneers of Boolean algebra: P. S. Poretski, E. Schröder, A. N. Whitehead (see Bibliography A).

Theorem 2. *Equation* (9) *is consistent if and only if*

(11) $$a\, b = 0.$$

Proof. Let x be a solution of (9). Then $a\, x = b\, \bar{x} = 0$, by (I.11′), hence $a \leqq \bar{x}$ and $\bar{x} \leqq \bar{b}$, by (I.23″), therefore $a \leqq \bar{b}$, by (I.86), or else $a\, b = 0$, again by (I.23″).

Conversely, suppose that relation (11) holds; then $x = b$ is a solution of (9).

The following remark will be frequently used in the sequel.

Remark 1. If $y = c\, x \cup d\, \bar{x}$, then $\bar{y} = \bar{c}\, x \cup \bar{d}\, \bar{x}$ (for $\bar{y} = (\bar{c} \cup \bar{x}) \cdot$ $\cdot\, (\bar{d} \cup x) = \bar{c}\, \bar{d} \cup \bar{c}\, x \cup \bar{d}\, \bar{x}$, by (I.14), (I.15), (I.8″) and (I.13″); but the term $\bar{c}\, \bar{d}$ may be written $\bar{c}\, \bar{d}(x \cup \bar{x}) = \bar{c}\, \bar{d}\, x \cup \bar{c}\, \bar{d}\, \bar{x}$, by (I.13′) and (I.10″), so that $\bar{y} = \bar{c}\, \bar{d}\, x \cup \bar{c}\, \bar{d}\, \bar{x} \cup \bar{c}\, x \cup \bar{d}\, \bar{x} = \bar{c}\, x \cup \bar{d}\, \bar{x}$, by absorption).

Theorem 3. *Assume that relation* (11) *holds. Then x is a solution of equation* (9) *if and only if*

(12) $$b \leqq x \leqq \bar{a},$$

or, equivalently, if and only if

(13) $$x = \bar{a}\, x \cup b\, \bar{x}.$$

Proof. If x satisfies (9), then $a\,x = 0$ and $b\,\bar{x} = 0$, by (I.11′), hence $x \leqq \bar{a}$ and $b \leqq \bar{\bar{x}} = x$, by (I.23″) and (I.15), thus proving (12).

Further, if x satisfies (12), then $x = \bar{a}\,x$, by (I.17″), and $b\,\bar{x} = 0$, by (I.23″); therefore $x = \bar{a}\,x = \bar{a}\,x \cup b\,\bar{x}$.

Finally, relation (13) implies $a\,x = a\,\bar{a}\,x \cup a\,b\,x = 0$, by (I.8″), (I.13″), (11), (I.9″) and (I.10′); further $\bar{x} = a\,x \cup \bar{b}\,\bar{x}$, by Remark 1, hence we deduce $b\,\bar{x} = 0$, as before. It follows that $a\,x \cup b\,\bar{x} = 0$, and the proof is completed.

Theorem 4. *Assume that relation* (11) *holds. Then the general solution of equation*

(9) $$a\,x \cup b\,\bar{x} = 0$$

is

(14) $$x = \bar{a}\,p \cup b\,\bar{p},$$

where p is an arbitrary parameter varying in B (that is, relation (14) *implies* (9) *whatever the value $p \in B$ may be, and, conversely, every solution of* (9) *may be written in the form* (14), *for a suitably chosen p).*

Proof. It follows from (14) and (11), taking also into account Remark 1, that

$$a\,x \cup b\,\bar{x} = a(\bar{a}\,p \cup b\,\bar{p}) \cup b(a\,p \cup \bar{b}\,\bar{p}) = 0.$$

Conversely, every solution x of equation (9) may be written in the form (14), with $p = x$, as shown by Theorem 3.

§ 3. Boolean Equations in n Unknowns: The Method of Successive Eliminations (Consistency and Solutions)

We consider now the general Boolean equation in n unknowns, that is

(15) $$f(x_1, \ldots, x_n) = 0,$$

where f is an arbitrary Boolean function; as we have shown, equation (15) may be written in the disjunctive canonical form

(16) $$\bigcup_{\alpha_1, \ldots, \alpha_n} c_{\alpha_1 \ldots \alpha_n}\, x_1^{\alpha_n} \ldots, x_n^{\alpha_n} = 0,$$

where

(17) $$c_{\alpha_1 \ldots \alpha_n} = f(\alpha_1, \ldots, \alpha_n) \qquad (\alpha_1, \ldots, \alpha_n = 0 \text{ and } 1).$$

Theorem 5 (A. N. Whitehead [A 1, A 2]). *Equation* (16) *is consistent if and only if*

(18) $$\prod_{\alpha_1, \ldots, \alpha_n} c_{\alpha_1 \ldots \alpha_n} = 0.$$

Proof. For $n = 1$, the above assertion reduces to Theorem 2. Now, assuming that Theorem 5 is valid for $n - 1$, we shall prove it for n.

Indeed, if $(x_1^*, \ldots, x_n^*) \in B^n$ is a solution of equation (16), then

$$(19) \quad \left(\bigcup_{\alpha_1, \ldots, \alpha_{n-1}} c_{\alpha_1 \ldots \alpha_{n-1} 1}\, x_1^{*\alpha_1} \ldots x_{n-1}^{*\alpha_{n-1}}\right) x_n^* \cup \cup \left(\bigcup_{\alpha_1, \ldots, \alpha_{n-1}} c_{\alpha_1, \ldots, \alpha_{n-1} 0}\, x_1^{*\alpha_1} \ldots x_{n-1}^{*\alpha_{n-1}}\right) \overline{x_n^*} = 0,$$

so that the single-unknown equation

$$(20) \quad \left(\bigcup_{\alpha_1, \ldots, \alpha_{n-1}} c_{\alpha_1 \ldots \alpha_{n-1} 1}\, x_1^{*\alpha_1} \ldots x_{n-1}^{*\alpha_{n-1}}\right) x_n \cup \cup \left(\bigcup_{\alpha_1, \ldots, \alpha_{n-1}} c_{\alpha_1, \ldots, \alpha_{n-1} 0}\, x_1^{*\alpha_1} \ldots x_{n-1}^{*\alpha_{n-1}}\right) \bar{x}_n = 0$$

has the solution $x_n = x_n^*$. It follows, according to Theoreme 2, that

$$(21) \quad \left(\bigcup_{\alpha_1, \ldots, \alpha_{n-1}} c_{\alpha_1 \ldots \alpha_{n-1} 1}\, x_1^{*\alpha_1} \ldots x_{n-1}^{*\alpha_{n-1}}\right) \cdot \cdot \left(\bigcup_{\alpha_1, \ldots, \alpha_{n-1}} c_{\alpha_1 \ldots \alpha_{n-1} 0}\, x_1^{*\alpha_1} \ldots x_{n-1}^{*\alpha_{n-1}}\right) = 0.$$

Since $\alpha \neq \beta$ implies $x^\alpha x^\beta = x^0 x^1 = \bar{x}\, x = 0$, relation (21) may also be written in the form

$$(22) \quad \bigcup_{\alpha_1, \ldots, \alpha_{n-1}} c_{\alpha_1 \ldots \alpha_{n-1} 1}\, c_{\alpha_1 \ldots \alpha_{n-1} 0}\, x_1^{*\alpha_1} \ldots x_{n-1}^{*\alpha_{n-1}} = 0,$$

showing that the $(n-1)$-unknown equation

$$(23) \quad \bigcup_{\alpha_1, \ldots, \alpha_{n-1}} c_{\alpha_1 \ldots \alpha_{n-1} 1}\, c_{\alpha_1 \ldots \alpha_{n-1} 0}\, x_1^{\alpha_1} \ldots x_{n-1}^{\alpha_{n-1}} = 0$$

has the solution $x_1 = x_1^*, \ldots, x_{n-1} = x_{n-1}^*$. Therefore, the inductive hypothesis shows that

$$(24) \quad \prod_{\alpha_1, \ldots, \alpha_{n-1}} c_{\alpha_1 \ldots \alpha_{n-1} 1}\, c_{\alpha_1 \ldots \alpha_{n-1} 0} = 0,$$

which is precisely the relation (18), in view of the associative and commutative laws.

Conversely, if relation (18) holds, we write it in the form (24), which shows that equation (23) has a solution $x_1 = x_1^*, \ldots, x_{n-1} = x_{n-1}^*$, by the inductive hypothesis. Hence $x_1^*, \ldots, x_{n-1}^*$ satisfy relation (22), which may also be written in the form (21). The latter relation means, in view of Theorem 2, that equation (20) has a solution $x_n = \bar{x}_n^*$, so that (19) holds, showing that equation (16) has the solution $x_1 = x_1^*, \ldots, x_{n-1}^* = x_{n-1}^*$, $x_n = x_n^*$. The theorem is thus established.

As a matter of fact, the above proof was based on Theorem 2 and on the remark that the n-unknown equation (16) is consistent if and only if there exist $x_1^*, \ldots, x_{n-1}^*$ so that the single-unknown equation (20) is consistent.

The same idea leads to the following procedure for *solving* equation (15).

Method of Successive Eliminations

For the necessity of the recursive procedure, we put

$$(25.1) \qquad f(x_1, \ldots, x_n) = f_1(x_1, \ldots, x_n);$$

then equation (15) becomes

$$(26.1) \qquad f_1(x_1, \ldots, x_n) = 0$$

and may be written in the form

$$(27.1) \qquad f_1(x_1, \ldots, x_{n-1}, 1)\, x_n \cup f_1(x_1, \ldots, x_{n-1}, 0)\, \bar{x}_n = 0;$$

If we set

$$(25.2) \qquad f_1(x_1, \ldots, x_{n-1}, 1)\, f_1(x_1, \ldots, x_{n-1}, 0) = f_2(x_1, \ldots, x_{n-1}),$$

the condition that equation (27.1) has a solution with respect to x_n becomes

$$(26.2) \qquad f_2(x_1, \ldots, x_{n-1}) = 0,$$

etc.

In the step i, we have to solve the equation

$$(26.\ i) \qquad f_i(x_1, \ldots, x_{n-i+1}) = 0;$$

we write it in the form

$$(27.\ i) \qquad f_i(x_1, \ldots, x_{n-i}, 1)\, x_{n-i+1} \cup f_i(x_1, \ldots, x_{n-i}, 0)\, \bar{x}_{n-i+1} = 0,$$

then we set

$$(25.\ i+1) \qquad f_i(x_1, \ldots, x_{n-i}, 1)\, f_i(x_1, \ldots, x_{n-i}, 0) = f_{i+1}(x_1, \ldots, x_{n-i})$$

and we form the equation

$$(26.\ i+1) \qquad f_{i+1}(x_1, \ldots, x_{n-i}) = 0,$$

etc.

In the step n, we obtain the equation

$$(26.\ n) \qquad f_n(x_1) = 0,$$

which may be written in the form

$$(27.\ n) \qquad f_n(1)\, x_1 \cup f_n(0)\, \bar{x}_1 = 0,$$

and we consider the constant

$$(25.\ n+1) \qquad f_n(1)\, f_n(0) = f_{n+1} \in B.$$

If $f_{n+1} \neq 0$, then equation (27. n)—or, equivalently, (26. n)—has no solution in view of Theorem 2. Since equation (26. n) is just the consistency condition $f_{n-1}(x_1, 1)\, f_{n-1}(x_1, 0) = 0$ of equation (27. $n-1$), it follows that the latter — which coincides with (26. $n-1$) — is also inconsistent. We deduce by induction that equation (26.1), which coincides with (15), has no solution.

If

$$(26.\,n+1) \qquad f_{n+1} = 0,$$

then equation (26. n) is consistent. By introducing its solutions into the equation

$$(26.\,n-1) \qquad f_{n-1}(x_1, x_2) = 0,$$

the latter becomes an equation with the single uknown x_2 and is consistent, for the reason explained above. We obtain thus the solutions (x_1, x_2) of equation (26. $n-1$), we introduce them into equation (26. $n-2$), etc. In the last step we introduce the solutions $(x_1, \ldots, x_{n-1})$ of equation (26.2) into equation (26.1), obtaining thus an equation with the single unknown x_n; after solving it, we have at hand the solutions $(x_1, \ldots, x_n)$ of equation (26.1), which coincides with (15).

Thus the method of successive eliminations consists of two stages. The first one, which lasts from the beginning to the finding of f_{n+1}, may also be considered as a way of deciding whether or not the given equation is consistent. This way becomes more efficient than the direct application of Theorem 5, in proportion as the number of variables increases.

If $f_{n+1} = 0$, then the second stage of the method leads to the determination of all the solutions of equation (15). We have intentionally described this stage in a somewhat vague form, because it may be performed in several ways (listing of all the solutions, recurrent inequalities, parametric form) which will be indicated in the sequel. Before doing this, we need a justification of the adequacy of the method. This is given by the following

Theorem 6*. *Equation* (15) *is equivalent to the system* (26) *obtained by the method of successive eliminations.*

Proof. Since equation (15) coincides with (26.1), each solution of (26) satisfies a fortiori (15).

Conversely, the above discussion has shown that every solution of equation (15) satisfies also equations (26.1, 26.2, ..., 26. n).

We shall now describe various ways of performing the second stage of the method of successive eliminations.

The first of them enables us to enumerate all the solutions. It can be described as follows:

Find all the solutions x_1^* of equation (26. n) as in § 2. For each x_1^*, introduce it into equation (26. $n-1$), obtaining thus a new equation $f_{n-1}(x_1^*, x_2) = 0$; find all its solutions x_2^* and form all the vectors (x_1^*, x_2^*), which may be called *partial solutions*. After performing these operations

---

* The method of successive eliminations has been used since the very beginning of the work on Boolean equations. For a formal proof of the adequacy of this method, see S. Rudeanu [A1]. Theorem 6 was explicitly stated in S. Rudeanu [A6].

on all x_1^*, construct the set of all partial solutions of the form (x_1^*, x_2^*); introduce, in turn, each of them into equation (26. n — 2), obtaining thus a set of new equations $f_{n-2}(x_1^*, x_2^*, x_3) = 0$ with the single unknown x_3, etc. In the last step we have at hand the set of all partial solutions of the form $(x_1^*, \ldots, x_{n-1}^*)$. For each partial solution $(x_1^*, \ldots, x_{n-1}^*)$, introduce the values $x_1^*, \ldots, x_{n-1}^*$ into equation (26.1), obtaining thus a new equation $f_1(x_1^*, \ldots, x_{n-1}^*, x_n) = 0$, find all its solutions x_n^* and form all the vectors $(x_1^*, \ldots, x_{n-1}^*, x_n^*)$, which are solutions of the given equation; each solution of (15) can be obtained in this way.

The conclusions of our discussion can be summarized as follows:

Theorem 7 (S. RUDEANU [A 11]). *The above described tree-like construction leads, when relation* (18) *is satisfied, to the list of all the solutions of the given equation* (15).

The following remarks prove to be useful for the practical application of Theorem 6.

Computational remarks. 1) If equation (15) is of the form

$$g(x_1, \ldots, x_{n-1}) \cup h(x_1, \ldots, x_{n-1})\, x_n \cup k(x_1, \ldots, x_{n-1})\, \bar{x}_n = 0, \tag{28}$$

then, by eliminating x_n we obtain the equation

$$g(x_1, \ldots, x_{n-1}) \cup h(x_1, \ldots, x_{n-1})\, k(x_1, \ldots, x_{n-1}) = 0, \tag{29}$$

[for $g = g\, x_n \cup g\, \bar{x}_n$, so that equation (28) may also be written in the form $(g \cup h)\, x_n \cup (g \cup k)\, \bar{x}_n = 0$, hence the result of the elimination of x_n is $(g \cup h)(g \cup k) = 0$, or else $g \cup h\, k = 0$, in view of (I.8')].

2) When we introduce a solution $(x_1^*, \ldots, x_{n-1}^*)$ of equation (29) into (28), we obtain simply the equation

$$h(x_1^*, \ldots, x_{n-1}^*)\, x_n \cup k(x_1^*, \ldots, x_{n-1}^*)\, \bar{x}_n = 0.$$

Example 2. Let us solve the equation

$$(b \cup c)\, \bar{x} \cup (c \cup a)\, \bar{y} \cup (a \cup b)\, \bar{z} \cup \bar{a}(\bar{b} \cup \bar{c})\, y\, z \cup \bar{b}(\bar{c} \cup \bar{a})\, z\, x \cup \bar{c}(\bar{a} \cup \bar{b})\, x\, y = 0 \tag{30}$$

in the Boolean algebra B_{2^8} of all 2^8 Boolean functions of a, b, c (Example I.2.3).

Taking into account the above computational remark 1, we can eliminate z from equation (30) and obtain thus

$$(b \cup c)\, \bar{x} \cup (c \cup a)\, \bar{y} \cup \bar{c}(\bar{a} \cup \bar{b})\, x\, y \cup [\bar{a}(\bar{b} \cup \bar{c})\, y \cup \bar{b}(\bar{c} \cup \bar{a})\, x]\, (a \cup b) = 0,$$

or else

$$(b \cup c)\, \bar{x} \cup (c \cup a)\, \bar{y} \cup \bar{c}(\bar{a} \cup \bar{b})\, x\, y \cup \bar{a}\, b\, \bar{c}\, y \cup a\, \bar{b}\, \bar{c}\, x = 0. \tag{31}$$

Now we eliminate y and obtain

$$(b \cup c)\, \bar{x} \cup a\, \bar{b}\, \bar{c}\, x \cup [\bar{c}(\bar{a} \cup \bar{b})\, x \cup \bar{a}\, b\, \bar{c}]\, (c \cup a) = 0,$$

or else

$$a\, \bar{b}\, \bar{c}\, x \cup (b \cup c)\, \bar{x} = 0. \tag{32}$$

Since $a\, \bar{b}\, \bar{c}(b \cup c) = 0$, equations (32), (31) and (30) are consistent.

Now according to Theorem 3, the solutions of equation (32) are given by

$$b \cup c \leqq x \leqq \bar{a} \cup b \cup c.$$

Using the canonical disjunctive forms of $b \cup c$ and $\bar{a} \cup b \cup c = \bar{a}\,\bar{b}\,\bar{c} \cup b \cup c$ [cf. (I.16′)], we can easily convince ourselves that in B_{2^8} there exist exactly two elements x satisfying the above inequalities, namely $b \cup c$ and $\bar{a} \cup b \cup c$.

We introduce the value $x = b \cup c$ into equation (31), and, taking into account the above computational remark 2, we obtain the new equation

$$\bar{a}\, b\, \bar{c}\, y \cup (c \cup a)\, \bar{y} = 0,$$

whose solutions are $y = a \cup c$ and $y = \bar{b} \cup a \cup c$, by the same arguments as before. We have thus obtained the partial solutions $(x = b \cup c, y = a \cup c)$ and $(x = b \cup c, y = a \cup \bar{b} \cup c)$.

Now, we take the other solution $x = \bar{a} \cup b \cup c$ of equation (32) and introduce it into equation (31), obtaining the equation

$$\bar{a}\, \bar{c}\, y \cup (c \cup a)\, \bar{y} = 0,$$

which has the single solution $y = a \cup c$. We have thus obtained the partial solution $(x = \bar{a} \cup b \cup c, y = a \cup c)$.

Finally, we have to introduce each of the above three partial solutions into equation (30).

For $x = b \cup c$, $y = a \cup c$, we get the equation

$$\bar{a}\, \bar{b}\, c\, z \cup (a \cup b)\, \bar{z} = 0,$$

whose solutions are $z = a \cup b$ and $z = a \cup b \cup \bar{c}$, by the same arguments as before. We have thus obtained the following two solutions of the given equation (30):

(33.1) $\qquad x = b \cup c, \quad y = a \cup c, \quad z = a \cup b,$

(33.2) $\qquad x = b \cup c, \quad y = a \cup c, \quad z = a \cup b \cup \bar{c}.$

For $x = b \cup c$, $y = a \cup \bar{b} \cup c$, we get the equation

$$\bar{a}\, \bar{b}\, z \cup (a \cup b)\, \bar{z} = 0,$$

whose unique solution is $z = a \cup b$; we have thus obtained the solution

(33.3) $\qquad x = b \cup c, \quad y = a \cup \bar{b} \cup c, \quad z = a \cup b$

of the given equation (30).

For $x = \bar{a} \cup b \cup c$, $y = a \cup c$, we get again the equation

$$\bar{a}\, \bar{b}\, z \cup (a \cup b)\, \bar{z} = 0,$$

with the unique solution $z = a \cup b$; hence the last solution of equation (30) is

(33.4) $\qquad x = \bar{a} \cup b \cup c, \quad y = a \cup c, \quad z = a \cup b.$

The method of successive eliminations was given before in a form appropriate for the construction of the list of all the solutions. If we desire to express them in a more concentrate form, we can apply the following

Theorem 8 (V. N. GREBENŠČIKOV [A 1], S. RUDEANU [A 2, A 6]). *Assume that relation* (18) *holds. Then the solutions of equation* (15) *are*

characterized by the following system of recurrent inequalities:

$$(34.1)\qquad f_n(0) \leqq x_1 \leqq \bar{f}_n(1),$$

$$(34.2)\qquad f_{n-1}(x_1, 0) \leqq x_2 \leqq \bar{f}_{n-1}(x_1, 1),$$

$$\cdots\cdots\cdots\cdots\cdots\cdots\cdots\cdots\cdots$$

$$(34.\,i)\qquad f_{n-i+1}(x_1, \ldots, x_{i-1}, 0) \leqq x_i \leqq \bar{f}_{n-i+1}(x_1, \ldots, x_{i-1}, 1),$$

$$\cdots\cdots\cdots\cdots\cdots\cdots\cdots\cdots\cdots$$

$$(34.\,n)\qquad f_1(x_1, \ldots, x_{n-1}, 0) \leqq x_n \leqq \bar{f}_1(x_1, \ldots, x_{n-1}, 1),$$

where $f_1, \ldots, f_n$ are the functions defined in the method of successive eliminations.

Proof. We have shown, in the proof of Theorem 7, that the solutions of equation (15) can be obtained in the following way: solve equation $f_n = 0$ with respect to x_1; introduce its solutions into equation $f_{n-1} = 0$ and solve the resulting equation(s) with respect to x_2, a. s. o. Now, solving the intermediate equations with the aid of Theorem 3, we obtain precisely formulas (34.1, 34.2, ..., 34. n), respectively [of course, in each formula (34. i) the variables $x_1, \ldots, x_{i-1}$ are supposed to satisfy the previous inequalities (34.1, ..., 34. $i - 1$)].

Example 3. In the case of equation (30) studied in Example 2, the equations (26) are (30), (31) and (32), so that

$$f_3(0) = b \cup c,$$

$$f_3(1) = a\,\bar{b}\,\bar{c},$$

$$f_2(x, 0) = (b \cup c)\,\bar{x} \cup (c \cup a) \cup a\,\bar{b}\,\bar{c}\,x = a \cup c \cup b\,\bar{x},$$

$$f_2(x, 1) = (b \cup c)\,\bar{x} \cup \bar{c}(\bar{a} \cup \bar{b})\,x \cup \bar{a}\,b\,\bar{c} \cup a\,\bar{b}\,\bar{c}\,x = \bar{c}(\bar{a} \cup \bar{b})\,x \cup (b \cup c)\,\bar{x},$$

$$f_1(x, y, 0) = (b \cup c)\,\bar{x} \cup (c \cup a)\,\bar{y} \cup (a \cup b) \cup \bar{c}(\bar{a} \cup \bar{b})\,x\,y = a \cup b \cup c\,\bar{x} \cup c\,\bar{y} \cup \bar{c}\,x\,y,$$

$$f_1(x, y, 1) = (b \cup c)\,\bar{x} \cup (c \cup a)\,\bar{y} \cup \bar{a}(\bar{b} \cup \bar{c})\,y \cup \bar{b}(\bar{c} \cup \bar{a})\,x \cup \bar{c}(\bar{a} \cup \bar{b})\,x\,y$$
$$= (b \cup c)\,\bar{x} \cup (c \cup a)\,\bar{y} \cup \bar{a}(\bar{b} \cup \bar{c})\,y \cup \bar{b}(\bar{c} \cup \bar{a})\,x;$$

hence the inequations (34) become (using Remark 1)

$$(35.1)\qquad b \cup c \leqq x \leqq \bar{a} \cup b \cup c,$$

$$(35.2)\qquad a \cup c \cup b\,\bar{x} \leqq y \leqq (a\,b \cup c)\,x \cup \bar{b}\,\bar{c}\,\bar{x},$$

$$(35.3)\qquad a \cup b \cup c(\bar{x} \cup \bar{y}) \cup \bar{c}\,x\,y \leqq z \leqq [(b \cup c\,a)\,x \cup \bar{b}\,\bar{c}\,\bar{x}]\,[(a \cup b\,c)\,y \cup \bar{a}\,\bar{c}\,\bar{y}].$$

Notice that equation (30) is equivalent to the system of recurrent inequalities (35), whatever the Boolean algebra B may be (not only for the Boolean algebra B_{2^8} considered in Example 2).

The method of successive eliminations enables us to express also the solutions in a *parametric form*; in other words, it enables us to find the *general solution*, in the sense of the following

Definition 4. A *particular solution* of the equation

$$(15)\qquad f(x_1, \ldots, x_n) = 0$$

is a vector $(x_1^*, \ldots, x_n^*)$ of elements $x_i^* \in B$ $(i = 1, \ldots, n)$ satisfying (15). The *general solution* of the same equation is a family $\{\varphi_i(p_1, \ldots, p_n)\}_{i=1,\ldots,n}$ of Boolean functions* satisfying the following two conditions:

(α) for every n values $p_1^*, \ldots, p_n^* \in B$ given to the *parameters* $p_1, \ldots, p_n$, the elements

$$x_i^* = \varphi_i(p_1^*, \ldots, p_n^*) \qquad (i = 1, \ldots, n) \tag{36}$$

form a particular solution of equation (15);

(β) conversely, if $(x_1^*, \ldots, x_n^*) \in B^n$ is particular solution of equation (15), then there exist n elements $p_1^*, \ldots, p_n^* \in B$ such that relations (36) hold.

We say also that formulas (36) *define* the general solution.

As it will be shown in § 4, there may exist several general solutions for the same equations, that is there may be several families of Boolean functions $\{\varphi_i\}$, $\{\varphi_i'\}$, $\{\varphi_i''\}, \ldots$ satisfying the above properties (α, β).

Lemma 3. *The double inequality*

$$c \leqq x \leqq d \tag{37}$$

is equivalent to relation

$$x = d\,p \cup c\,\bar{p}, \tag{38}$$

where p is an arbitrary parameter.

Proof. The given inequality is equivalent to equation $c\,\bar{x} \cup \bar{d}\,x = 0$, by (I.23″) and (I.11′), therefore it is equivalent to relation (38), by Theorem 4.

Taking into account Lemma 3, we see that Theorem 8 implies the following

Corollary 1. Assume that relation (18) holds. Then the solutions of equation (15) are given by the following recurrent equalities:

$$x_1 = \bar{f}_n(1)\,p_1 \cup f_n(0)\,\bar{p}_1, \tag{39.1}$$

$$x_2 = \bar{f}_{n-1}(x_1, 1)\,p_2 \cup f_{n-1}(x_1, 0)\,\bar{p}_2, \tag{39.2}$$

$$\cdots\cdots\cdots\cdots\cdots\cdots$$

$$x_i = \bar{f}_{n-i+1}(x_1, \ldots, x_{i-1}, 1)\,p_i \cup f_{n-i+1}(x_1, \ldots, x_{i-1}, 0)\,\bar{p}_i, \tag{39.i}$$

$$\cdots\cdots\cdots\cdots\cdots\cdots$$

$$x_n = \bar{f}_1(x_1, \ldots, x_{n-1}, 1)\,p_n \cup f_1(x_1, \ldots, x_{n-1}, 0)\,\bar{p}_n, \tag{39.n}$$

where $p_1, \ldots, p_n$ are arbitrary parameters.

Hence we deduce the following

Theorem 9. *Assume that relation* (18) *holds. Then, after performing successively the substitutions indicated by relations* (39), *we obtain the*

---

* It may happen that some (or all) of the functions φ_i do not depend essentially on all the variables $p_1, \ldots, p_n$. However, any function of $m \leqq n$ variables $p_{i_1}, \ldots, p_{i_m}$, may be considered as a function of n variables $p_1, \ldots, p_n$.

general solution of equation (15) *in the following* "triangular" *form*:

$$x_1 = \varphi_1(p_1), \tag{40.1}$$

$$x_2 = \varphi_2(p_1, p_2), \tag{40.2}$$

$$\dots\dots\dots\dots\dots$$

$$x_i = \varphi_i(p_1, \dots, p_i), \tag{40. i}$$

$$\dots\dots\dots\dots\dots$$

$$x_n = \varphi_n(p_1, \dots, p_n). \tag{40. n}$$

Example 4. Let us resume equation (30), studied in Examples 2 and 3. According to Lemma 3, the recurrent inequalities (35) may be translated in the following parametric form:

$$x = (\bar{a} \cup b \cup c)\, p \cup (b \cup c)\, \bar{p}, \tag{41.1}$$

$$y = [(a\, b \cup c)\, x \cup \bar{b}\, \bar{c}\, \bar{x}]\, q \cup (a \cup c \cup b\, \bar{x})\, \bar{q}, \tag{41.2}$$

$$z = [(b \cup c\, a)\, x \cup \bar{b}\, \bar{c}\, \bar{x}]\, [(a \cup b\, c)\, y \cup \bar{a}\, \bar{c}\, \bar{y}]\, r \cup [a \cup b \cup c (\bar{x} \cup \bar{y}) \cup \bar{c}\, x\, y]\, \bar{r}. \tag{41.3}$$

Now, we introduce the expression (41.1) of x into relations (41.2), (41.3) and the expression (41.2) of y into (41.3), as indicated in Theorem 9. We obtain the following form of the general solution:

$$x = b \cup c \cup \bar{a}\, p, \tag{42.1}$$

$$y = a \cup c \cup \bar{b}\, \bar{p}\, q, \tag{42.2}$$

$$z = a \cup b \cup \bar{c}\, \bar{p}\, \bar{q}\, r. \tag{42.3}$$

Here, as in Example 3, the results are valid in any Boolean algebra.

Notice that if the Boolean algebra B has $|B|$ elements ($|B|$ being finite or not) and the general solution of equation (15) depends upon n parameters, then we have to give $|B|^n$ systems of values to the parameters in order to obtain all the particular solutions. However, it may happen that the number of solutions be far smaller than $|B|^n$, so that this procedure for finding all the particular solutions may be very redundant.

M. Carvallo [*A* 1, *A* 2] suggests that this redundancy be reduced by determining a form of the general solution with less than n parameters. Another possibility is that of finding an "irredundant form" of the general solution. This means that we wish to impose some constraints upon the parameters $p_1, \dots, p_n$, in such a way that there exists a one-to-one correspondence between the vectors $(p_1, \dots, p_n)$ satisfying these constraints and the particular solutions of equation (15). More precisely:

Definition 5. An *irredundant form of the general solution* of equation (15) is a family $\{\varphi_i(p_1, \dots, p_n)\}_{i=1,\dots,n}$ of Boolean functions, together with a system of constraints C imposed upon the parameters $p_1, \dots, p_n$, so that:

(γ) if $(p_1^*, \ldots, p_n^*) \in B^n$ is a vector satisfying the constraints C, then the vector $(x_1^*, \ldots, x_n^*)$ defined by

$$(36) \qquad x_i^* = \varphi_i(p_1^*, \ldots, p_n^*) \qquad (i = 1, \ldots, n)$$

is a particular solution of equation (15);

(δ) conversely, each particular solution $(x_1^*, \ldots, x_n^*)$ of equation (15) can be written in the form (36), where $(p_1^*, \ldots, p_n^*) \in B^n$ is a suitably chosen vector satisfying the contraints C;

( ) if the vectors $(p_1^*, \ldots, p_n^*)$ and $(p_1^{**}, \ldots, p_n^{**})$ satisfy the constraints C and $p_i^* \neq p_i^{**}$ for at least one i, then the corresponding solutions, $(x_1^*, \ldots, x_n^*)$ and $(x_1^{**}, \ldots, x_n^{**})$, are different, i.e. $x_j^* \neq x_j^{**}$ for at least one j.

We finish this section by giving an irredundant form of the general solution.

Lemma 4. *If relation* $a\,b = 0$ *holds, then an irredundant form of the general solution of equation*

$$(9) \qquad a\,x \cup b\,\bar{x} = 0$$

is given by formula

$$(43) \qquad x = b \cup p,$$

together with the constraint

$$(44) \qquad p \leqq \bar{a}\,\bar{b}.$$

Proof. Notice first that $a\,b = 0$ means $b \leqq \bar{a}$ or $b \cup \bar{a} = \bar{a}$. Now, relations (43) and (44) imply $b \leqq b \cup p = x \leqq b \cup \bar{a}\,\bar{b} = b \cup \bar{a} = \bar{a}$, hence they imply that x is a solution of equation (9), by Theorem 3.

Conversely, if x is a solution of equation (9), then the same Theorem 3, shows that $x = b \cup x = b \cup x\,\bar{a} = b \cup \bar{b}\,x\,\bar{a}$, so that, choosing $p = \bar{a}\,\bar{b}\,x$, relations (43) and (44) are satisfied.

Finally, let p', p'' be two elements of B satisfying (44); we have to prove that the equality

$$(45) \qquad b \cup p' = b \cup p''$$

implies $p' = p''$. But relation (45) may be written in the form $(b \cup p') \cdot \cdot \bar{b}\,\bar{p}'' \cup \bar{b}\,\bar{p}'(b \cup p'') = 0$, or else

$$(46) \qquad \bar{b}\,(p'\,\bar{p}'' \cup \bar{p}'\,p'') = 0.$$

On the other hand, relations $p' \leqq \bar{a}\,\bar{b}$ and $p'' \leqq \bar{a}\,\bar{b}$ imply $p' \leqq \bar{b}$ and $p'' \leqq \bar{b}$, respectively, hence a fortiori $p'\,\bar{p}'' \leqq \bar{b}$ and $\bar{p}'\,p'' \leqq \bar{b}$, therefore $p'\,\bar{p}'' \cup \bar{p}'\,p'' \leqq \bar{b}$, or else $b\,(p'\,\bar{p}'' \cup \bar{p}'\,p'') = 0$. By comparing with (46), we deduce $p'\,\bar{p}'' \cup \bar{p}'\,p'' = 0$, or else $p' = p''$.

Theorem 10 (S. Rudeanu [A 7, A 11]). *Assume that relation* (18) *holds and let* $f_1(x_1, \ldots, x_n)$, $f_2(x_1, \ldots, x_{n-1}), \ldots, f_n(x_1)$ *be the functions* *defined in the method of successive eliminations. Then the formulas*

(47.1) $$x_1 = f_n(0) \cup p_1,$$

(47.2) $$x_2 = f_{n-1}(x_1, 0) \cup p_2,$$

. . . .

(47. i) $$x_i = f_{n-i+1}(x_1, \ldots, x_{i-1}, 0) \cup p_i,$$

. . . .

(47. n) $$x_n = f_1(x_1, \ldots, x_{n-1}, 0) \cup p_n,$$

ogether with the constraints

(48.1) $$p_1 \leqq \bar{f}_n(0)\, \bar{f}_n(1),$$

(48.2) $$p_2 \leqq \bar{f}_{n-1}(x_1, 0)\, \bar{f}_{n-1}(x_1, 1),$$

. . . .

(48. i) $$p_i \leqq \bar{f}_{n-i+1}(x_1, \ldots, x_{i-1}, 0)\, \bar{f}_{n-i+1}(x_1, \ldots, x_{i-1}, 1),$$

. . . .

(48. n) $$p_n \leqq \bar{f}_1(x_1, \ldots, x_{n-1}, 0)\, \bar{f}_1(x_1, \ldots, x_{n-1}, 1)$$

define an irredundant form of the general solution of equation (15).

Proof. Immediate from Theorem 6 and Lemma 4.

Computational remark. It seems to be preferable to apply successively Lemma 4, instead of a direct application of the above theorem.

Example 5. Let us resume equation (30), which was studied in Examples 2, 3 and 4. We have shown that, after eliminating z and y, we obtain the equations (31) and (32), respectively.

According to Lemma 4, the irredundant solution of equation (32) is

(49.1) $$x = b \cup c \cup p,$$

where

(50.1) $$p \leqq \bar{a}\, \bar{b}\, \bar{c}.$$

Now, equation (31) becomes

$$\bar{c}(\bar{a} \cup \bar{b})\,(b \cup p)\, y \cup (c \cup a)\, \bar{y} = 0$$

and it has the irredundant solution

(49.2) $$y = a \cup c \cup q,$$

where

(50.2) $$q \leqq \bar{a}\, \bar{b}\, \bar{c}\, \bar{p}.$$

Finally, we obtain from (30) the equation

$$[\bar{a}(\bar{b} \cup \bar{c})\,(c \cup q) \cup \bar{b}(\bar{a} \cup \bar{c})\,(c \cup p)]\, z \cup (a \cup b)\, \bar{z} = 0,$$

whose irredundant general solution is

(49.3) $$z = a \cup b \cup r,$$

where

(50.3) $$r \leqq \bar{a}\, \bar{b}\, \bar{c}\, \bar{p}\, \bar{q}.$$

Formulas (49) together, with the constraints (50), define the irredundant general solution of equation (30) in an arbitrary Boolean algebra B.

In the particular case when B is the Boolean algebra B_{2^8} of all 2^8 Boolean functions of a, b, c, there exist exactly two elements p satisfying (50.1), namely 0 and $\bar{a}\,\bar{b}\,\bar{c}$. This remark enables us to construct immediately all the particular solutions and the reader may easily convince himself that we obtain precisely the four solutions (33) which were already found in Example 2. Notice that a vector $(p_1, p_2, p_3) \in B_{2^8}^3$ not subject to constraints, has 2^{24} possible values.

§ 4. The Löwenheim Form of the General Solution

When a particular solution of equation

$$f(x_1, \ldots, x_n) = 0 \tag{15}$$

is known, then the general solution can be immediately written, in view of the following

Theorem 11 (L. Löwenheim [A 1, A 2]). *If* $(\xi_1, \ldots, \xi_n) \in B^n$ *is a particular solution of equation* (15), *then formulas*

$$x_i = \xi_i f(p_1, \ldots, p_n) \cup p_i \bar{f}(p_1, \ldots, p_n) \qquad (i = 1, \ldots, n) \tag{51}$$

define the general solution. The solution (51) *is reproductive, i.e. for every particular solution* $(x_1^*, \ldots, x_n^*)$ *of equation* (15), *we have*

$$x_i^* = \xi_i f(x_1^*, \ldots, x_n^*) \cup x_i^* \bar{f}(x_1^*, \ldots, x_n^*) \qquad (i = 1, \ldots, n). \tag{52}$$

The proof will use the disjunctive canonical form of the function f and Lemma 5 below, which has an intrinsic interest.

Lemma 5. *The following identities hold:*

$$\Big(\bigcup_{\gamma_1, \ldots, \gamma_n} a_{\gamma_1 \ldots \gamma_n} x_1^{\gamma_1} \ldots x_n^{\gamma_n}\Big) \cup \Big(\bigcup_{\gamma_1, \ldots, \gamma_n} b_{\gamma_1 \ldots \gamma_n} x_1^{\gamma_1} \ldots x_n^{\gamma_n}\Big) = \bigcup_{\gamma_1, \ldots, \gamma_n} (a_{\gamma_1 \ldots \gamma_n} \cup b_{\gamma_1 \ldots \gamma_n}) x_1^{\gamma_1} \ldots x_n^{\gamma_n}, \tag{53}$$

$$\Big(\bigcup_{\gamma_1, \ldots, \gamma_n} a_{\gamma_1 \ldots \gamma_n} x_1^{\gamma_1} \ldots x_n^{\gamma_n}\Big) \Big(\bigcup_{\gamma_1, \ldots, \gamma_n} b_{\gamma_1 \ldots \gamma_n} x_1^{\gamma_1} \ldots x_n^{\gamma_n}\Big) = \bigcup_{\gamma_1, \ldots, \gamma_n} a_{\gamma_1 \ldots \gamma_n} b_{\gamma_1 \ldots \gamma_n} x_1^{\gamma_1} \ldots x_n^{\gamma}, \tag{54}$$

$$\overline{\bigcup_{\gamma_1, \ldots, \gamma_n} a_{\gamma_1 \ldots \gamma_n} x_1^{\gamma_1} \ldots x_n^{\gamma_n}} = \bigcup_{\gamma_1, \ldots, \gamma_n} \bar{a}_{\gamma_1 \ldots \gamma_n} x_1^{\gamma_1} \ldots x_n^{\gamma_n}. \tag{55}$$

Proof. Relation (53) is obvious. Identity (54) follows from the remark that $\gamma' \neq \gamma''$ implies $x^{\gamma'} x^{\gamma''} = x^0 x^1 = \bar{x}\, x = 0$, while $x^\gamma x^\gamma = x^\gamma$. Finally (55) follows by the De Morgan laws and the same remark.

Proof of Theorem 11. Let

$$f(x_1, \ldots, x_n) = \bigcup_{\alpha_1, \ldots, \alpha_n} c_{\alpha_1 \ldots \alpha_n} x_1^{\alpha_1} \ldots x_n^{\alpha_n} \tag{56}$$

be the disjunctive canonical form of the function f and let x_i have the expression (51). Then

$$\bar{x}_i = \bar{\xi}_i f(p_1, \ldots, p_n) \cup \bar{p}_i \bar{f}(p_1, \ldots, p_n) \qquad (i = 1, \ldots, n) \tag{57}$$

by (55). Relations (51) and (57) may be written in the concentrate form

$$x_i^{\alpha_1} = \xi_i^{\alpha_1} f(p_1, \ldots, p_n) \cup p_i^{\alpha_1} \bar{f}(p_1, \ldots, p_n) \qquad (i = 1, \ldots, n) \tag{58}$$

where $\alpha_i = 0$ or 1. By multiplying the equalities (58), we obtain:

(59) $x_1^{\alpha_1} \ldots x_n^{\alpha_n} = \xi_1^{\alpha_1} \ldots \xi_n^{\alpha_n} f(p_1, \ldots, p_n) \cup p_1^{\alpha_1} \ldots p_n^{\alpha_n} \bar{f}(p_1, \ldots, p_n)$.

We deduce from (51), via (56) and (59), that

$$\begin{aligned} f(x_1, \ldots, x_n) &= \bigcup_{\alpha_1, \ldots, \alpha_n} [c_{\alpha_1 \ldots \alpha_n} \xi_1^{\alpha_1} \ldots \xi_n^{\alpha_n} f(p_1, \ldots, p_n) \cup \\ &\quad \cup c_{\alpha_1 \ldots \alpha_n} p_1^{\alpha_1} \ldots p_n^{\alpha_n} \bar{f}(p_1, \ldots, p_n)] \\ &= \Big(\bigcup_{\alpha_1, \ldots, \alpha_n} c_{\alpha_1 \ldots \alpha_n} \xi_1^{\alpha_1} \ldots \xi_n^{\alpha_n} \Big) f(p_1, \ldots, p_n) \cup \\ &\quad \cup \Big(\bigcup_{\alpha_1, \ldots, \alpha_n} c_{\alpha_1 \ldots \alpha_n} p_1^{\alpha_1} \ldots p_n^{\alpha_n} \Big) \bar{f}(p_1, \ldots, p_n) \\ &= f(\xi_1, \ldots, \xi_n) f(p_1, \ldots, p_n) \cup f(p_1, \ldots, p_n) \bar{f}(p_1, \ldots, p_n) = 0. \end{aligned}$$

Conversely, if $(x_1^*, \ldots, x_n^*)$ is a particular solution of equation (15), then $f(x_1^*, \ldots, x_n^*) = 0$, $\bar{f}(x_1^*, \ldots, x_n^*) = 1$, so that relations (52) are satisfied; this shows that x_i^* are of the form (51), with $p_i = x_i^*$ $(i = 1, \ldots, n)$. The proof is complete.

S. Rudeanu [*A* 4, *A* 6] has shown that Löwenheim's result can be used in order to determinate *all* possible forms of the general solution; this may be useful for the finding of a simple form of the general solution.

Example 6. Let a_{ij} be n^2 elements of the Boolean algebra B and consider the equation

$$\bigcup_{h=1}^{n} \bigcup_{j=1}^{n} a_{hj} x_h x_j = 0,$$

which has the particular solution $\xi_1 = \cdots = \xi_n = 0$. Hence its general solution (51) becomes

$$x_i = p_i \bar{f}(p_1, \ldots, p_n) \qquad (i = 1, \ldots, n),$$

that is

$$x_i = p_i \prod_{h=1}^{n} \prod_{j=1}^{n} (\bar{a}_{hj} \cup \bar{x}_h \cup \bar{x}_j) \qquad (i = 1, \ldots, n)$$

where $p_1, \ldots, p_n$ are arbitrary parameters in B.

B. Equations in the Two-Element Boolean Algebra

§ 5. The Method of Bifurcations

Many applications require the solving of equations with the unknowns in the two-element Boolean algebra B_2, or, more generally, the solving of Boolean equations whose unknowns are Boolean functions (in the sense of Definition I.2) of some given variables. Notice also that every functional equation in B_2 reduces to a system of ordinary equations, whose unknowns are the coefficient of the disjunctive canonical form(s) of the unknown function(s) (See, for instance, S. Rudeanu [A 3], M. Carvallo [A 1], etc.)

Equations of the above described types may be solved by special methods, using the peculiarities of the (very simple!) two-element

Boolean algebra B_2 (for instance, tabular methods and other procedures suitable for electronic digital computers have been widely employed. See, for example, H. ANGSTL [A 1], BALAKRAN [A 1], O. BEAUFAYS [A 1], E. CARRUCCIO [A 1], M. CARVALLO [A 1, A 2], M. JA. EĬNGORIN [A 1, A 2, A 4], M. GOTŌ [A 1], JU. I. GRIGORIAN [A 1], M. ITOH [A 4], J. KLÍR [A 1], V. LALAN [A 1], R. S. LEDLEY [A 1—A 3], K. K. MAITRA [A 1], M. NADLER and B. ELSPAS [A 1], S. NORDIO [A 1], J. POSTLEY [A 1], N. ROUCHE [A 1], E. L. SCHUBERT [A 1], W. SEMON [A 1], A. SVOBODA [A 1] (or A. SVOBODA and K. ČULIK [A 1]), Y. TOHMA [A 1], R. M. TOMS [A 1], A. ŽELEZNIKAR [A 1, A 2], H. ZEMANEK [A 1].

A thorough presentation of the above works would be very useful, but, unfortunately, would augment too much the dimensions of this book.

Thus, we shall describe only two methods for solving equations in B_2.

The first of them is very suitable for hand computation, but also workable with an electronic digital computer. It may be called:

The Method of Bifurcations. Let $x_1, \ldots, x_n$ be the unknowns of the given (system of) equation(s). Choose an index i_1 $(1 \leqq i_1 \leqq n)$, put $x_{i_1} = 1$ (or $x_{i_1} = 0$) and solve the resulting system with $n - 1$ unknowns. Then choose another index i_2 $(1 \leqq i_2 \leqq n)$, put $x_{i_1} = 0$, $x_{i_2} = 1$ (respectively, $x_{i_1} = 1$, $x_{i_2} = 0$) and solve the resulting system with $n - 2$ unknowns, a. s. o. Thus the discussion can be divided into several pairwise disjoint cases, including all the possibilities. If the (system of) equation(s) obtained in one of these cases, say $x_{i_1} = \cdots = x_{i_{k-1}} = 0$, $x_{i_k} = 1$ (respectively, $x_{i_1} = \cdots = x_{i_{k-1}} = 1$, $x_{i_k} = 0$), is too complicated, then divide it into several subcases, following the same procedure described above. If the complementary case $x_{i_1} = \cdots = x_{i_{k-1}} = x_{i_k} = 0$ (respectively, $x_{i_1} = \cdots = x_{i_{k-1}} = x_{i_k} = 1$) can be solved without introducing a new condition $x_{i_{k+1}} = 1$ (respectively, $x_{i_{k+1}} = 0$), then this case will be the last one to be discussed (this situation occurs surely for $k = n$, if not before).

Example 7*. Let us solve the system

$$x_a = \bar{x}_b \bar{x}_h \cup \bar{x}_b \bar{x}_c \bar{x}_d \cup \bar{x}_b \bar{x}_d \bar{x}_e \bar{x}_f \cup \bar{x}_d \bar{x}_f \bar{x}_g \cup \bar{x}_d \bar{x}_h \bar{x}_i, \tag{60.1}$$

$$x_b = \bar{x}_a \bar{x}_h \cup \bar{x}_c \bar{x}_d \cup \bar{x}_d \bar{x}_e \bar{x}_f \cup \bar{x}_d \bar{x}_f \bar{x}_g \cup \bar{x}_d \bar{x}_h \bar{x}_i, \tag{60.2}$$

$$x_c = \bar{x}_a \bar{x}_b \bar{x}_h \cup \bar{x}_b \bar{x}_d \cup \bar{x}_d \bar{x}_f \bar{x}_g \cup \bar{x}_d \bar{x}_h \bar{x}_i, \tag{60.3}$$

$$x_d = \bar{x}_a \bar{x}_b \bar{x}_h \cup \bar{x}_b \bar{x}_c \cup \bar{x}_b \bar{x}_e \bar{x}_f \cup \bar{x}_f \bar{x}_g \cup \bar{x}_h \bar{x}_i, \tag{60.4}$$

$$x_e = \bar{x}_a \bar{x}_b \bar{x}_h \cup \bar{x}_b \bar{x}_c \bar{x}_d \cup \bar{x}_b \bar{x}_d \bar{x}_f \cup \bar{x}_d \bar{x}_f \bar{x}_g \cup \bar{x}_d \bar{x}_h \bar{x}_i, \tag{60.5}$$

$$x_f = \bar{x}_a \bar{x}_b \bar{x}_h \cup \bar{x}_b \bar{x}_c \bar{x}_d \cup \bar{x}_b \bar{x}_d \bar{x}_e \cup \bar{x}_d \bar{x}_g \cup \bar{x}_d \bar{x}_h \bar{x}_i, \tag{60.6}$$

$$x_g = \bar{x}_a \bar{x}_b \bar{x}_h \cup \bar{x}_b \bar{x}_c \bar{x}_d \cup \bar{x}_d \bar{x}_f \cup \bar{x}_d \bar{x}_h \bar{x}_i, \tag{60.7}$$

$$x_h = \bar{x}_a \bar{x}_b \cup \bar{x}_b \bar{x}_c \bar{x}_d \cup \bar{x}_b \bar{x}_d \bar{x}_e \bar{x}_f \cup \bar{x}_d \bar{x}_f \bar{x}_g \cup \bar{x}_d \bar{x}_i, \tag{60.8}$$

$$x_i = \bar{x}_a \bar{x}_b \bar{x}_h \cup \bar{x}_b \bar{x}_c \bar{x}_d \cup \bar{x}_b \bar{x}_d \bar{x}_e \bar{x}_f \cup \bar{x}_d \bar{x}_f \bar{x}_g \cup \bar{x}_d \bar{x}_h. \tag{60.9}$$

* See also Example X.10.

We begin the discussion with the most frequent unknown, i.e. x_d.

First case: $x_d = 1$. The equations (60.1), (60.2), (60.4) and (60.8) reduce to

$$(60'.1) \qquad x_a = \bar{x}_b \bar{x}_h,$$

$$(60'.2) \qquad x_b = \bar{x}_a \bar{x}_h,$$

$$(60'.4) \qquad 1 = \bar{x}_a \bar{x}_b \bar{x}_h \cup \bar{x}_b \bar{x}_c \cup \bar{x}_b \bar{x}_e \bar{x}_f \cup \bar{x}_f \bar{x}_g \cup \bar{x}_h \bar{x}_i,$$

$$(60'.8) \qquad x_h = \bar{x}_a \bar{x}_b,$$

respectively, while the other equations (60) give $x_c = x_e = x_f = x_g = x_i = \bar{x}_a \bar{x}_b \bar{x}_h$.

But equation (60'.1) implies $\bar{x}_a \bar{x}_b \bar{x}_h = 0$, so that $x_c = x_e = x_f = x_g = x_i = 0$ and these values satisfy equation (60'.4), whatever x_a, x_b, x_h may be. The system (60') reduces thus to the system consisting of equations (60'.1), (60'.2) and (60'.8). The last one has obviously three solutions: $(x_a = 0, x_b = 0, x_h = 1)$, $(x_a = 0, x_b = 1, x_h = 0)$ and $(x_a = 1, x_b = 0, x_h = 0)$.

We have thus obtained the following solutions of the given system (60):

$$(61.1) \quad x_a = x_b = x_c = 0, \quad x_d = 1, \quad x_e = x_f = x_g = 0, \quad x_h = 1, \quad x_i = 0,$$

$$(61.2) \quad x_a = 0, \quad x_b = 1, \quad x_c = 0, \quad x_d = 1, \quad x_e = x_f = x_g = x_h = x_i = 0,$$

$$(61.3) \quad x_a = 1, \quad x_b = x_c = 0, \quad x_d = 1, \quad x_e = x_f = x_g = x_h = x_i = 0.$$

Now, the most frequent unknown, after x_d, is x_b.

Second case: $x_d = 0$, $x_b = 1$. Equations (60.1), (60.3), (60.5) and (60.4) give

$$x_a = x_c = x_e = \bar{x}_f \bar{x}_g \cup \bar{x}_h \bar{x}_i = 0,$$

hence equations (60.6) and (60.7) reduce both to $x_f = \bar{x}_g$, while equations (60.8) and (60.9) become $x_h = \bar{x}_i$; these values transform equation (60.2) into an identity.

Giving to x_g and x_i all possible values, we obtain the following solutions of the system (60):

$$(61.4) \quad x_a = 0, \quad x_b = 1, \quad x_c = x_d = x_e = 0, \quad x_f = 1, \quad x_g = 0, \quad x_h = 1, \quad x_i = 0,$$

$$(61.5) \quad x_a = 0, \quad x_b = 1, \quad x_c = x_d = x_e = 0, \quad x_f = 1, \quad x_g = x_h = 0, \quad x_i = 1,$$

$$(61.6) \quad x_a = 0, \quad x_b = 1, \quad x_c = x_d = x_e = x_f = 0, \quad x_g = x_h = 1, \quad x_i = 0,$$

$$(61.7) \quad x_a = 0, \quad x_b = 1, \quad x_c = x_d = x_e = x_f = 0, \quad x_g = 1, \quad x_h = 0, \quad x_i = 1.$$

Third (and last) *case:* $x_d = 0$, $x_b = 0$. Equation (60.3) reduces to $x_c = 1$, hence both equations (60.2) and (60.4) reduce to $\bar{x}_a \bar{x}_h \cup \bar{x}_e \bar{x}_f \cup \bar{x}_f \bar{x}_g \cup \bar{x}_h \bar{x}_i = 0$. It follows that equations (60.1), (60.5), (60.7) and (60.9) become $x_a = \bar{x}_h$, $x_e = \bar{x}_f$, $x_g = \bar{x}_f$ and $x_i = \bar{x}_h$, respectively. These values transform equations (60.2), (60.4), (60.6) and (60.8) into identities.

Giving to x_f and x_h all possible values, we obtain the last solutions of the system (60):

$$(61.8) \quad x_a = 1, \quad x_b = 0, \quad x_c = 1, \quad x_d = 0, \quad x_e = 1, \quad x_f = 0, \quad x_g = 1, \quad x_h = 0, \quad x_i = 1,$$

$$(61.9) \quad x_a = x_b = 0, \quad x_c = 1, \quad x_d = 0, \quad x_e = 1, \quad x_f = 0, \quad x_g = x_h = 1, \quad x_i = 0,$$

$$(61.10) \quad x_a = 1, \quad x_b = 0, \quad x_c = 1, \quad x_d = x_e = 0, \quad x_f = 1, \quad x_g = x_h = 0, \quad x_i = 1,$$

$$(61.11) \quad x_a = 0, \quad x_b = 0, \quad x_c = 1, \quad x_d = x_e = 0, \quad x_f = 1, \quad x_g = 0, \quad x_h = 1, x_i = 0.$$

Example 8*. Let us solve the system

(62.1) $$x_a = \bar{x}_b \bar{x}_h,$$

(62.2) $$x_b = \bar{x}_a \bar{x}_c \bar{x}_d \bar{x}_e \bar{x}_h,$$

(62.3) $$x_c = \bar{x}_b \bar{x}_d,$$

(62.4) $$x_d = \bar{x}_b \bar{x}_c \bar{x}_e \bar{x}_f \bar{x}_g \bar{x}_h \bar{x}_i,$$

(62.5) $$x_e = \bar{x}_b \bar{x}_d \bar{x}_f,$$

(62.6) $$x_f = \bar{x}_d \bar{x}_e \bar{x}_g,$$

(62.7) $$x_g = \bar{x}_d \bar{x}_f,$$

(62.8) $$x_h = \bar{x}_a \bar{x}_b \bar{x}_d \bar{x}_i,$$

(62.9) $$x_i = \bar{x}_d \bar{x}_h.$$

We begin the discussion with the unknown corresponding to the "longest" equation.

First case: $x_d = 1$. Equation (62.4) gives $x_b = x_c = x_e = x_f = x_g = x_h = x_i = 0$, hence $x_a = 1$, by (62.1); these values satisfy all the equations (62). We have obtained the solution

(63.1) $$x_a = 1, \quad x_b = x_c = 0, \quad x_d = 1, \quad x_e = x_f = x_g = x_h = x_i = 0.$$

Now, the longest equation, after (62.4), is (62.2).

Second case: $x_d = 0, x_b = 1$. Equation (62.2) gives $x_a = x_c = x_d = x_e = x_h = 0$. These values satisfy equations (62.1), (62.3), (62.4), (62.5), (62.8), reduce equations (62.6) and (62.7) to $x_g = \bar{x}_f$, and transform equation (62.9) into $x_i = 1$.

Giving to x_f the values 0, 1, we obtain the solutions

(63.2) $$x_a = 0, \quad x_b = 1, \quad x_c = x_d = x_e = x_f = 0, \quad x_g = 1, \quad x_h = 0, \quad x_i = 1,$$

(63.3) $$x_a = 0, \quad x_b = 1, \quad x_c = x_d = x_e = 0, \quad x_f = 1, \quad x_g = x_h = 0, \quad x_i = 1.$$

Third case: $x_d = x_b = 0, x_h = 1$. Equation (62.8) gives $x_a = x_b = x_d = x_i = 0$. These values satisfy equations (62.1), (62.2), (62.4), (62.9), reduce equation (62.3) to $x_c = 1$ and transform equations (62.5) and (62.7) into $x_e = \bar{x}_f$ and $x_g = \bar{x}_f$, respectively, so that (62.6) is satisfied.

The variable x_f remains arbitrary; we get the solutions

(63.4) $$x_a = x_b = 0, \quad x_c = 1, \quad x_d = 0, \quad x_e = 1, \quad x_f = 0, \quad x_g = x_h = 1, \quad x_i = 0,$$

(63.5) $$x_a = x_b = 0, \quad x_c = 1, \quad x_d = x_e = 0, \quad x_f = 1, \quad x_g = 0, \quad x_h = 1, \quad x_i = 0.$$

Fourth case: $x_d = x_b = x_h = 0, x_e = 1$. Equation (62.5) gives $x_b = x_d = x_f = 0$. Now, equations (62.2), (62.4) and (62.6) are satisfied, while (62.1), (62.3), (62.7) and (62.9) reduce to $x_a = 1$, $x_c = 1$, $x_g = 1$ and $x_i = 1$, respectively. Hence equation (62.8) is also verified and we have obtained the solution

(63.6) $$x_a = 1, \; x_b = 0, \; x_c = 1, \; x_d = 0, \; x_e = 1, \; x_f = 0, \; x_g = 1, \; x_h = 0, \; x_i = 1.$$

Fifth case: $x_d = x_b = x_h = x_e = 0$. Equations (62.1), (62.3), (62.5) and (62.9) reduce to $x_a = 1$, $x_c = 1$, $0 = \bar{x}_f$, and $x_i = 1$, respectively. Hence: $x_f = 1$,

* See also Example X.5.1.

equations (62.2), (62.4) and (62.8) are satisfied, while (62.6) and (62.7) reduce to $x_f = \bar{x}_g$, showing that $x_g = 0$. Thus the last solution is

(63.7) $x_a = 1, \quad x_b = 0, \quad x_c = 1, \quad x_d = x_e = 0, \quad x_f = 1, \quad x_g = x_h = 0, \quad x_i = 1.$

Example 9*. Let us solve (in B_2) the following system of 20 equations with the 20 unknowns $W, L, R, Y, S, N, P, K, F, M, \check{S}, T, Z, V, B, G, D, H, \check{Z}, \underset{,}{T}$:

(64.1) $W = \bar{L}\ \bar{R}\ \bar{S}\ \bar{N}\ P\ \bar{K}\ \bar{F}\ \bar{M}\ \bar{\check{S}}\ \bar{T}\ \bar{Z}\ \bar{V}\ \bar{B}\ \bar{G}\ \bar{D}\ \bar{H}\ \bar{\check{Z}}\ \bar{\underset{,}{T}}$

(64.2) $L = \bar{W}\ \bar{Y}\ \bar{S}\ \bar{P}\ \bar{K}\ \bar{F}\ \bar{M}\ \bar{\check{S}}\ \bar{Z}\ \bar{V}\ \bar{B}\ \bar{G}\ \bar{H},$

(64.3) $R = \bar{W}\ \bar{P}\ K\ \bar{F}\ \bar{M}\ \bar{\check{S}}\ \bar{T}\ \bar{V}\ \bar{B}\ \bar{G}\ \bar{D}\ \bar{H},$

(64.4) $Y = \bar{L}\ S\ \bar{N}\ \bar{P}\ \bar{F}\ \bar{M}\ \bar{T}\ \bar{Z}\ \bar{V}\ \bar{B}\ \bar{D}\ \bar{\underset{,}{T}},$

(64.5) $S = \bar{W}\ \bar{L}\ \bar{Y}\ \bar{N}\ \bar{P}\ \bar{K}\ \bar{F}\ \bar{M}\ \bar{T},$

(64.6) $N = \bar{W}\ \bar{Y}\ \bar{S}\ \bar{P}\ \bar{K}\ \bar{M}\ \bar{\check{S}}\ \bar{G}\ \bar{\check{Z}},$

(64.7) $P = \bar{W}\ \bar{L}\ \bar{R}\ \bar{Y}\ \bar{S}\ \bar{N}\ \bar{\check{S}}\ \bar{T},$

(64.8) $K = \bar{W}\ \bar{L}\ \bar{R}\ \bar{S}\ \bar{N}\ \bar{F}\ \bar{T}\ \bar{V},$

(64.9) $F = \bar{W}\ \bar{L}\ \bar{R}\ \bar{Y}\ S\ \bar{K}\ \bar{\check{S}}\ \bar{T},$

(64.10) $M = \bar{W}\ \bar{L}\ \bar{R}\ \bar{Y}\ \bar{S}\ \bar{N}\ \bar{\check{S}}\ \bar{Z},$

(64.11) $\check{S} = \bar{W}\ \bar{L}\ \bar{R}\ \bar{N}\ \bar{P}\ \bar{F}\ \bar{M}\ \bar{T},$

(64.12) $T = \bar{W}\ \bar{R}\ \bar{Y}\ \bar{S}\ \bar{P}\ \bar{K}\ \bar{F}\ \bar{\check{S}},$

(64.13) $Z = \bar{W}\ \bar{L}\ \bar{Y}\ \bar{M}\ \bar{B}\ \bar{G}\ \bar{D},$

(64.14) $V = \bar{W}\ \bar{L}\ \bar{R}\ \bar{Y}\ \bar{K}\ \bar{\check{Z}},$

(64.15) $B = \bar{W}\ \bar{L}\ \bar{R}\ \bar{Y}\ \bar{Z},$

(64.16) $G = \bar{W}\ \bar{L}\ \bar{R}\ \bar{N}\ \bar{Z},$

(64.17) $D = \bar{W}\ \bar{R}\ \bar{Y}\ \bar{Z},$

(64.18) $H = \bar{W}\ \bar{L}\ \bar{R},$

(64.19) $\check{Z} = \bar{W}\ \bar{N}\ \bar{V},$

(64.20) $\underset{,}{T} = \bar{W}\ \bar{Y}.$

Case I. We seek the solutions with $W = 1$.

It follows from (64.1) that $L = R = S = N = P = K = F = M = \hat{S} = T = Z = V = B = G = D = H = \check{Z} = \underset{,}{T} = 0$, hence $Y = 1$ by (64.4).

Case II. We seek the solutions with $W = 0$, $L = 1$.

It follows from (64.2) that $W = Y = S = P = K = F = M = \check{S} = Z = V = B = G = H = 0$. Therefore relations (64.20), (64.19), (64.17) and (64.12)

---

* See also Example X.5.2.

imply $\underset{,}{T} = 1$, $\check{Z} = \bar{N}$, $D = \bar{R}$, $T = \bar{R}$. The other equations are satisfied, N and R being arbitrary.

Casse III. Solutions with $W = L = 0$, $R = 1$.

It follows from (64.3) that $W = P = K = F = M = \check{S} = T = V = B = G = D = H = 0$. Therefore relations (64.20), (64.19), (64.13) and (64.5) imply $\underset{,}{T} = \bar{Y}$, $\check{Z} = \bar{N}$, $Z = \bar{Y}$, $S = \bar{N}\,\bar{Y}$. Now relation (64.6) becomes $N = \bar{Y}N$ or else $N\,Y = 0$. This condition upon the parameters N, Y assures that the other equations are satisfied.

In the following case, $W = L = R = 0$, implying $H = 1$, by (64.18).

Case IV. Solutions with $H = 1$, $W = L = R = 0$, $Y = 1$.

It follows from (64.2) that $B = P = V = F = M = D = N = S = T = \underset{,}{T} = Z = 0$. Equations (64.19), (64.16), (64.11) and (64.8) become $\check{Z} = 1$, $G = 1$, $\check{S} = 1$, $K = 1$. The other equations are satisfied.

In the following cases, $W = Y = 0$, implying $\underset{,}{T} = 1$, by (64.20).

Case V. Solutions with $H = \underset{,}{T} = 1$, $W = L = R = Y = 0$, $S = 1$.

It follows from (64.5) that $N = P = K = F = M = T = 0$. Therefore relations (64.19), (64.17), (64.16), (64.15) and (64.11) imply $\check{Z} = \bar{V}$, $D = \bar{Z}$, $G = \bar{Z}$, $B = \bar{Z}$, $\check{S} = 1$. The other equations are satisfied, V and Z being arbitrary.

Case VI. Solutions with $H = \underset{,}{T} = 1$, $W = L = R = Y = S = 0$, $N = 1$.

It follows from (64.6) that $P = K = M = \check{S} = G = \check{Z} = 0$. Therefore relations (64.17), (64.15), (64.14) and (64.12) imply $D = \bar{Z}$, $B = \bar{Z}$, $V = 1$, $T = \bar{F}$. The other equations are satisfied, F and Z being arbitrary.

Case VII. Solutions with $H = \underset{,}{T} = 1$, $W = L = R = Y = S = N = 0$, $P = 1$.

It follows from (64.7) that $\check{S} = T = 0$. Therefore relations (64.19), (64.17), (64.16), (64.15), (64.10) and (64.9) imply $\check{Z} = \bar{V}$, $D = \bar{Z}$, $G = \bar{Z}$, $B = \bar{Z}$, $M = \bar{Z}$, $F = \bar{K}$. Now relation (64.14) becomes $V = V\bar{K}$, or else $V K = 0$. This condition upon the parameters K, V, assures that the other equations are satisfied, Z being arbitrary.

Case VIII. Solutions with $H = \underset{,}{T} = 1$, $W = L = R = Y = S = P = N = 0$, $K = 1$.

It follows from (64.8) that $F = T = V = 0$. Therefore, relations (64.19), (64.17), (64.16), (64.15) and (64.12) imply $\check{Z} = 1$, $D = \bar{Z}$, $G = \bar{Z}$, $B = \bar{Z}$, $\check{S} = 1$. Hence $M = 0$, by (64.10). The other equations are satisfied, Z being arbitrary.

In the following cases ($K = 0$), relations (64.7) and (64.9) imply $0 = \bar{\check{S}}\,\bar{T}$, $F = \bar{\check{S}}\,\bar{T}$, hence $F = 0$.

Case IX. Solutions with $H = \underset{,}{T} = 1$, $W = L = R = Y = S = P = N = K = F = 0$, $M = 1$.

It follows from (64.10) that $\check{S} = Z = 0$. Therefore relations (64.19), (64.17) and (64.12) imply $\check{Z} = \bar{V}$, $D = L$, $G = 1$, $B = 1$, $T = 1$. The other equations are satisfied, V being arbitrary.

Case X. Solutions with $H = \underset{,}{T} = 1$, $W = L = R = Y = S = P = N = K = F = M = 0$.

Relation (64.5) implies $T = 1$, hence (64.12) becomes $\check{S} = 0$. It follows, from (64.10), that $Z = 1$, hence $B = G = D = 0$, by (64.13). This implies, on the account of (64.6), that $\check{Z} = 1$, hence $V = 0$, by (64.14). This is the last solution of our system.

Giving to the parameters occurring in the cases II, III, V—IX all possible values, we obtain immediately the following solutions:

Table 1

	W	L	R	Y	S	N	P	K	F	M	$Š$	T	Z	V	B	G	D	H	$Ž$	$Ţ$
1	1	0	0	1	0	0	0	0	0	0	0	0	0	0	0	0	0	0	0	0
2	0	1	1	0	0	1	0	0	0	0	0	0	0	0	0	0	0	0	0	1
3	0	1	0	0	0	1	0	0	0	0	0	1	0	0	0	0	1	0	0	1
4	0	1	1	0	0	0	0	0	0	0	0	0	0	0	0	0	0	0	1	1
5	0	1	0	0	0	0	0	0	0	0	0	1	0	0	0	0	1	0	1	1
6	0	0	1	1	0	0	0	0	0	0	0	0	0	0	0	0	0	0	1	0
7	0	0	1	0	0	1	0	0	0	0	0	0	1	0	0	0	0	0	0	1
8	0	0	1	0	1	0	0	0	0	0	0	0	1	0	0	0	0	0	1	1
9	0	0	0	1	0	0	0	1	0	0	1	0	0	0	0	1	0	1	1	0
10	0	0	0	0	1	0	0	0	0	0	1	0	1	1	0	0	0	1	0	1
11	0	0	0	0	1	0	0	0	0	0	1	0	0	1	1	1	1	1	0	1
12	0	0	0	0	1	0	0	0	0	0	1	0	1	0	0	0	0	1	1	1
13	0	0	0	0	1	0	0	0	0	0	1	0	0	0	1	1	1	1	1	1
14	0	0	0	0	0	1	0	0	1	0	0	0	1	1	0	0	0	1	0	1
15	0	0	0	0	0	1	0	0	1	0	0	0	0	1	1	0	1	1	0	1
16	0	0	0	0	0	1	0	0	0	0	0	1	1	1	0	0	0	1	0	1
17	0	0	0	0	0	1	0	0	0	0	0	1	0	1	1	0	1	1	0	1
18	0	0	0	0	0	0	1	1	0	0	0	0	1	0	0	0	0	1	1	1
19	0	0	0	0	0	0	1	1	0	1	0	0	0	0	1	1	1	1	1	1
20	0	0	0	0	0	0	1	0	1	0	0	0	1	1	0	0	0	1	0	1
21	0	0	0	0	0	0	1	0	1	0	0	0	1	0	0	0	0	1	1	1
22	0	0	0	0	0	0	1	0	1	1	0	0	0	1	1	1	1	1	0	1
23	0	0	0	0	0	0	1	0	1	1	0	0	0	0	1	1	1	1	1	1
24	0	0	0	0	0	0	0	1	0	0	1	0	1	0	0	0	0	1	1	1
25	0	0	0	0	0	0	0	1	0	0	1	0	0	0	1	1	1	1	1	1
26	0	0	0	0	0	0	0	0	0	1	0	1	0	1	1	1	1	1	0	1
27	0	0	0	0	0	0	0	0	0	1	0	1	0	0	1	1	1	1	1	1
28	0	0	0	0	0	0	0	0	0	0	0	1	1	0	0	0	0	1	1	1

§ 6. Families of Solutions

We describe below a second method for solving Boolean equations in B_2. This procedure is based on the notion of *families of solutions.* The procedure consists of two steps.

In the first one, the given equation (or system of equations) is expressed as a single equation of the form

$$f(x_1, \ldots, x_n) = 1, \tag{65}$$

the function f being written in a disjunctive form, i.e.

$$C_1 \cup \cdots \cup C_m = 1, \tag{65'}$$

where $C_1, \ldots, C_m$ are elementary conjunctions.

In the second step, we consider, in turn, each conjunction C_i and putting

$$C_i = x_{i_1}^{\alpha_{i_1}} x_{i_2}^{\alpha_{i_2}} \ldots x_{i_{p(i)}}^{\alpha_{i_{p(i)}}} = 1 \qquad (i = 1, \ldots, m), \tag{66}$$

we get

$$(67)\qquad x_{i_1}^{\alpha_{i_1}} = x_{i_2}^{\alpha_{i_2}} = \cdots = x_{i_{p(i)}}^{\alpha_{i_{p(i)}}} = 1 \qquad (i = 1, \ldots, m).$$

Hence, taking into account that $x^\alpha = 1$ if and only if $x = \alpha$ (for $x^1 = 1$ means $x = 1$, while $x^0 = 1$ means $\bar{x} = 1$, i.e. $x = 0$), we obtain the solutions grouped into the so-called *"families of solutions"*, i.e.

$$(68)\qquad x_{i_1} = \alpha_{i_1}, \quad x_{i_2} = \alpha_{i_2}, \ldots, x_{i_{p(i)}} = \alpha_{i_{p(i)}} \qquad (i = 1, \ldots, m),$$

which obviously cover all the solutions of (65).

We shall make an extensive use of this method in Chapter IV.

Example 10. Let us solve the equation

$$(69)\qquad x_2\,\bar{x}_4 \cup x_1\,x_2\,x_5 \cup \bar{x}_3\,\bar{x}_4 \cup \bar{x}_2\,\bar{x}_3 \cup \bar{x}_3\,x_5 \cup \bar{x}_1\,\bar{x}_4\,\bar{x}_5 = 0.$$

We have

$$(\bar{x}_2 \cup x_4)\,(\bar{x}_1 \cup \bar{x}_2 \cup \bar{x}_5)\,(x_3 \cup x_4)\,(x_2 \cup x_3)\,(x_3 \cup \bar{x}_5)\,(x_1 \cup x_4 \cup x_5) = 1,$$

or else, after performing all the multiplications and absorptions,

$$(69')\qquad x_1\,\bar{x}_2\,x_3 \cup \bar{x}_1\,x_3\,x_4 \cup x_2\,x_4\,\bar{x}_5 \cup \bar{x}_2\,x_3\,x_5 = 1.$$

Hence we obtain the following families of solutions:

(70.1) $x_1 = 1,\quad x_2 = 0,\quad x_3 = 1,\quad x_4$ and x_5 arbitrary;

(70.2) $x_1 = 0,\quad x_3 = 1,\quad x_4 = 1,\quad x_2$ and x_5 arbitrary;

(70.3) $x_2 = 1,\quad x_4 = 1,\quad x_5 = 0,\quad x_1$ and x_3 arbitrary;

(70.4) $x_2 = 0,\quad x_3 = 1,\quad x_5 = 1,\quad x_1$ and x_4 arbitrary.

These families may be represented in Table 2 below, where the dashes denote the arbitrary variables:

Table 2

x_1	x_2	x_3	x_4	x_5
1	0	1	—	—
0	—	1	1	—
—	1	—	1	0
—	0	1	—	1

If, in each family, we give all possible values to the arbitrary variables, we obtain the explicit list of all the solutions.

Table 3

x_1	x_2	x_3	x_4	x_5
1	0	1	0	0
1	0	1	0	1
1	0	1	1	0
1	0	1	1	1
0	0	1	1	0
0	0	1	1	1
0	1	1	1	0
0	1	1	1	1
0	1	0	1	0
1	1	0	1	0
1	1	1	1	0
0	0	1	0	1

C. Addendum

§ 7. Other Researches Concerning Boolean Equations and their Generalizations

In order to keep our book within reasonable dimensions, we have limited the presentation of Boolean equations to those problems and methods which are actually used in the next chapters.

We give below, for the reader having a special interest in this field, a few references to other researches (see also the previous section).

The first basic works are due to G. Boole [A 1, A 2], S. Jevons [A 1], E. Schröder [A 1, A 2], P. S. Poretski [A 1—A 10], A. N. Whitehead [A 1, A 2]; comprehensive and clear accounts of their results were given by A. del Re [A 1] and L. Couturat [A 1].

L. Löwenheim [A 1—A 4] has developed a theory of Boolean matrices and determinants, using them in order to solve systems of Boolean equations.

Symmetric Boolean equations were studied by E. Schröder [A 2], W. E. Johnson [A 1], Th. Skolem [A 1], and M. Ja. Èĭngorin [A 2].

A. N. Whitehead [A 2], B. A. Bernstein [A 1], W. L. Parker and B. A. Bernstein [A 1], and R. Martelotta [A 1] have studied Boolean equations having unique solutions.

The Boolean ring structure may also be used for solving Boolean equations, as it was shown by G. Birkhoff [A 1], W. L. Parker and B. A. Bernstein [A 1], V. Lalan [A 1] and M. Barr [A 1].

Various forms of the parametric general solution were given by M. Itoh [A 7], M. Gotō [A 1—A3] and S. Rudeanu [A 6].

S. B. Akers Jr. [A 1] has created a Boolean "differential calculus", with applications to Boolean equations.

Certain particular equations were solved by S. Ćetković [A 1], C. Stanojević [A 1], V. Sedmak [A 1], G. Andreoli [A 1], Y. Tohma [A 1] and S. Rudeanu [A 9, A 10].

Equations in certain particular Boolean algebras were studied by E. Stamm [A 1, A 2] (in the Boolean algebra of all the intervals of a given Boolean algebra), M. Gotō [A 4] (in finite Boolean algebras), G. Andreoli [A 2] (in atomic Boolean algebras), H. D. Sprinkle [A 1] and M. Barr [A 1] (in complete Boolean algebras), S. Ćetković [A 1] and A. W. Goodman [A 1] (set equations).

The so-called "sequential" Boolean equations (or, equations with delays) were studied by H. Wang [A 1], M. Gotō and Y. Komamiya [A 1], Ju. Ja. Bazilevskiĭ [A 1] and M. Ja. Èĭngorin [A 4].

M. Itoh [A 1—A 6] has studied equations in n-valued lattices (instead of the two-element Boolean algebra B_2), while A. Tauts [A 1—A 3] was concerned with equations in propositional calculus and in the first-order predicate calculus.

Recently, R. L. Goodstein [A 1] has studied equations in distributive lattices with 0 and 1.

Another generalization, the theory of pseudo-Boolean equations, will be presented in the next chapter.

Chapter III

Linear Pseudo-Boolean Equations and Inequalities

Pseudo-Boolean equations, that is equations of the form

$$f(x_1, \ldots, x_n) = 0, \tag{1}$$

where f is a pseudo-Boolean function, have been studied by R. FORTET [1—3], P. CAMION [2], and P. L. HAMMER [3, 4]. These authors have suggested various methods for expressing the general solution in a parametric form.

The procedure described below (P. L. HAMMER and S. RUDEANU [4—7]), offers the sought solutions either completely listed, or grouped into "families" of solutions (as in § II.6), each family being characterized by the fact that for certain fixed indices $i_1, \ldots, i_p$ the corresponding variables have fixed values: $x_{i_1} = \alpha_{i_1}, \ldots, x_{i_p} = \alpha_{i_p}$, while the other variables: $x_{i_{p+1}}, \ldots, x_{i_n}$, remain arbitrary. This way of expressing the solutions is very advantageous for our method of minimizing pseudo-Boolean functions (because that method uses certain pseudo-Boolean equations and inequalities; see Chapters V, VI, VII).

In this Chapter we study only (systems of) linear equations and inequalities. The nonlinear case will be dealt with in Chapter IV.

The treatment of the linear case is, in its essence, a systematic accelerated search of the solutions within an associated tree. This tree-like construction was used by S. RUDEANU [12] for solving Boolean equations; in the same paper, he conjectured that the method can be adapted to the case of pseudo-Boolean equations and inequalities.

Our procedure has certain similarities with some techniques, used in integer linear programming, which involve the setting up of tree-like schemes. One class of such techniques is that of the so-called branching methods ("branch and bound", "branch and exclude") developed by R. BELLMAN [1], R. BELLMAN and S. E. DREYFUS [1], G. F. BENDERS, A. R. CATCHPOLE and C. KUIKEN [1], W. C. HEALY JR [1], A. H. LAND and A. G. DOIG [1], J. D. C. LITTLE, K. G. MURTY, D. W. SWEENEY and K. CAROLINE [1]; see also the surveys of M. L. BALINSKI [1], E. M. L. BEALE [1], A. BEN-ISRAEL and A. CHARNES [1]. Another class of such techniques is that of the "SEP procedures" (séparation et évaluation progressives) developed by P. BERTIER and Ph. T. NGHIEM [1], P. BERTIER, Ph. T. NGHIEM and B. ROY [1], P. BERTIER and B. ROY [1], B. ROY, Ph. T. NGHIEM and P. BERTIER [1], B. ROY and B. SUSSMANN [1]. The methods of R. FAURE and Y. MALGRANGE [1, 2], and that of E. BALAS [1—3] belong to the same type.

A different method, due to P. CAMION, will be described in the last section.

§ 1. Linear Pseudo-Boolean Equations

Let

$$a_1 z_1 + b_1 \bar{z}_1 + a_2 z_2 + b_2 \bar{z}_2 + \cdots + a_n z_n + b_n \bar{z}_n = k, \tag{2}$$

where a_i, b_i $(i = 1, \ldots, n)$ and k are given constants, be the general form of a linear pseudo-Boolean equation with the unknowns $z_1, \ldots, z_n$. Of course, we may assume, without loss of generality, that $a_i \neq b_i$ for all i (if not, the term $a_i z_i + b_i \bar{z}_i$ in the left-hand side of (2) is simply the constant a_i).

For each i, let us set

$$x_i = \begin{cases} z_i, & \text{if } a_i > b_i, \\ \bar{z}_i, & \text{if } a_i < b_i. \end{cases} \tag{3}$$

Then the terms $a_i z_i + b_i \bar{z}_i$ may be transformed as follows:

$$a_i z_i + b_i \bar{z}_i = \begin{cases} (a_i - b_i) x_i + b_i, & \text{if } a_i > b_i, \\ (b_i - a_i) x_i + a_i, & \text{if } a_i < b_i. \end{cases} \tag{4}$$

Thus, equation (2) becomes

$$c_1 x_1 + c_2 x_2 + \cdots + c_n x_n = d, \tag{5}$$

where $c_1, \ldots, c_n$, d are constants, $c_i > 0$ $(i = 1, \ldots, n)$, and where (after re-indexing the unknowns), we can suppose that

$$c_1 \geqq c_2 \geqq \cdots \geqq c_n > 0. \tag{6}$$

Now, we are concentrating our attention on a procedure for solving the "canonical" form (5) under the assumption (6).

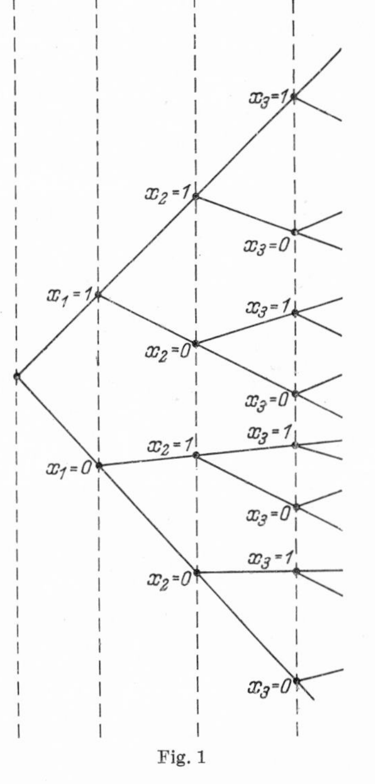

Fig. 1

We shall track down the solutions of equation (5) along the branches of the tree in Fig. 1. This tree has $n + 1$ levels $0, 1, \ldots, n$.

Each level r contains 2^r nodes. Each node of the r-th level is characterized by the fact that the values of the variables $x_1, \ldots, x_r$ are fixed $(x_1 = \xi_1, \ldots, x_r = \xi_r)$, while the variables $x_{r+1}, \ldots, x_n$ are subject to the equation

$$\sum_{j=r+1}^{n} c_j x_j = d' \tag{5. r}$$

$\left(\text{where } d' = d - \sum_{k=1}^{r} c_k \xi_k\right)$, which is of the type (5).

Of course, it would be unreasonable to follow all the 2^n possible paths. Fortunately, most of the blind alleys can be avoided by a systematic use of the following:

Table 1

No	Case	Conclusions
1	$d < 0$	No solutions
2	$d = 0$	The unique solution is $x_1 = x_2 = \cdots = x_n = 0$
3	$d > 0$ and $c_1 \geqq \cdots \geqq c_p > d \geqq c_{p+1} \geqq \cdots \geqq c_n$	The solutions (if any) satisfy $x_1 = \cdots = x_p = 0$ and $\sum_{j=p+1}^{n} c_j x_j = d$
4	$d > 0$ and $c_1 = \cdots = c_p = d > c_{p+1} \geqq \cdots \geqq c_n$	α) For every $k = 1, 2, \ldots, p$: $x_k = 1$, $x_1 = \cdots = x_{k-1} = x_{k+1} = \cdots = x_n = 0$ is a solution β) The other solutions (if any) satisfy $x_1 = \cdots x_p = 0$ and $\sum_{j=p+1}^{n} c_j x_j = d$
5	$d > 0$, $c_i < d$ $(i = 1, 2, \ldots, n)$ and $\sum_{i=1}^{n} c_i < d$	No solutions
6	$d > 0$, $c_i < d$ $(i = 1, 2, \ldots, n)$ and $\sum_{i=1}^{n} c_i = d$	The unique solution is $x_1 = x_2 = \cdots = x_n = 1$
7	$d > 0$, $c_i < d$ $(i = 1, 2, \ldots, n)$, $\sum_{i=1}^{n} c_i > d$ and $\sum_{j=2}^{n} c_i < d$	The solutions (if any) satisfy $x_1 = 1$ and $\sum_{j=2}^{n} c_j x_j = d - c_1$
8	$d > 0$, $c_i < d$ $(i = 1, 2, \ldots, n)$, $\sum_{i=1}^{n} c_i > d$ and $\sum_{j=2}^{n} c_j \geqq d$	The solutions (if any) satisfy either $x_1 = 1$ and $\sum_{j=2}^{n} c_j x_j = d - c_1$, or $x_1 = 0$ and $\sum_{j=2}^{n} c_j x_j = d$

Table 1 discusses 8 mutually exclusive cases concerning equation (5) and covering all the situations; for each case, an obvious conclusion is drawn. We see that the following circumstances may occur:

— equation (5) is inconsistent (cases 1 and 5);

— equation (5) has a unique solution (cases 2 and 6);

— equation (5) is replaced by an equation of the same type, but with less variables (cases 3, 4 and 7);

— equation (5) is replaced by two equations of the same type, but with less variables; each of these equations is to be discussed separately (case 8).

Therefore, unless equation (5) is inconsistent or it has a unique solution, we have to continue the investigation by applying the conclusions in Table 1 to the new equation(s) that resulted at the first step. This process is continued until we have exhausted all the possibilities.

We have thus proved

Theorem 1 (i). *The above described procedure leads to* all *the solutions of the canonical equation* (5). (ii) *If T is the transformation from* (2) *to* (5), *then the solutions of* (2) *are obtained by applying T^{-1} to the solutions of* (5).

Example 1. Let us solve the linear pseudo-Boolean equation

$$4z_1 + \bar{z}_1 - 3z_2 + \bar{z}_2 + 5z_3 - 2z_4 + 5z_5 + 2z_6 - z_7 = 7. \tag{7}$$

Applying the transformation (3), we set

$$y_1 = z_1, \quad y_2 = \bar{z}_2, \quad y_3 = z_3, \quad y_4 = \bar{z}_4, \quad y_5 = z_5, \quad y_6 = z_6, \quad y_7 = \bar{z}_7; \tag{8}$$

hence equation (7) becomes

$$(3y_1 + 1) + (4y_2 - 3) + 5y_3 + (2y_4 - 2) + 5y_5 + 2y_6 + (y_7 - 1) = 7,$$

or

$$3y_1 + 4y_2 + 5y_3 + 2y_4 + 5y_5 + 2y_6 + y_7 = 12,$$

or else, ordering the unknowns so as condition (6) be fulfilled, we get

$$5x_1 + 5x_2 + 4x_3 + 3x_4 + 2x_5 + 2x_6 + x_7 = 12, \tag{9}$$

where

$$\begin{cases} x_1 = y_3 = z_3, \\ x_2 = y_5 = z_5, \\ x_3 = y_2 = \bar{z}_2, \\ x_4 = y_1 = z_1, \\ x_5 = y_4 = \bar{z}_4, \\ x_6 = y_6 = z_6, \\ x_7 = y_7 = \bar{z}_7. \end{cases} \tag{10}$$

We begin now the tree-like construction of the solutions. The reader is advised to follow simultaneously the computations and Fig. 2 on p. 54.

Since we are in Case 8, we have simply to consider the two equations obtained from (9) by making $x_1 = 1$ and $x_1 = 0$. Equation (9) becomes

$$5x_2 + 4x_3 + 3x_4 + 2x_5 + 2x_6 + x_7 = 7 \tag{9.1}$$

and

$$5x_2 + 4x_3 + 3x_4 + 2x_5 + 2x_6 + x_7 = 12, \tag{9.0}$$

respectively. The number (9.1) indicates that the corresponding equation was obtained from (9) by making $x_1 = 1$; similarly for (9.0) etc. So, for instance, the label (9.1100) below indicates the equation obtained from (9) by making $x_1 = 1$, $x_2 = 1$, $x_3 = 0$, $x_4 = 0$.

We begin by following the branch $x_1 = 1$ corresponding to (9.1). We are again in Case 8, so that we continue the splitting with respect to x_2:

$$4x_3 + 3x_4 + 2x_5 + 2x_6 + x_7 = 2, \tag{9.11}$$

$$4x_3 + 3x_4 + 2x_5 + 2x_6 + x_7 = 7. \tag{9.10}$$

Applying the conclusion of the Case **3** to (9.11), we obtain $x_3 = x_4 = 0$ and the equation

$$2x_5 + 2x_6 + x_7 = 2; \tag{9.1100}$$

by 4, we get the solutions

$$x_1 = 1,\quad x_2 = 1,\quad x_3 = 0,\quad x_4 = 0,\quad x_5 = 1,\quad x_6 = 0,\quad x_7 = 0, \tag{9.1100100}$$

$$x_1 = 1,\quad x_2 = 1,\quad x_3 = 0,\quad x_4 = 0,\quad x_5 = 0,\quad x_6 = 1,\quad x_7 = 0 \tag{9.1100010}$$

and the equation

$$x_7 = 2, \tag{9.110000}$$

which has no solutions (by 5).

Now we come back to equation (9.10), which is in case 8, so that we have to consider separately the cases $x_3 = 1$ and $x_3 = 0$:

$$3x_4 + 2x_5 + 2x_6 + x_7 = 3, \tag{9.101}$$

$$3x_4 + 2x_5 + 2x_6 + x_7 = 7. \tag{9.100}$$

Using the conclusions of the case 4, we see that equation (9.101) leads to the solution

$$x_1 = 1,\quad x_2 = 0,\quad x_3 = 1,\quad x_4 = 1,\quad x_5 = 0,\quad x_6 = 0,\quad x_7 = 0 \tag{9.1011000}$$

and to the equation

$$2x_5 + 2x_6 + x_7 = 3, \tag{9.1010}$$

which, in its turn, leads to the equations

$$2x_6 + x_7 = 1, \tag{9.10101}$$

$$2x_6 + x_7 = 3. \tag{9.10100}$$

The conclusion 3, applied to (9.10101), shows that $x_6 = 0$, hence $x_7 = 1$, therefore

$$x_1 = 1,\quad x_2 = 0,\quad x_3 = 1,\quad x_4 = 0,\quad x_5 = 1,\quad x_6 = 0,\quad x_7 = 1, \tag{9.1010101}$$

while equation (9.10100) can be solved with the aid of 6:

$$x_1 = 1,\quad x_2 = 0,\ x_3 = 1,\quad x_4 = 0,\quad x_5 = 0,\quad x_6 = 1,\quad x_7 = 1. \tag{9.1010011}$$

Now we come back to equation (9.100), which, by a repeated application of the conclusion of the case 7, implies first that $x_4 = 1$ and

$$2x_5 + 2x_6 + x_7 = 4, \tag{9.1001}$$

then $x_5 = 1$ and

(9.10011) $$2x_6 + x_7 = 2,$$

hence, by 4 and 5, we deduce the solution

(9.1001110) $$x_1 = 1, \quad x = 0, \quad x_3 = 0, \quad x_4 = 1, \quad x_5 = 1, \quad x_6 = 1, \quad x_7 = 0.$$

We have thus found all the solutions of equation (9.1), so that it remains to determine the solutions of (9.0). We must split again, obtaining thus the equations

(9.01) $$4x_3 + 3x_4 + 2x_5 + 2x_6 + x_7 = 7,$$

(9.00) $$4x_3 + 3x_4 + 2x_5 + 2x_6 + x_7 = 12.$$

Equation (9.01) coincides with (9.10), whose solutions were determined before. Therefore, the solutions of (9.01) can simply be obtained from those of (9.10) by taking $x_1 = 0$ and $x_2 = 1$, instead of $x_1 = 1$ and $x_2 = 0$:

(9.0111000) $$x_1 = 0, \quad x_2 = 1, \quad x_3 = 1, \quad x_4 = 1, \quad x_5 = 0, \quad x_6 = 0, \quad x_7 = 0,$$

(9.0110101) $$x_1 = 0, \quad x_2 = 1, \quad x_3 = 1, \quad x_4 = 0, \quad x_5 = 1, \quad x_6 = 0, \quad x_7 = 1,$$

(9.0110011) $$x_1 = 0, \quad x_2 = 1, \quad x_3 = 1, \quad x_4 = 0, \quad x_5 = 0, \quad x_6 = 1, \quad x_7 = 1,$$

(9.0101110) $$x_1 = 0, \quad x_2 = 1, \quad x_3 = 0, \quad x_4 = 1, \quad x_5 = 1, \quad x_6 = 1, \quad x_7 = 0.$$

As to equation (9.00), 6 shows that it has the unique solutions $x_3 = \cdots = x_7 = 1$, so that we obtain the following solution of the initial equation (9):

(9.0011111) $$x_1 = 0, \quad x_2 = 0, \quad x_3 = 1, \quad x_4 = 1, \quad x_5 = 1, \quad x_6 = 1, \quad x_7 = 1.$$

We have thus found all the solution of (9); taking into account the transformation formulas (10), we obtain the solution of the given equation (7):

Table 2

z_1	z_2	z_3	z_4	z_5	z_6	z_7
0	1	1	0	1	0	1
0	1	1	1	1	1	1
1	0	1	1	0	0	1
0	0	1	0	0	0	0
0	0	1	1	0	1	0
1	1	1	0	0	1	1
1	0	0	1	1	0	1
0	0	0	0	1	0	0
0	0	0	1	1	1	0
1	1	0	0	1	1	1
1	0	0	0	0	1	0

Comments. 1) The above described method determines all the solutions, and no solution is found twice.

2) In the above example, we have tested 12 paths, 11 of which have led to (all the) solutions. In other words, only one path was unfruitful; the other 116 paths which correspond to non-solutions were avoided.

3) Moreover, the fact that we have obtained the same equation at two distinct stages of the process (corresponding to the points 10 and 01) contributed also to the reduction of the amount of computations.

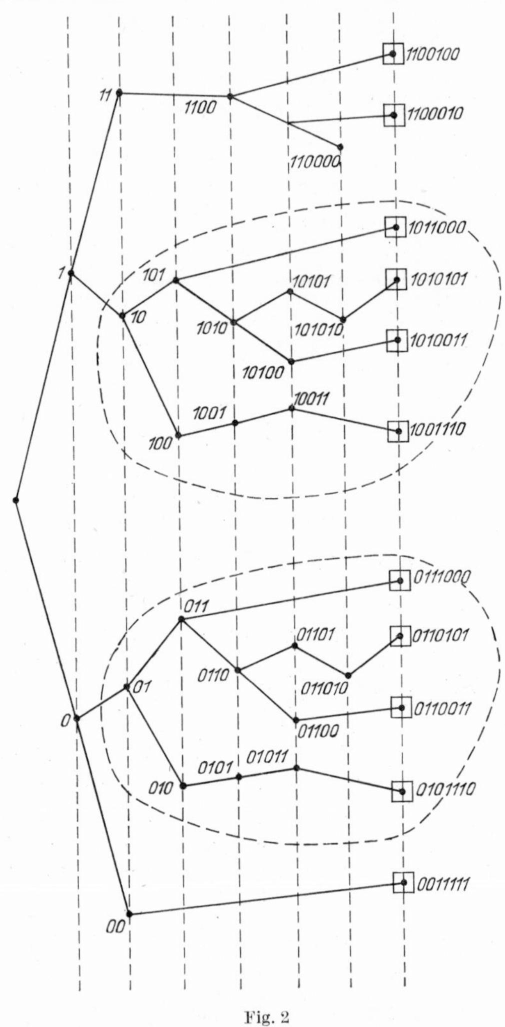

Fig. 2

§ 2. Linear Pseudo-Boolean Inequalities

The most general form of a linear pseudo-Boolean inequality is either

$$a_1 z_1 + b_1 \bar{z}_1 + a_2 z_2 + b_2 \bar{z}_2 + \cdots + a_n z_n + b_n \bar{z}_n > h, \tag{11}$$

or

$$a_1 z_1 + b_1 \bar{z}_1 + a_2 z_2 + b_2 \bar{z}_2 + \cdots + a_n z_n + b_n \bar{z}_n \geqq k, \tag{12}$$

where a_i, b_i, h and k are constants and we may assume that $a_i \neq b_i$ for all i (if we have the sign $<$ or $\leq$ instead of $>$ or $\geqq$, respectively, we multiply the whole inequality by -1). If the constants a_i, b_i and h are *integers* (and this is the usual case in applications), then the *strict* inequality (11) may also be written in the form (12), if we take $k = h + 1$. Therefore, we shall confine our attention to inequalities of the form (12). As a matter of fact, the method developed in this section for solving the inequality (12), will directly offer the solutions of the equation

(2) $$a_1 z_1 + b_1 \bar{z}_1 + a_2 z_2 + b_2 \bar{z}_2 + \cdots + a_n z_n + b_n \bar{z}_n = k,$$

as well as those of the strict inequality

(13) $$a_1 z_1 + b_1 \bar{z}_1 + a_2 z_2 + b_2 \bar{z}_2 + \cdots + a_n z_n + b_n \bar{z}_n > k.$$

We shall prove that the solutions of the inequality (12) (if any) can be grouped into "families of solutions", in the sense of the following

Definition 1. Let $S = (z_1^*, \ldots, z_n^*)$ be a solution of (12) and let I be a set of indices: $I \subseteqq \{1, 2, \ldots, n\}$. Let $\Sigma(S, I)$ be the set of all vectors $(z_1, \ldots, z_n) \in B_2^n$ satisfying

$$z_i = z_i^* \quad \text{for all} \quad i \in I,$$

the other variables z_j ($j \notin I$) being arbitrary. If all the vectors $(z_1, \ldots, z_n) \in \Sigma(S, I)$ satisfy the inequality (12), then $\Sigma(S, I)$ is said to be a *family of solutions* of (12). We say also that this family is *generated by the pair* (S, I); the variables z_i for which $i \in I$ are called the *fixed variables* of the family.

Notice that relation $S \in \Sigma(S, I)$ holds for every pair (S, I). If $I = \{1, 2, \ldots, n\}$, then $\Sigma(S, I)$ is a degenerate family containing a single solution, namely S. More generally, if the set I consists of r indices, then $\Sigma(S, I)$ contains 2^{n-r} elements; if $r < n$, the family may be called *non-degenerate*.

We want to obtain the solutions grouped into families of solutions, so that the number of these families should be as small as possible. Therefore we are interested in obtaining, whenever possible, non-degenerate families of solutions. Moreover, it will be shown that it is possible to obtain the solutions grouped into set-theoretically disjoint families.

Since it is not easy to detect non-degenerate families of solutions directly on the inequality (12), we shall first reduce it to a standard form. Namely, applying to each z_i the transformation (3), and re-ordering the unknowns, as in § 1, we see that the inequality (12) may be brought to the canonical form

(14) $$c_1 x_1 + c_2 x_2 + \cdots + c_n x_n \geqq d,$$

where $c_1, \ldots, c_n$, d are constants and

(15) $$c_1 \geqq c_2 \geqq \cdots \geqq c_n > 0.$$

We give below a procedure which enables us to obtain the solutions of (14) grouped into several non-degenerate and pairwise disjoint families of solutions; after this has been done, we apply the inverse transformation (from (14) to (12) and obtain immediately the families of solutions of (12).

To this end, let us introduce the following

Definition 2. A vector $(x_1^*, \ldots, x_n^*)$ satisfying the inequality (14) is called a *basic solution* of (14), if for each index i such that $x_i^* = 1$, the vector $(x_1^*, \ldots, x_{i-1}^*, 0, x_{i+1}^*, \ldots, x_n^*)$ is not a solution of (14).

Remark 1. The solutions of the equation

$$c_1 x_1 + c_2 x_2 + \cdots + c_n x_n = d \tag{5}$$

(if any) are basic solutions of the inequality (14).

We shall prove that the solutions of (14) may be found by a process involving two steps:

a) We determine all the basic solutions of (14).

b) To each basic solution S_k we associate a certain set of indices I_k in such a way that $\Sigma(S_k, I_k)$ should be a family of solutions and that the system $\{\Sigma(S_k, I_k)\}_{k=1,\ldots,m}$ should be "complete" [i.e., it should include all the solutions of (14)].

We proceed now to the first step:

a) Determination of the basic solutions

The basic solutions of (14) will be determined by a tree-like construction similar to that used for solving linear equations.

We need the following three lemmas:

Lemma 1. *Let* $(x_1^*, \ldots, x_p^*, x_{p+1}^*, \ldots, x_n^*)$ *be a basic solution of* (14); *then* $(x_{p+1}^*, \ldots, x_n^*)$ *is a basic solution of the inequality*

$$\sum_{j=p+1}^{n} c_j x_j \geqq d - \sum_{k=1}^{p} c_k x_k^* . \tag{16}$$

Proof. Obvious.

Lemma 2. *If* $(x_{p+1}^*, \ldots, x_n^*)$ *is a basic solution of the inequality*

$$\sum_{j=p+1}^{n} c_j x_j \geqq d, \tag{17}$$

then $(\underbrace{0, \ldots, 0}_{p \text{ times}}, x_{p+1}^*, \ldots, x_n^*)$ *is a basic solution of* (14).

Proof. Obvious.

Lemma 3. *If* $d > 0$ *and* $(x_2^*, \ldots, x_n^*)$ *is a basic solution of*

$$\sum_{j=2}^{n} c_j x_j \geqq d - c_1, \tag{18}$$

then $(1, x_2^*, \ldots, x_n^*)$ *is a basic solution of* (14).

Proof. If $x_2^* = \ldots = x_n^* = 0$, the lemma is obvious; hence we can assume that $\sum_{j=2}^{n} x_j^* > 0$. Since $(1, x_2^*, \ldots, x_n^*)$ is clearly a solution of (14), it remains to prove it is a basic one.

Since $(x_2^*, \ldots, x_n^*)$ is a basic solution of (18), it follows that

$$\sum_{\substack{j=2 \\ j \neq k}}^{n} c_j x_j^* < d - c_1 \tag{19. k}$$

for every k with $2 \leqq k \leqq n$ and $x_k^* = 1$. Hence

$$c_1 \cdot 1 + \sum_{\substack{j=2 \\ j \neq k}}^{n} c_j x^* < c_1 + d - c_1 = d,$$

i.e. $(1, x_2^*, \ldots, x_{k-1}^*, 0, x_{k+1}^*, \ldots, x_n^*)$ does not satisfy (14). Further,

$$\sum_{j=2}^{n} c_j x_j^* = c_k + \sum_{\substack{j=2 \\ j \neq k}}^{n} c_j x_j^* < c_k + d - c_1 \leqq d,$$

i.e. $(0, x_2^*, \ldots, x_n^*)$ is not a solution of (14). This completes the proof.

Lemmas 1, 2 and 3 enable us to build up Table 3 on p. 58, which is the analogue of Table 1 in § 1.

We conclude:

Theorem 2. *The above described method determines all the basic solutions of* (14).

We proceed now to the step

b) Determination of a complete system of families of solutions of (14)

To each basic solution $S = (x_1^*, \ldots, x_n^*)$ we associate a family of solutions $\sum(S, J_S)$ defined as follows. Let i_0 be the last index for which $x_i^* = 1$, (i.e. $x_{i_0}^* = 1$ and $x_i^* = 0$ for all $i > i_0$), and let J_S be the set of all indices $i \leqq i_0$. Then $\sum(S, J_S)$ (see Definition 1) is the set of all vectors $(x_1, \ldots, x_n)$ satisfying.

$$x_i = \begin{cases} x_i^*, & \text{for } i \leqq i_0, \\ \text{arbitrary}, & \text{for } i > i_0. \end{cases} \tag{20}$$

Theorem 3. *Let $S_1, \ldots, S_m$ be all the basic solutions of* (14) *and let $\sum_k = \sum(S_k, J_{S_k})$ $(k = 1, \ldots, m)$ be constructed as above. Then every solution $(x_1, \ldots, x_n)$ of* (14) *belongs to exactly one of these families of solutions.*

Proof. Let $(x_1, \ldots, x_n)$ be a vector in $\sum_k$; since $x_i^* = 0$ for all $i > i_0$, relation (20) shows that $x_i^* \leqq x_i$ for all i, hence $d \leqq \sum_{i=1}^{n} c_i x_i^* \leqq \sum_{i=1}^{n} c_i x_i$, i.e. every $\sum_k$ is a family of solutions. Further, if $S' = (x_1', \ldots, x_n')$ and $S'' = (x_1'', \ldots, x_n'')$ are distinct basic solutions, then there is an

Table 3

No.	Case	Conclusions	Valid
1	$d \leqq 0$	The unique basic solution is $x_1 = x_2 = \cdots = x_n = 0$	obviously
2	$d > 0$ and $c_1 \geqq \cdots \geqq c_p \geqq d >$ $> c_{p+1} \geqq \cdots \geqq c_n$	α) For every $k = 1, 2, \ldots, p$: $x_k = 1$, $x_1 = \cdots = x_{k-1} = x_{k+1} = \cdots = x_n = 0$ is a basic solution. β) The other basic solutions (if any) are characterized by the property: $x_1 = \cdots = x_p = 0$, and $(x_{p+1}, \ldots, x_n)$ is a basic solution of $\sum_{j=p+1}^{n} c_j x_j \geqq d$	obviously by lemmas 1, 2
3	$d > 0$, $c_i < d$ $(i = 1, 2, \ldots, n)$ and $\sum_{i=1}^{n} c_i < d$	No solutions	obviously
4	$d > 0$, $c_i < d$ $(i = 1, 2, \ldots, n)$ and $\sum_{i=1}^{n} c_i = d$	The unique (basic) solution is $x_1 = x_2 = \cdots = x_n = 1$	obviously
5	$d > 0$, $c_i < d$ $(i = 1, 2, \ldots, n)$ $\sum_{i=1}^{n} c_i > d$ and $\sum_{j=2}^{n} c_j < d$	The basic solutions (if any) are characterized by the property: $x_1 = 1$, and $(x_2, \ldots, x_n)$ is a basic solution of $\sum_{j=2}^{u} c_j x_j \geqq d - c_1$	by lemmas 1, 3
6	$d > 0$, $c_i < d$ $(i = 1, 2, \ldots, n)$ $\sum_{i=1}^{n} c_i > d$ and $\sum_{j=2}^{n} c_j \geqq d$	The basic solutions (if any) are characterized by the property: either $x_1 = 1$ and $(x_2, \ldots, x_n)$ is a basic solution of $\sum_{j=2}^{n} c_j x_j \geqq d - c_1$, or: $x_1 = 0$ and $(x_2, \ldots, x_n)$ is a basic solution of $\sum_{j=2}^{n} c_j x_j \geqq d$	by lemmas 1, 3 and 2

$i \in J_{S'} \cap J_{S''}$ such that $x'_i \neq x''_i$; hence the families $\sum_k$ are pairwise disjoint. It remains to prove that each solution belongs to one of the $\sum_k$.

Denote by $(x_1, \ldots, x_n)$ an arbitrary solution of (14). Let us consider the sequence of vectors: $(x_1, \ldots, x_n)$, $(x_1, \ldots, x_{n-1}, 0)$ $(x_1, \ldots, x_{n-2}, 0, 0), \ldots,$ $(x_1, \ldots, x_i, 0, \ldots, 0), \ldots$ Let $S = (x_1, \ldots, x_p, 0, \ldots, 0)$ be last vector of this sequence which is a solution of (14). Then obviously $x_p = 1$ and $(x_1, \ldots, x_{p-1}, 0, \ldots, 0)$ is not a solution, i.e. $c_1 x_1 + \cdots + c_{p-1} x_{p-1} < d$. Let us consider an index q with $x_q = 1$ $(q \leqq p)$. Then $(x_1, \ldots, x_{q-1}, 0, x_{q+1}, \ldots, x_p, 0, \ldots, 0)$ is not a solution, for $c_q \geqq c_p$ and hence

$$c_1 x_1 + \cdots + c_{q-1} x_{q-1} + c_{q+1} x_{q+1} + \cdots + c_{p-1} x_{p-1} + c_p x_p$$
$$= c_1 x_1 + \cdots + c_q x_q + \cdots + c_{p-1} x_{p-1} + (c_p - c_q) \leqq$$
$$\leqq c_1 x_1 + \cdots + c_{p-1} x_{p-1} < d.$$

We have thus shown that S is a basic solution of (14) and we see that the original solution $(x_1, \ldots, x_n)$ belongs to $\sum(S, J_S)$.

Concluding this discussion, we come to

Theorem 4. *The procedure summarized in Theorems* 2 *and* 3 *gives all the solutions of the canonical inequality* (14).

Corollary 1. Let T be the transformation leading from (12) to (14); we apply the inverse transformation T^{-1} to the solutions of (14) and obtain the solutions of the original inequality (12).

Example 2. Let us solve the linear pseudo-Boolean inequality

$$2\bar{z}_1 - 5z_2 + 3z_3 + 4\bar{z}_4 - 7z_5 + 16z_6 - z_7 \geqq -4. \tag{21}$$

We set

$$z_1 = \bar{y}_1, \quad z_2 = \bar{y}_2, \quad z_3 = y_3, \quad z_4 = \bar{y}_4, \quad z_5 = \bar{y}_5, \quad z_6 = y_6, \quad z_7 = \bar{y}_7. \tag{22}$$

Hence

$$2y_1 + 5y_2 + 3y_3 + 4y_4 + 7y_5 + 16y_6 + y_7 \geqq 9,$$

or else

$$16x_1 + 7x_2 + 5x_3 + 4x_4 + 3x_5 + 2x_6 + x_7 \geqq 9, \tag{23}$$

where

$$\begin{cases} x_1 = y_6 = z_6, \\ x_2 = y_5 = \bar{z}_5, \\ x_3 = y_2 = \bar{z}_2, \\ x_4 = y_4 = \bar{z}_4, \\ x_5 = y_3 = z_3, \\ x_6 = y_1 = \bar{z}_1, \\ x_7 = y_7 = \bar{z}_7. \end{cases} \tag{24}$$

The first coefficient being >9, we apply 2 and obtain the basic solution

(23.1000000) $x_1 = 1, \quad x_2 = 0, \quad x_3 = 0, \quad x_4 = 0, \quad x_5 = 0, \quad x_6 = 0, \quad x_7 = 0;$

the other basic solutions satisfy $x_1 = 0$ and

(23.0) $$7x_2 + 5x_3 + 4x_4 + 3x_5 + 2x_6 + x_7 \geqq 9$$

(here again, as in the previous section, the label (23.0) indicates the inequality obtained from (23) by making $x_1 = 0$, etc.).

As we are now in case 6, we shall examine distinctly the inequalities

(23.01) $$5x_3 + 4x_4 + 3x_5 + 2x_6 + x_7 \geqq 2,$$

and

(23.00) $$5x_3 + 4x_4 + 3x_5 + 2x_6 + x_7 \geqq 9,$$

corresponding to $x_2 = 1$ and $x_2 = 0$, respectively.

Applying now 2 to (23.01), we obtain the following basic solutions:

(23.0110000) $x_1 = 0, \quad x_2 = 1, \quad x_3 = 1, \quad x_4 = 0, \quad x_5 = 0, \quad x_6 = 0, \quad x_7 = 0,$

(23.0101000) $x_1 = 0, \quad x_2 = 1, \quad x_3 = 0, \quad x_4 = 1, \quad x_5 = 0, \quad x_6 = 0, \quad x_7 = 0,$

(23.0100100) $x_1 = 0, \quad x_2 = 1, \quad x_3 = 0, \quad x_4 = 0, \quad x_5 = 1, \quad x_6 = 0, \quad x_7 = 0,$

(23.0100010) $x_1 = 0, \quad x_2 = 1, \quad x_3 = 0, \quad x_4 = 0, \quad x_5 = 0, \quad x_6 = 1, \quad x_7 = 0,$

and the inequality

(23.010000) $$x_7 \geqq 2,$$

which has no solutions (see 3).

We come back now to the inequality (23.00), which satisfies 6. We consider the two subcases $x_3 = 1$ and $x_3 = 0$:

(23.001) $$4x_4 + 3x_5 + 2x_6 + x_7 \geqq 4,$$

(23.000) $$4x_4 + 3x_5 + 2x_6 + x_7 \geqq 9.$$

The inequality (23.001) has the basic solution (see 2) $x_4 = 1, x_5 = x_6 = x_7 = 0$, leading to

(23.0011000) $x_1 = 0, \quad x_2 = 0, \quad x_3 = 1, \quad x_4 = 1, \quad x_5 = 0, \quad x_6 = 0, \quad x_7 = 0,$

while the other basic solutions satisfy $x_4 = 0$ and

(23.0010) $$3x_5 + 2x_6 + x_7 \geqq 4;$$

the conclusion 5 shows that $x_5 = 1$ and

(23.00101) $$2x_6 + x_7 \geqq 1,$$

which, in view of 2, admits the basic solutions $x_6 = 1, x_7 = 0$ and $x_6 = 0, x_7 = 1$, leading to:

(23.0010110) $x_1 = 0, \quad x_2 = 0, \quad x_3 = 1, \quad x_4 = 0, \quad x_5 = 1, \quad x_6 = 1, \quad x_7 = 0$

and to

(23.0010101) $x_1 = 0, \quad x_2 = 0, \quad x_3 = 1, \quad x_4 = 0, \quad x_5 = 1, \quad x_6 = 0, \quad x_7 = 1,$

respectively.

We have now to consider the inequality (23.000) which, falling into the case 5, implies $x_4 = 1$ and

(23.0001) $$3x_5 + 2x_6 + x_7 \geqq 5.$$

This inequality, by the same arguments, gives $x_5 = 1$ and

(23.00011) $$2x_6 + x_7 \geqq 2;$$

then, case 2 shows that we have the basic solution

(23.0001110) $$x_1 = 0, \quad x_2 = 0, \quad x_3 = 0, \quad x_4 = 1, \quad x_5 = 1, \quad x_6 = 1, \quad x_7 = 0$$

and the inequality

(23.000110) $$x_7 \geqq 2,$$

which has no solutions (by 3).

The tree-like construction having come to an end, we have obtained all the basic solutions of (23), which we group together in Table 4 below, where we indicate by a label the solutions of the equation

(25) $$16x_1 + 7x_2 + 5x_3 + 4x_4 + 3x_5 + 2x_6 + x_7 = 9$$

(see Remark 1).

Table 4

No.	x_1^*	x_2^*	x_3^*	x_4^*	x_5^*	x_6^*	x_7^*	(25)?
1	1	0	0	0	0	0	0	
2	0	1	1	0	0	0	0	
3	0	1	0	1	0	0	0	
4	0	1	0	0	1	0	0	
5	0	1	0	0	0	1	0	V
6	0	0	1	1	0	0	0	V
7	0	0	1	0	1	1	0	
8	0	0	1	0	1	0	1	V
9	0	0	0	1	1	1	0	V

The corresponding families of solutions $\sum(S, J_S)$ of (23) are given in Table 5 below, where the dashes indicate the arbitrary variables*.

Table 5

No.	x_1	x_2	x_3	x_4	x_5	x_6	x_7
1	1	—	—	—	—	—	—
2	0	1	1	—	—	—	—
3	0	1	0	1	—	—	—
4	0	1	0	0	1	—	—
5	0	1	0	0	0	1	—
6	0	0	1	1	—	—	—
7	0	0	1	0	1	1	—
8	0	0	1	0	1	0	1
9	0	0	0	1	1	1	—

* After some practice, the reader can obtain directly the table of the families of solutions from the canonical form of an inequality.

Taking into account the transformation formulas (24), we obtain the families of solutions of the original inequality (21):

Table 6

No.	z_1	z_2	z_3	z_4	z_5	z_6	z_7
1	—	—	—	—	—	1	—
2	—	0	—	—	0	0	—
3	—	1	—	0	0	0	—
4	—	1	1	1	0	0	—
5	0	1	0	1	0	0	—
6	—	0	—	0	1	0	—
7	0	0	1	1	1	0	—
8	1	0	1	1	1	0	0
9	0	1	1	0	1	0	—

As we have announced at the beginning of this section, we can solve simultaneously* the inequality (12), the equation (2) and the strict inequality (13). It suffices to know how to solve simultaneously the inequality (14), the strict inequality

(26) $$c_1 x_1 + c_2 x_2 + \cdots + c_n x_n > d,$$

and the equation

(5) $$c_1 x_1 + c_2 x_2 + \cdots + c_n x_n = d.$$

We have already noticed (Remark 1) that the solutions of (5) are to be sought among the basic solutions of (14).

The knowledge of the families of solutions of the inequality (14) enables us to determine the families of solutions of the strict inequality (26).

We notice first, that the concept of basic solution applies to inequalities of the form (26) too. Therefore, in order to solve the inequality (26), it suffices to find its basic solutions.

To do this, we examine the basic solutions of (14); they satisfy either (i) the strict inequality (26) or (ii) the equation (5). The solutions (i) are obviously basic solutions of the strict inequality (26). As to the case (ii), consider a solution $(x_1^*, \ldots, x_n^*)$ of the equation (5), and let p be the place of the last 1 in this solution i.e.: $x_p^* = 1, x_{p+1}^* = \cdots = x_n^* = 0$. We change, in turn, each of the last $n - p$ zeros into 1, obtaining thus $n - p$ vectors which will prove to be basic solutions of the strict inequality (26).

The above described procedure provides us with all the basic solutions of (26) as proved below.

* A (perhaps easier) way is suggested in Chap. IV, § 2, note added in proofs.

More exactly, the procedure runs as follows:

Let B be the set of all the basic solutions of (14). Let M' be the set of those basic solutions of (14) which are not solutions of the equation (5). Let $S^* = (x_1^*, \ldots, x_n^*)$ be an element of $B - M'$ [i.e., a solution of (5)] and let p be the greatest index for which $x_p^* = 1$. We associate to S^* the vectors $R_j^* = (y_{j1}^*, \ldots, y_{jn}^*)$ $(j = p + 1, \ldots, n)$ defined as follows

$$y_{ji}^* = \begin{cases} x_i^* & \text{if } i \neq j, \\ 1 = \overline{x_j^*}, & \text{if } i = j. \end{cases} \tag{27}$$

The set of all the vectors R_j^* $(j = p + 1, \ldots, n)$ associated to the different elements of $B - M'$ will be denoted by M''.

Let us denote by M the set of all basic solutions of (26). We have:

Theorem 5. $M = M' \cup M''$.

Proof. (α) If S^* is a basic solution of (14) which does not satisfy (5), then S^* is obviously a basic solution of (26), so that $M' \subseteqq M$.

(β) Let $S^* = (x_1^*, \ldots, x_n^*)$ be a solution of (5) where $x_p^* = 1$, and $x_{p+1}^* = \cdots = x_n^* = 0$, and let $R_j^* = (y_{j1}^*, \ldots, y_{jn}^*)$ be defined by the above formula (27). Then R_j^* is obviously a solution of (26); we shall prove it is a basic one. Indeed, in the opposite case, reasoning as in the proof of Theorem 3, we would obtain a basic solution of (26) of the form $(y_{j1}^*, \ldots, y_{jq}^*, 0, \ldots, 0)$, with $q < j$, and $y_{jq}^* = x_q^* = 1$, but this implies $q \leqq p$ and

$$\sum_{h=1}^{q} c_h y_{jh}^* \leqq \sum_{h=1}^{p} c_h y_{jh}^* = \sum_{h=1}^{p} c_h x_h^* = \sum_{h=1}^{n} c_h x_h^* = d,$$

a contradiction.

(γ) We have proved at α) and β) that $M' \subseteqq M$ and $M'' \subseteqq M$, respectively; hence $M' \cup M'' \subseteqq M$ and it remains to demonstrate the converse inclusion.

Let $R^* = (y_1^*, \ldots, y_n^*)$ be an element of M, that is a basic solution of (26); let j be the last index for which $y_j^* = 1$, and let $S^* = (x_1^*, \ldots, x_n^*)$ be defined by

$$x_h^* = \begin{cases} y_h^*, & \text{if } h \neq j, \\ 0, & \text{if } h = j. \end{cases}$$

Since R^* was a basic solution of (26), it follows that $\sum_{h=1}^{n} c_h x_h^* \leqq d$. If $\sum_{h=1}^{n} c_h x_h^* = d$, then $R^* \in M''$. If $\sum_{h=1}^{n} c_h x_h^* < d$, we shall prove that $R^* \in M'$, i.e. that R^* is a basic solution of (14). To prove this, it suffices to consider a vector $(z_1^*, \ldots, z_n^*) \neq (x_1^*, \ldots, x_n^*)$ satisfying

$$z_h^* = \begin{cases} y_h^*, & \text{if } h \neq k, \\ 0, & \text{if } h = k, \end{cases}$$

where k is an index for which $y_k^* = 1$. Hence $k < j$, therefore $z_j^* = y_j^* = 1$, $x_k^* = y_k^* = 1$, $c_k \geqq c_j$ and

$$\sum_{h=1}^{n} c_h z_h^* = c_j + \sum_{\substack{h=1\\ h\neq k\\ h\neq j}}^{n} c_h z_h^* = c_j - c_k + c_k x_k^* + \sum_{\substack{h=1\\ h\neq k\\ h\neq j}}^{n} c_h x_h^*$$

$$= c_j - c_k + \sum_{h=1}^{n} c_h x_h^* \leqq \sum_{h=1}^{n} c_h x_h^* < d,$$

completing the proof.

Thus $R^* \in M$ implies $R^* \in M'$ or $R^* \in M''$, i.e. $M \subseteqq M' \cup M''$, hence $M = M' \cup M''$, by α) and β), thus completing the proof.

Corollary 2. The solutions of the strict inequality (26) may be determined as follows: **a)** Find the basic solutions of (26) as indicated by Theorem 5. b) Find the families of solutions as indicated by Theorem 3.

Example 3. In Example 2 we have solved the inequality

$$(21) \qquad 2\bar{z}_1 - 5z_2 + 3z_3 + 4\bar{z}_4 - 7z_5 + 16z_6 - z_7 \geqq -4,$$

which has the canonical form

$$(23) \qquad 16x_1 + 7x_2 + 5x_3 + 4x_4 + 3x_5 + 2x_6 + x_7 \geqq 9,$$

the basic solutions of which were given in Table 4.

Table 7 below gives the solutions of the equation

$$(28) \qquad 2\bar{z}_1 - 5z_2 + 3z_3 + 4\bar{z}_4 - 7z_5 + 16z_6 - z_7 = -4$$

associated to the inequality (21); they are simply the transforms of the solutions of (25), which were labeled in Table 4.

Table 7

No.	z_1	z_2	z_3	z_4	z_5	z_6	z_7
5	0	1	0	1	0	0	1
6	1	0	0	0	1	0	1
8	1	0	1	1	1	0	0
9	0	1	1	0	1	0	1

Further, in order to solve the strict inequality

$$(29) \qquad 2\bar{z}_1 - 5z_2 + 3z_3 + 4\bar{z}_4 - 7z_5 + 16z_6 - z_7 > -4,$$

we have to find the basic solutions of the canonical strict inequality

$$(30) \qquad 16x_1 + 7x_2 + 5x_3 + 4x_4 + 3x_5 + 2x_6 + x_7 > 9.$$

According to Theorem 5, these solutions are: (i) those basic solutions of (23) which do not satisfy (25) (these solutions are simply the non-labeled solutions in Table 4); (ii) the solutions of (30) associated to the solutions of (25) (the latter are the labeled solutions in Table 4). We obtain thus the table of all the basic solutions of the inequality (30):

Table 8

No.	x_1^*	x_2^*	x_3^*	x_4^*	x_5^*	x_6^*	x_7^*
1	1	0	0	0	0	0	0
2	0	1	1	0	0	0	0
3	0	1	0	1	0	0	0
4	0	1	0	0	1	0	0
5′	0	1	0	0	0	1	1
6′	0	0	1	1	1	0	0
6″	0	0	1	1	0	1	0
6‴	0	0	1	1	0	0	1
7	0	0	1	0	1	1	0
9′	0	0	0	1	1	1	1

In Tables 4 and 8, we have denoted by the same number a solution of (25) and the associated solution of (30). Notice that there is no solution of (30) associated to the solution No. 8.

From the basic solutions in Table 8 we obtain the families of solutions of (30), and then those of (29), by the same procedure as in Example 2. The result is given in Table 9:

Table 9

No.	z_1	z_2	z_3	z_4	z_5	z_6	z_7
1	—	—	—	—	—	1	—
2	—	0	—	—	0	0	—
3	—	1	—	0	0	0	—
4	—	1	1	1	0	0	—
5′	0	1	0	1	0	0	0
6′	—	0	1	0	1	0	—
6″	0	0	0	0	1	0	—
6‴	1	0	0	0	1	0	0
7	0	0	1	1	1	0	—
9′	0	1	1	0	1	0	0

The families 1, 2, 3, 4 and 7 are, of course, the same as in Table 6 while the families 5, 6 and 9 in that table were replaced by the families 5′, 6′, 6″, 6‴ and 9′ generated by the associated solutions of (30). Although Table 9 contains no family corresponding to the family 8 of Table 6, Theorem 5 assures us that the solutions of (29) which belong to the family 8 in Table 6 are not lost: they are contained in various families of Table 9.

§ 3. Systems of Linear Pseudo-Boolean Equations and/or Inequalities

The method exposed in the preceding two sections for solving a linear pseudo-Boolean equation or inequality can easily be adapted to the more general case of a system of linear equations and/or inequalities (with real coefficients).

The algorithm for solving linear systems will comprise three stages.

Stage 1. Replacing: (α)* each inequality of the form $f > 0$ by $f - 1 \geqq 0$, each inequality $g < 0$ by $-g - 1 \geqq 0$, and (β) each inequality $h \leqq 0$ by $-h \geqq 0$, we obtain a system containing inequalities of the form $F \geqq 0$, or equations $G = 0$, or both.

Stage 2. Let $x_1, \ldots, x_n$ be the unknowns of the system. Using the relations $\bar{x}_i = 1 - x_i$ and $x_j = 1 - \bar{x}_j$, we can write, for each i, the i-th inequality, in the form

$$(31) \qquad c^i_{i_1} \tilde{x}_{i_1} + c^i_{i_2} \tilde{x}_{i_2} + \cdots + c^i_{i_m} \tilde{x}_{i_m} \geqq d^i,$$

where: $x_{i_1}, \ldots, x_{i_m}$ are those variables the corresponding inequality depends effectively on, $\tilde{x}$ is either x or $\bar{x}$, so that $c^i_{i_1} \geqq c^i_{i_2} \geqq \cdots \geqq c^i_{i_m} > 0$. The equations of the system are to be written in a similar way. In other words, we bring each equation and inequality to the canonical form with respect to the variables occuring effectively in it, but without changing the notation.

Stage 3. We apply now the following idea. Each equation (inequality), considered separately, is written in the canonical form with respect to the variables $\tilde{x}$ contained in it; therefore a certain conclusion can be drawn from Table 1 (respectively, from Table 3); this deduction leads to another conclusion refering to the whole system.

For instance, when a certain inequality or equation of the system has no solutions, then the whole system is inconsistent. In the same way, if the equation $f(x_{i_1}, \ldots, x_{i_m}) = 0$ has the unique solution $x_{i_1} = x^*_{i_1}, \ldots, x_{i_m} = x^*_{i_m}$, then each solution of the system (if any!) must satisfy the remaining relations, the variables $x_{i_1}, \ldots, x_{i_m}$ (which are not necessarily exhausting the set of all the variables of the system) having the above fixed values.

Further, we cannot transpose the notion of basic solution to the case of an arbitrary system of linear inequalities; therefore, the conclusions in Table 3 are to be re-formulated so as to indicate the corresponding families of solutions. For instance, assume that the inequality (31) is in the case 2, that is $c^i_{i_1} \geqq c^i_{i_2} \geqq \cdots \geqq c^i_{i_p} \geqq d^i > c^i_{i_{p+1}} \geqq \cdots \geqq c^i_{i_m}$. Then, instead of the basic solutions

$$(32.\,k) \qquad \begin{cases} \tilde{x}_{i_k} = 1, \\ \tilde{x}_{i_1} = \cdots = \tilde{x}_{i_{k-1}} = \tilde{x}_{i_{k+1}} = \cdots = \tilde{x}_{i_m} = 0 \quad (k = 1, \ldots, p) \end{cases}$$

we have to consider simply the p branches

$$(33.\,k) \qquad \tilde{x}_{i_k} = 1, \quad \tilde{x}_{i_j} \text{ arbitrary for } j \neq k \qquad (k = 1, \ldots, p).$$

---

* (α) is applicable only when the coefficients are integer; if they are not, then rules similar to those in § 2 are to be established for solving strict inequalties.

No.	Case	Informations		
		Conclusions	Fixed variables	Remaining equation
1	$d^i < 0$	No solutions	—	—
2	$d^i = 0$	All of appearing variables fixed	$\tilde{x}_{i_1} = \cdots = \tilde{x}_{i_m} = 0$	—
3	$d^i > 0$ and $c^i_{i_1} \geqq \cdots \geqq c^i_{i_p} > d^i \geqq c^i_{i_{p+1}} \geqq \cdots \geqq c^i_{i_m}$	Part of appearing variables fixed	$\tilde{x}_{i_1} = \cdots = \tilde{x}_{i_p} = 0$	$\sum_{j=p+1}^{m} c^i_{i_j} \tilde{x}_{i_j} = d^i$
4	$d^i > 0$ and $c^i_{i_1} = \cdots = c^i_{i_p} = d^i \geqq c^i_{i_{p+1}} \geqq \cdots \geqq c^i_{i_m}$	There are $p+1$ possibilities $\alpha_1, \ldots, \alpha_p, \beta$	α_k: $\tilde{x}_{i_k} = 1, \tilde{x}_{i_1} = \cdots = \tilde{x}_{i_{k-1}} = \tilde{x}_{i_{k+1}} = \cdots = \tilde{x}_{i_m} = 0$ $(k = 1, \ldots, p)$	—
			β: $\tilde{x}_{i_1} = \cdots = \tilde{x}_{i_p} = 0$	$\sum_{j=p+1}^{m} c^i_{i_j} \tilde{x}_{i_j} = d^i$
5	$d^i > 0, \quad c^i_{i_j} < d^i \quad (j = 1, 2, \ldots, m)$ and $\sum_{j=1}^{m} c^i_{i_j} < d^i$	No solutions	—	—
6	$d^i > 0, \quad c^i_{i_j} < d^i \quad (j = 1, 2, \ldots, m)$ and $\sum_{j=1}^{m} c^i_{i_j} = d^i$	All of appearing variables fixed	$\tilde{x}_{i_1} = \cdots = \tilde{x}_{i_m} = 1$	—
7	$d^i > 0, \quad c^i_{i_j} < d^i \quad (j = 1, 2, \ldots, m)$ $\sum_{j=1}^{m} c^i_{i_j} > d^i$ and $\sum_{j=2}^{m} c^i_{i_j} < d^i$	One variable fixed	$\tilde{x}_{i_1} = 1$	$\sum_{j=2}^{m} c^i_{i_j} \tilde{x}_{i_j} = d^i - c^i_{i_1}$
8	$d^i > 0, \quad c^i_{i_j} < d^i \quad (j = 1, \ldots, m)$ $\sum_{j=1}^{m} c^i_{i_j} > d^i$ and $\sum_{j=2}^{m} c^i_{i_j} \geqq d^i$	There are two possibilities γ_1, γ_2	γ_1: $\tilde{x}_{i_1} = 1$	$\sum_{j=2}^{m} c^i_{i_j} \tilde{x}_{i_j} = d^i - c^i_{i_1}$
			γ_2: $\tilde{x}_{i_1} = 0$	$\sum_{j=2}^{m} c^i_{i_j} \tilde{x}_{i_j} = d^i$

Table 10 (Continued) B. Inequality

No.	Case	Informations		
		Conclusions	Fixed variables	Remaining inequality
1	$d^i \leqq 0$	Redundant inequality	—	—
2	$d^i > 0$ and $c^i_{i_1} \geqq \cdots \geqq c^i_{i_p} \geqq d^i > c^i_{i_{p+1}} \geqq \cdots \geqq c^i_{i_m}$	There are $p+1$ possibilities $\alpha_1, \ldots, \alpha_p, \beta$	α_k: $\tilde{x}_{i_1} = \cdots = \tilde{x}_{i_{k-1}} = 0$, $\tilde{x}_{i_k} = 1$ $(k = 1, \ldots, p)$	—
			β: $\tilde{x}_{i_1} = \cdots = \tilde{x}_{i_p} = 0$	$\sum_{j=p+1}^{m} c^i_{i_j} \tilde{x}_{i_j} \geqq d^i$
3	$d^i > 0$, $c^i_{i_j} < d^i$ $(j = 1, \ldots, m)$ and $\sum_{j=1}^{m} c^i_{i_j} < d^i$	No solutions	—	—
4	$d^i > 0$, $c^i_{i_j} < d^i$ $(j = 1, \ldots, m)$ $\sum_{j=1}^{m} c^i_{i_j} = d^i$	All of appearing variables fixed	$\tilde{x}_{i_1} = \cdots = \tilde{x}_{i_m} = 1$	—
5	$d^i > 0$, $c^i_{i_j} < d^i$ $(j = 1, \ldots, m)$ $\sum_{j=1}^{m} c^i_{i_j} > d^i$ and $\sum_{j=2}^{m} c^i_{i_j} < d^i$	One variable fixed	$\tilde{x}_{i_1} = 1$	$\sum_{j=2}^{m} c^i_{i_j} \tilde{x}_{i_j} \geqq d^i - c^i_{i_1}$
6	$d^i > 0$, $c^i_{i_j} < d^i$ $(j = 1, 2, \ldots, m)$ $\sum_{j=1}^{m} c^i_{i_j} > d^i$ and $\sum_{j=2}^{m} c^i_{i_j} \geqq d^i$	There are two possibilities γ_1, γ_2	γ_1: $\tilde{x}_{i_1} = 1$	$\sum_{j=2}^{m} c^i_{i_j} \tilde{x}_{i_j} \geqq d^i - c^i_{i_1}$
			γ_2: $\tilde{x}_{i_1} = 0$	$\sum_{j=2}^{m} c^i_{i_j} \tilde{x}_{i_j} \geqq d^i$

Of course, it is convenient to consider set-theoretically disjoint families of solutions, so that we shall follow the branches

$$(34.\,k)\qquad \begin{cases} \tilde{x}_{i_1} = \cdots = \tilde{x}_{i_{k-1}} = 0, \\ \tilde{x}_{i_k} = 1, \quad \tilde{x}_{i_j} \text{ arbitrary for } j > k \qquad (k = 1, \ldots, p), \end{cases}$$

instead of (33. k).

We give in Table 10 on pp. 67 – 68 the complete list of these conclusions.

As we see, there are cases in which some variables are fixed, or in which there are no solutions, or in which the considered equation (inequality) is redundant; we call these cases "determinate". There are other cases when we have practically no information and we are obliged to split the discussion into two cases; we call these cases "undeterminate". Finally, there are cases when the discussion is to be split into $p + 1$ cases with increased informations; we call them "partially determinate". This classification is given in Table 11 below:

Table 11

Preferential order	Equation (Table 10 A)	Inequality (Table 10 B)	Characterization
First	1, 5 2, 6 3, 7	3 1, 4 5	"Determinate"
Second	4	2	"Partially determinate"
Third	8	6	"Undeterminate"

Now the 3-rd stage of the procedure for solving a system of linear equation and/or inequalities may be continued as follows:

If some equations and inequalities belong to "determinate" cases we draw all the corresponding conclusions and collate them. Two situations may arise. If at least one equation or inequality has no solutions, or if two distinct equations or inequalities lead to conclusions of the form $\tilde{x}_i = 1$ and $\tilde{x}_i = 0$ respectively, then the system has no solutions. In the other cases, the values of certain variables are determined and this leads to a smaller system which is to be examined.

If none of the equations and inequalities is in a determinate case, but there are equations or inequalities in partially determinate cases, then we follow the conclusion corresponding to one of these cases. It seems covenient to choose the equation or inequality corresponding to the greatest p (see Table 10, the cases $4A$ and $2B$).

Finally, if all the equations and inequalities are in undeterminate cases, we split the discussion with respect to one of the variables; it seems convenient to choose (one of) the variable (s) appearing with the greatest coefficient in the system.

We have the following

Theorem 6. *The above described procedure leads to all the solutions of the considered system of linear pseudo-Boolean equations and/or inequalities.*

Of course, the above method may be enriched by adding several supplementary rules for speeding up the computations. In Example 4 below, we have deliberately abstained from using such accelerating remarks, in order to illustrate only the essence of the procedure.

Remark 2. The families of solutions obtained by the method described in this section correspond to different branches of the associated tree, so that these families are pairwise disjoint.

Example 4. Let us solve the system

$$2x_1 - 4x_2 + 8x_3 + 3x_4 - 6x_5 = -2, \tag{35.1}$$

$$5x_1 - 4x_3 + 3x_5 + 2x_6 - x_7 + 9x_8 \leqq 5, \tag{35.2}$$

$$4x_1 + 6x_2 + 4x_4 - 5x_5 - 9x_6 + 8x_7 > -1, \tag{35.3}$$

$$2x_2 - 4x_4 - x_6 + 3x_8 \geqq 1. \tag{35.4}$$

Performing the transformations indicated at the stages 1 and 2, we obtain the following equivalent system

$$8x_3 + 6\bar{x}_5 + 4\bar{x}_2 + 3x_4 + 2x_1 = 8, \tag{36.1}$$

$$9\bar{x}_8 + 5\bar{x}_1 + 4x_3 + 3\bar{x}_5 + 2\bar{x}_6 + x_7 \geqq 14, \tag{36.2}$$

$$9\bar{x}_6 + 8x_7 + 6x_2 + 5\bar{x}_5 + 4x_1 + 4x_4 \geqq 14. \tag{36.3}$$

$$4\bar{x}_4 + 3x_8 + 2x_2 + \bar{x}_6 \geqq 6. \tag{36.4}$$

Equation (36.1) is in the "partially determinate" case 4 A, while the other relations are in "undeterminate" cases. Here we have $p = 1$, therefore there are two alternatives:

α) $x_3 = 1, \quad \bar{x}_5 = \bar{x}_2 = x_4 = x_1 = 0$

and

β) $x_3 = 0$ and $6\bar{x}_5 + 4\bar{x}_2 + 3x_4 + 2x_1 = 8$.

In the alternative α, equation (36.1) vanishes and the system (36) reduces to

$$x_1 = x_4 = 0, \quad x_2 = x_3 = x_5 = 1, \tag{37.0}$$

$$9\bar{x}_8 + 2\bar{x}_6 + x_7 \geqq 5, \tag{37.2}$$

$$9\bar{x}_6 + 8x_7 \geqq 8, \tag{37.3}$$

$$3x_8 + \bar{x}_6 \geqq 0; \tag{37.4}$$

we have introduced the supplementary relation (37.0), in order to indicate the fixed variables which have led to the system (37).

In the above system, the single inequality in a determinate case is (37.4): as it is in case 1 B, it is redundant and so our system reduces to (37.2) and (37.3).

These inequalities are both in the "partially determinate" case 2 B; the p for (37.2) is 1, while for (37.3) it is 2; therefore we split the discussion into the three alternatives

α_1') $\bar{x}_6 = 1$,

α_2') $\bar{x}_6 = 0, \quad x_7 = 1$,

β') $\bar{x}_6 = x_7 = 0$ and the remaining inequality would be the absurdity $0 \geqq 8$; hence this alternative can be dropped.

In the case α_1', we obtain a single inequality to be solved:

(38.0) $$x_1 = x_4 = x_6 = 0, \quad x_2 = x_3 = x_5 = 1,$$

(38.2) $$9\bar{x}_8 + x_7 \geqq 3.$$

According to the case 2 B, we have two possibilities:

α'') $\bar{x}_8 = 1$,
β'') $\bar{x}_8 = 0$ and $x_7 \geqq 3$, which is inconsistent; hence this alternative can be dropped.

In the case α'' we have no more conditions, so that we have come to the solutions:

(39) $$x_1 = 0, \; x_2 = 1, \; x_3 = 1, \; x_4 = 0, \; x_5 = 1, \; x_6 = 0, \; x_7 \text{ arbitrary}, \; x_8 = 0.$$

Now we come back to the alternative α_2', in which (37.3) is also verified; we have

(40.0) $$x_1 = x_4 = 0, \quad x_2 = x_3 = x_5 = x_6 = x_7 = 1,$$

(40.2) $$9\bar{x}_8 \geqq 4,$$

hence $\bar{x}_8 = 1$ and we have obtained the following solution of the system (36):

(41) $$x_1 = 0, \; x_2 = 1, \; x_3 = 1, \; x_4 = 0, \; x_5 = 1, \; x_6 = 1, \; x_7 = 1, \; x_8 = 0.$$

It remains to examine the case β. We have:

(42.0) $$x_3 = 0,$$

(42.1) $$6\bar{x}_5 + 4\bar{x}_2 + 3x_4 + 2x_1 = 8,$$

(42.2) $$9\bar{x}_8 + 5\bar{x}_1 + 3\bar{x}_5 + 2\bar{x}_6 + x_7 \geqq 14,$$

(42.3) $$9\bar{x}_6 + 8x_7 + 6x_2 + 5\bar{x}_5 + 4x_1 + 4x_4 \geqq 14,$$

(42.4) $$4\bar{x}_4 + 3x_8 + 2x_2 + \bar{x}_6 \geqq 6.$$

The single relation in a "determinate" case is (42.2) which, according to the case 5 B, implies $\bar{x}_8 = 1$. Therefore:

(43.0) $$x_3 = x_8 = 0,$$

(43.1) $$6\bar{x}_5 + 4\bar{x}_2 + 3x_4 + 2x_1 = 8,$$

(43.2) $$5\bar{x}_1 + 3\bar{x}_5 + 2\bar{x}_6 + \bar{x}_7 \geqq 5,$$

(43.3) $$9\bar{x}_6 + 8x_7 + 6x_2 + 5\bar{x}_5 + 4x_1 + 4x_4 \geqq 14,$$

(43.4) $$4\bar{x}_4 + 2x_2 + \bar{x}_6 \geqq 6.$$

The single relation in a "determinate" case is (43.4), implying $\bar{x}_4 = 1$. Hence:

(44.0) $$x_3 = x_4 = x_8 = 0,$$

(44.1) $$6\bar{x}_5 + 4\bar{x}_2 + 2x_1 = 8,$$

(44.2) $$5\bar{x}_1 + 3\bar{x}_5 + 2\bar{x}_6 + \bar{x}_7 \geqq 5,$$

(44.3) $$9\bar{x}_6 + 8x_7 + 6x_2 + 5\bar{x}_5 + 4x_1 \geqq 14,$$

(44.4) $$2x_2 + \bar{x}_6 \geqq 2.$$

Reasoning as above, the equation (44.1) implies $\bar{x}_5 = 1$, hence:

(45.0) $$x_3 = x_4 = x_5 = x_8 = 0,$$

(45.1) $$4\bar{x}_2 + 2x_1 = 2,$$

(45.2) $$5\bar{x}_1 + 2\bar{x}_6 + \bar{x}_7 \geqq 2,$$

(45.3) $$9\bar{x}_6 + 8x_7 + 6x_2 + 4x_1 \geqq 9,$$

(45.4) $$2x_2 + \bar{x}_6 \geqq 2.$$

By the same reasoning, the first equation (case 3 A) implies $\bar{x}_2 = 0$, hence:

(46.0) $$x_3 = x_4 = x_5 = x_8 = 0, \quad x_2 = 1,$$

(46.1) $$2x_1 = 2,$$

(46.2) $$5\bar{x}_1 + 2\bar{x}_6 + \bar{x}_7 \geqq 2,$$

(46.3) $$9\bar{x}_6 + 8x_7 + 4x_1 \geqq 3,$$

(46.4) $$\bar{x}_6 \geqq 0.$$

The last inequality is identically satisfied (case 1 B), while the equation (46.1) has the solution $x_1 = 1$. The system reduces to:

(47.0) $$x_3 = x_4 = x_5 = x_8 = 0, \quad x_1 = x_2 = 1,$$

(47.2) $$2\bar{x}_6 + \bar{x}_7 \geqq 2,$$

(47.3) $$9\bar{x}_6 + 8x_7 \geqq -1.$$

The second inequality is identically satisfied (case 1 B), while the first one leads to the cases

α''') $\bar{x}_6 = 1$,

β''') $\bar{x}_6 = 0$, and $\bar{x}_7 \geqq 2$, which is inconsistent; hence this alternative can be dropped.

The alternative α''' corresponds to the solutions

(48) $x_1 = 1, \quad x_2 = 1, \quad x_3 = 0, \quad x_4 = 0, \quad x_5 = 0, \quad x_6 = 0, \quad x_7$ arbitrary, $\quad x_8 = 0.$

We have thus found all the families of solutions of the system (36):

Table 12

x_1	x_2	x_3	x_4	x_5	x_6	x_7	x_8
0	1	1	0	1	0	—	0
0	1	1	0	1	1	1	0
1	1	0	0	0	0	—	0

Example 5. Let us solve the system

(49.1) $$x_1 - 3x_2 + 12x_3 + x_5 - 7x_6 + x_7 - 3x_{10} + 5x_{11} + x_{12} - 6 \geqq 0,$$

(49.2) $$-3x_1 + 7x_2 - x_4 - 6x_5 + 1 \geqq 0,$$

(49.3) $$-11x_1 - x_3 + 7x_4 + x_6 - 2x_7 - x_8 + 5x_9 - 9x_{11} - 4 \geqq 0$$

(49.4) $$-5x_2 - 6x_3 + 12x_5 - 7x_6 - 3x_8 - x_9 + 8x_{10} - 5x_{12} + 8 \geqq 0,$$

(49.5) $$7x_1 + x_2 + 5x_3 - 3x_4 - x_5 + 8x_6 + 2x_8 - 7x_9 - x_{10} + 7x_{12} - 7 \geqq 0,$$

(49.6) $$2x_1 + 4x_4 + 3x_7 + 5x_8 + x_9 - x_{11} - x_{12} - 4 \geqq 0,$$

which can be brought to the equivalent form

(50.1) $12x_3 + 7\bar{x}_6 + 5x_{11} + 3\bar{x}_2 + 3\bar{x}_{10} + x_1 + x_5 + x_7 + x_{12} \geqq 19,$

(50.2) $7x_2 + 6\bar{x}_5 + 3\bar{x}_1 + \bar{x}_4 \geqq 9,$

(50.3) $11\bar{x}_1 + 9\bar{x}_{11} + 7x_4 + 5x_9 + 2\bar{x}_7 + \bar{x}_3 + x_6 + \bar{x}_8 \geqq 28,$

(50.4) $12x_5 + 8x_{10} + 7\bar{x}_6 + 6\bar{x}_3 + 5\bar{x}_2 + 5\bar{x}_{12} + 3\bar{x}_8 + \bar{x}_9 \geqq 19,$

(50.5) $8x_6 + 7x_1 + 7\bar{x}_9 + 7x_{12} + 5x_3 + 3\bar{x}_4 + 2x_8 + x_2 + \bar{x}_5 + \bar{x}_{10} \geqq 19,$

(50.6) $5x_8 + 4x_4 + 3x_7 + 2x_1 + x_9 + \bar{x}_{11} + \bar{x}_{12} \geqq 6.$

The inequality (50.3) is in the case 5 B, implying $\bar{x}_1 = 1$, that is $x_1 = 0$. We nitroduce this value in the system and we see that there is no inequality in a "determinate" case; relation (50.2) reduces to $7x_2 + 6\bar{x}_5 + \bar{x}_4 \geqq 6$, which is in the case 2 B. We have to consider the three alternatives:

α_1) $x_2 = 1$,

α_2) $x_2 = 0$, $\bar{x}_5 = 1$,

β) $x_2 = \bar{x}_5 = 0$ and $\bar{x}_4 \geqq 6$, which is inconsistent; hence this alternative can be dropped.

We begin with the alternative α_1, which means that (50.2) is verified and $x_1 = 0$ (see above), $x_2 = 1$; these values reduce the inequality (50.1) to $12x_3 + 7\bar{x}_6 + 5x_{11} + 3\bar{x}_{10} + x_5 + x_7 + x_{12} \geqq 19$, which is in the case 5 B, implying $x_3 = 1$. The values $x_1 = 0$, $x_2 = 1$, $x_3 = 1$ reduce (50.3) to the inequality $9\bar{x}_{11} + 7x_4 + 5x_9 + 2\bar{x}_7 + x_6 + \bar{x}_8 \geqq 17$, which is again in case 5 B, implying $\bar{x}_{11} = 1$. Now the inequality (50.1) is reduced to $7\bar{x}_6 + 3\bar{x}_{10} + x_5 + x_7 + x_{12} \geqq 7$, which, belonging to the case 5 B, implies $\bar{x}_6 = 1$. The inequality (50.1) is transformed into an identity, so that the system (50) reduces to

(51.0) $x_1 = x_6 = x_{11} = 0, \quad x_2 = x_3 = 1,$

(51.3) $7x_4 + 5x_9 + 2\bar{x}_7 + \bar{x}_8 \geqq 8,$

(51.4) $12x_5 + 8x_{10} + 5\bar{x}_{12} + 3\bar{x}_8 + \bar{x}_9 \geqq 12,$

(51.5) $7\bar{x}_9 + 7x_{12} + 3\bar{x}_4 + 2x_8 + \bar{x}_5 + \bar{x}_{10} \geqq 13,$

(51.6) $5x_8 + 4x_4 + 3x_7 + x_9 + \bar{x}_{12} \geqq 5.$

The inequalities (51.4) and (51.6) are in case 2 B, while (51.3) and (51.5) belong to 6 B. We perform the splitting resulting from (51.6):

α') $x_8 = 1$,

β') $x_8 = 0$, and $4x_4 + 3x_7 + x_9 + \bar{x}_{12} \geqq 5$.

In the alternative α', the inequality (51.3) reduces to $7x_4 + 5x_9 + 2\bar{x}_7 \geqq 8$, which is in the case 5 B, implying $x_4 = 1$. Hence (51.5) becomes $7\bar{x}_9 + 7x_{12} + \bar{x}_5 + \bar{x}_{10} \geqq 11$, therefore (case 5 B) $\bar{x}_9 = 1$ and $7x_{12} + \bar{x}_5 + \bar{x}_{10} \geqq 4$, implying $x_{12} = 1$, again by the conclusion 5 B. Now (51.3) becomes $2\bar{x}_7 \geqq 1$, hence $\bar{x}_7 = 1$; further, (51.4) reduces to $12x_5 + 8x_{10} \geqq 11$, implying $x_5 = 1$. These values satisfy the system (51), so that we have found the following solutions of (50):

(52) $x_1 = 0,\ x_2 = 1,\ x_3 = 1,\ x_4 = 1,\ x_5 = 1,\ x_6 = 0,\ x_7 = 0,\ x_8 = 1,\ x_9 = 0,$

x_{10} arbitrary, $\quad x_{11} = 0, \quad x_{12} = 1.$

In the alternative β', the inequality (51.5) becomes $7\bar{x}_9 + 7x_{12} + 3\bar{x}_4 + \bar{x}_5 + \bar{x}_{10} \geqq 13$, implying $\bar{x}_9 = 1$ and $7x_{12} + 3\bar{x}_4 + \bar{x}_5 + \bar{x}_{10} \geqq 6$. We must take $x_{12} = 1$, otherwise we would obtain an inconsistent inequality. Further, (51.6) becomes $4x_4 + 3x_7 \geqq 5$, implying $x_4 = 1$ and $3x_7 \geqq 1$, hence $x_7 = 1$. Now the inequalities (51.3), (51.5) and (51.6) are verified, so that the system (51) reduces to (51.4), which becomes $12x_5 + 8x_{10} \geqq 8$. This inequality is solved taking either $x_5 = 1$, or $x_5 = 0$ and $x_{10} = 1$, leading to the following solutions of the system (50):

(53) $x_1 = 0,\quad x_2 = 1,\quad x_3 = 1,\quad x_4 = 1,\quad x_5 = 1,\quad x_6 = 0,\quad x_7 = 1,\quad x_8 = 0,$

$x_9 = 0,\quad x_{10}$ arbitrary, $\quad x_{11} = 0,\quad x_{12} = 1$

and

(54) $x_1 = 0,\quad x_2 = 1,\quad x_3 = 1,\quad x_4 = 1,\quad x_5 = 0,\quad x_6 = 0,\quad x_7 = 1,$

$x_8 = 0,\quad x_9 = 0,\quad x_{10} = 1,\quad x_{11} = 0,\quad x_{12} = 1$

respectively.

Now it remains the alternative α_2:

(55.0) $$x_1 = x_2 = x_5 = 0,$$

(55.1) $$12x_3 + 7\bar{x}_6 + 5x_{11} + 3\bar{x}_{10} + x_7 + x_{12} \geqq 16,$$

(55.3) $$9\bar{x}_{11} + 7x_4 + 5x_9 + 2\bar{x}_7 + \bar{x}_3 + x_6 + \bar{x}_8 \geqq 17,$$

(55.4) $$8x_{10} + 7\bar{x}_6 + 6\bar{x}_3 + 5\bar{x}_{12} + 3\bar{x}_8 + \bar{x}_9 \geqq 14,$$

(55.5) $$8x_6 + 7\bar{x}_9 + 7x_{12} + 5x_3 + 3\bar{x}_4 + 2x_8 + \bar{x}_{10} \geqq 18,$$

(55.6) $$5x_8 + 4x_4 + 3x_7 + x_9 + \bar{x}_{11} + \bar{x}_{12} \geqq 6.$$

All these inequalities are in the case 6 B. We shall split the discussion with respect to the variable x_8, i.e.:

γ_1) $x_8 = 1$

and

γ_2) $x_8 = 0$.

In the alternative γ_1, we have $x_8 = 1$ and (55.3) is reduced to $9\bar{x}_{11} + 7x_4 + 5x_9 + 2\bar{x}_7 + \bar{x}_3 + x_6 \geqq 17$, which is in case 5 B and implies $\bar{x}_{11} = 1$. Then (55.1) is transformed into $12x_3 + 7\bar{x}_6 + 3\bar{x}_{10} + x_7 + x_{12} \geqq 16$, implying $x_3 = 1$. Now the inequality (55.4) becomes $8x_{10} + 7\bar{x}_6 + 5\bar{x}_{12} + \bar{x}_9 \geqq 14$, implying thus $x_{10} = 1$, hence (55.1) becomes $7\bar{x}_6 + x_7 + x_{12} \geqq 4$. Since $\bar{x}_6 = 0$ would imply $x_7 + x_{12} \geqq 4$, we must take $\bar{x}_6 = 1$; the inequality (55.1) is thus verified. Now (55.3) becomes $7x_4 + 5x_9 + 2\bar{x}_7 \geqq 8$ and implies $x_4 = 1$; further, the inequality (55.5) is reduced to $7\bar{x}_9 + 7x_{12} \geqq 11$, implying $\bar{x}_9 = 1$ and $7x_{12} \geqq 4$, hence $x_{12} = 1$. Also (55.1) reduces to $2\bar{x}_7 \geqq 1$, i.e. to $\bar{x}_7 = 1$. The values found above satisfy the system (55), so that we have found the following solution of the system (50):

(56) $x_1 = 0,\quad x_2 = 0,\quad x_3 = 1,\quad x_4 = 1,\quad x_5 = 0,\quad x_6 = 0,\quad x_7 = 0,$

$x_8 = 1,\quad x_9 = 0,\quad x_{10} = 1,\quad x_{11} = 0,\quad x_{12} = 1.$

In the alternative γ_2, all the inequalities (55) are in the case 6 B; we shall split the discussion with respect to x_{11}:

γ_1') $x_8 = 0,\ x_{11} = 1$

and

γ_2') $x_8 = x_{11} = 0$.

In the alternative γ_1', the inequality (55.3) becomes $7x_4 + 5x_9 + 2\bar{x}_7 + \bar{x}_3 + x_6 \geqq 16$, which implies $x_4 = x_9 = \bar{x}_7 = \bar{x}_3 = x_6 = 1$, so that (55.1) reduces to the inequality $3\bar{x}_{10} + x_{12} \geqq 11$, which is inconsistent.

It remains the alternative γ_2', in which (55.1) reduces to $12x_3 + 7\bar{x}_6 + 3\bar{x}_{10} + x_7 + x_{12} \geqq 16$ and implies $x_3 = 1$. Hence the system (55) becomes

(57.0) $$x_1 = x_2 = x_5 = x_8 = x_{11} = 0, \quad x_3 = 1,$$

(57.1) $$7\bar{x}_6 + 3\bar{x}_{10} + x_7 + x_{12} \geqq 4,$$

(57.3) $$7x_4 + 5x_9 + 2\bar{x}_7 + x_6 + \bar{x}_8 \geqq 8,$$

(57.4) $$8x_{10} + 7\bar{x}_6 + 5\bar{x}_{12} + \bar{x}_9 \geqq 11,$$

(57.5) $$8x_6 + 7\bar{x}_9 + 7x_{12} + 3\bar{x}_4 + \bar{x}_{10} \geqq 13,$$

(57.6) $$4x_4 + 3x_7 + x_9 + \bar{x}_{11} + \bar{x}_{12} \geqq 6.$$

If $x_6 = 1$, then (57.1) becomes $3\bar{x}_{10} + x_7 + x_{12} \geqq 4$, hence $\bar{x}_{10} = 1$, while (57.4) reduces to $8x_{10} + 5\bar{x}_{12} + \bar{x}_9 \geqq 11$, implying $x_{10} = 1$, a contradiction.

If $x_6 = 0$, then the inequality (57.5) reduces to $7\bar{x}_9 + 7x_{12} + 3\bar{x}_4 + \bar{x}_{10} \geqq 13$, implying $\bar{x}_9 = 1$ and $7x_{12} + 3\bar{x}_4 + \bar{x}_{10} \geqq 6$, hence $x_{12} = 1$. Now (57.3) becomes $7x_4 + 2\bar{x}_7 \geqq 7$, hence $x_4 = 1$, while (57.4) reduces to $8x_{10} \geqq 3$, implying thus $x_{10} = 1$. Further, (57.6) is transformed into $3x_7 \geqq 1$, i.e., $x_7 = 1$. These values satisfy the system (57) so that we have found the last solution of the system (50):

(58) $$x_1 = 0, \quad x_2 = 0, \quad x_3 = 1, \quad x_4 = 1, \quad x_5 = 0, \quad x_6 = 0, \quad x_7 = 1,$$
$$x_8 = 0, \quad x_9 = 0, \quad x_{10} = 1, \quad x_{11} = 0, \quad x_{12} = 1.$$

Thus the table of all the solutions of (50) is the following:

Table 13

x_1	x_2	x_3	x_4	x_5	x_6	x_7	x_8	x_9	x_{10}	x_{11}	x_{12}
0	1	1	1	1	0	0	1	0	—	0	1
0	1	1	1	1	0	1	0	0	—	0	1
0	1	1	1	0	0	1	0	0	1	0	1
0	0	1	1	0	0	0	1	0	1	0	1
0	0	1	1	0	0	1	0	0	1	0	1

The above procedures were tested on several problems solved by hand computation, and the results seem to be satisfactory. So, for instance, Example 4 of § 3, having 6 inequalities with 12 unknowns, was solved in less than one hundred minutes. (We recall that by direct inspection we should be faced with the checking of $2^{12} = 4096$ variants).

The programming of the method for a MECIPT-1 computer is in progress.

§ 4. The Methods of R. Fortet and P. Camion

The first systematic approaches to problems of bivalent programming are contained in the papers of R. FORTET [1—3] and P. CAMION [2]. These approaches include a way for solving pseudo-Boolean equations and inequalities with integer coefficients, and this will be described below.

Let us consider, for this sake, two non-negative integers A and B, and let

(59) $$A = \sum_{h=0}^{m} 2^h a_h \qquad (a_h = 0 \text{ or } 1, \quad \text{for } h = 0, \ldots, m),$$

(60) $$B = \sum_{h=0}^{m} 2^h b_h \qquad (b_h = 0 \text{ or } 1, \quad \text{for } h = 0, \ldots, m)$$

be their binary developments. If

(61) $$A + B = \sum_{k=0}^{m+1} 2^k s_k \qquad (s_k = 0 \text{ or } 1, \quad \text{for } k = 0, \ldots, m+1)$$

is the binary development of $A + B$, then the coefficients s_k depend on the a_h's and b_h's, i.e.

(62) $$s_k = s_k(a_1, \ldots, a_m, b_1, \ldots, b_m) \qquad (k = 0, \ldots, m+1).$$

where s_k are Boolean functions.

More generally, if $A_1, \ldots, A_p$ are m non-negative integers having the binary developments

(63) $$A_i = \sum_{h=0}^{m_i} 2^h a_{i_h} \qquad (i = 1, \ldots, p),$$

then the binary coefficients s_k of the sum

(64) $$\sum_{i=1}^{p} A_i = \sum_{k=0}^{r} 2^k s_k$$

are Boolean functions of the coefficients a_{ih}:

(65) $$s_k = s_k(a_{11}, \ldots, a_{1m_1}, \ldots, a_{p1}, \ldots, a_{pm_p}) \qquad (k = 0, \ldots, r).$$

Further, if A is given by (59) and x is a binary (0, 1) variable, then the product $A\,x$ has the binary development

$$A\,x = \sum_{h=0}^{m} 2^h a_h\,x \qquad (a_h\,x = 0 \text{ or } 1, \quad \text{for } h = 0, \ldots, m)$$

The above remarks allow us to obtain the binary development of an expression of the form

(66) $$\sum_{i=1}^{p} A_i\,x_i,$$

where A_i are non-negative integers, while x_i are binary variables, i.e.

(67) $$\sum_{i=1}^{p} A_i\,x_i = \sum_{k=0}^{r} 2^k t_k,$$

where t_k are Boolean functions of the a_{ih}'s and x_i's:

(68) $$t_k = t_k(a_{11}, \ldots, a_{1m_1}, \ldots, a_{p1} \ldots, a_{pm_p}, x_1, \ldots, x_p) \quad (k = 0, \ldots, r)$$

Now, every linear pseudo-Boolean equation with integer coefficients may be written in the form

$$\sum_{i=1}^{p} A_i \tilde{x}_i = \sum_{j=1}^{q} B_j \tilde{x}_{p+j} \tag{69}$$

where A_i and B_j are positive coefficients, while each $\tilde{x}$ is either x or $\bar{x}$. In view of the above discussion, the right-hand side of (69) has the form

$$\sum_{j=1}^{q} B_j \tilde{x}_{p+j} = \sum_{k=0}^{r} 2^k u_k(b_{11}, \ldots, b_{q n_q}, \tilde{x}_{p+1}, \ldots, \tilde{x}_{p+q}), \tag{70}$$

so that the pseudo-Boolean equation (69) is equivalent to the system of *Boolean equations*

$$t_k(a_{11}, \ldots, a_{p m_p}, \tilde{x}_1, \ldots, \tilde{x}_p) = u_k(b_{11}, \ldots, b_{q n_q}, \tilde{x}_{p+1}, \ldots, \tilde{x}_{p+q}) \quad (k = 0, \ldots, r) \tag{71}$$

Further, if

$$C = \sum_{k=0}^{r} 2^k c_k \tag{72}$$

$$D = \sum_{k=0}^{r} 2^k d_k \tag{73}$$

are the binary developments of the non-negative integers C, D, then, as it will be shown below, the inequality $C \leqq D$ is equivalent to a Boolean equation of the form

$$R_r(c_0, \ldots, c_r, d_0, \ldots, d_r) = 1 \tag{74}$$

where R_r is a certain recursively defined Boolean function.

On the other hand, every linear pseudo-Boolean inequality with integer coefficients may be written in the form

$$\sum_{i=1}^{p} A_i \tilde{x}_i \leqq \sum_{j=1}^{q} B_j \tilde{x}_{p+j}, \tag{75}$$

so that it is equivalent to the *Boolean equation*

$$\begin{aligned} R_r[&t_0(a_{11}, \ldots, a_{p m_p}, \tilde{x}_1, \ldots, \tilde{x}_p), \ldots, t_r(a_{11}, \ldots, a_{p m_p}, \tilde{x}_1, \ldots, \tilde{x}_p), \\ &u_0(b_{11}, \ldots, b_{q n_q}, \tilde{x}_{p+1}, \ldots, \tilde{x}_{p+q}), \ldots, \\ &u_r(b_{11}, \ldots, b_{q n_q}, \tilde{x}_{p+1}, \ldots, \tilde{x}_{p+q})] = 1. \end{aligned} \tag{76}$$

The above discussion shows that:

(i) the knowledge of the functions t_k allows the reduction of any linear pseudo-Boolean equation with integer coefficients to a system of Boolean equations.

(ii) the knowledge of the functions t_k and R_r allows the reduction of any linear pseudo-Boolean inequality with integer coefficients to a Boolean equation.

Methods for finding the functions t_k and R_k were given by R. FORTET in [1—3] and by P. CAMION in [2].

P. CAMION suggests the use of the sum modulo 2 (for the properties of $\oplus$ see § 5 Chapter I)

$$x \oplus y = x \bar{y} \cup \bar{x} y. \tag{I.87}$$

Namely, it is easy to deduce that the binary coefficients s_k of the sum (61) are given by

$$s_0 = a_0 \oplus b_0, \tag{77.0}$$

$$s_1 = a_1 \oplus b_1 \oplus r_1, \tag{77.1}$$

$$s_2 = a_2 \oplus b_2 \oplus r_2, \tag{77.2}$$

$$\cdots\cdots\cdots\cdots$$

$$s_m = a_m \oplus b_m \oplus r_m, \tag{77. p}$$

$$s_{m+1} = r_m, \tag{77. p + 1}$$

where

$$r_1 = a_0 b_0, \tag{78.1}$$

$$r_2 = (a_1 \oplus b_1) r_1 \oplus a_1 b_1, \tag{78.2}$$

$$r_3 = (a_2 \oplus b_2) r_2 \oplus a_2 b_2, \tag{78.3}$$

$$\cdots\cdots\cdots\cdots$$

$$r_m = (a_{m-1} \oplus b_{m-1}) r_{m-1} \oplus a_{m-1} b_{m-1}. \tag{78. p}$$

By iteration, we can obtain the binary coefficients of the sum (64).

Variants of this method may be found in P. CAMION [2].

R. FORTET [3] suggests a procedure based on the use of symmetric Boolean functions.

As concerns the inequality

$$C \leqq D, \tag{79}$$

where C and D have the binary developments (72) and (73), the function R_r is obviously given by the recurrence formulas

$$R_r = \bar{c}_r d_r \cup (\bar{c}_r \cup d_r) R_{r-1}, \tag{80. r}$$

$$\cdots\cdots\cdots\cdots$$

$$R_2 = \bar{c}_2 d_2 \cup (\bar{c}_2 \cup d_2) R_1, \tag{80.2}$$

$$R_1 = \bar{c}_1 d_1 \cup (\bar{c}_1 \cup d_1) R_0, \tag{80.1}$$

$$R_0 = \bar{c}_0 \cup d_0. \tag{80.0}$$

A similar treatment can be applied to the strict inequality

$$C < D; \tag{81}$$

in this case, the function R_0 is to be replaced by

$$R_0' = \bar{c}_0 d_0. \tag{82}$$

Example 6. Let us compute the binary development of the pseudo-Boolean function

(83) $$f = 3\bar{x}_1 + 2x_2 + 4x_3.$$

We have

$$\begin{aligned} f &= [(\bar{x}_1 + 2x_1) + (0 + 2x_2)] + 4x_3 \\ &= [(\bar{x}_1 \oplus 0) + 2(\bar{x}_1 \oplus x_2 \oplus 0) + 2^2(0 \oplus \bar{x}_1 x_2)] + 4x_3 \\ &= (\bar{x}_1 + 2(\bar{x}_1 \oplus x_2) + 2^2 \bar{x}_1 x_2) + (0 + 2.0 + 2^2 x_3) \\ &= (\bar{x}_1 \oplus 0) + 2(\bar{x}_1 \oplus x_2 \oplus 0 \oplus 0) + 2^2(\bar{x}_1 x_2 \oplus x_3 \oplus 0) + 2^3(0 \oplus \bar{x}_1 x_2 x_3) \end{aligned}$$

hence

(84) $$f = \bar{x}_1 + 2(\bar{x}_1 \oplus x_2) + 2^2(\bar{x}_1 x_2 \oplus x_3) + 2^3 \bar{x}_1 x_2 x_3.$$

Example 7. In order to illustrate a variant of the above method, also suggested by P. Camion, let us find the binary development of

(85) $$g = 5x_4 + 2\bar{x}_5 + 6.$$

We have

$$\begin{aligned} g &= (x_4 + 2^2 x_4) + 2\bar{x}_5 + (2 + 2^2) = x_4 + 2(\bar{x}_5 + 1) + 2^2(x_4 + 1) \\ &= x_4 + [2(\bar{x}_5 \oplus 1) + 2^2 \bar{x}_5] + [2^2(x_4 \oplus 1) + 2^3 x_4] \\ &= x_4 + 2x_5 + 2^2(\bar{x}_5 + \bar{x}_4) + 2^3 x_4 \\ &= x_4 + 2x_5 + 2^2(\bar{x}_5 \oplus \bar{x}_4) + 2^3 \bar{x}_5 \bar{x}_4 + 2^3 x_4 \\ &= x_4 + 2x_5 + 2^2(x_5 \oplus x_4) + 2^3(\bar{x}_5 \bar{x}_4 + x_4) \\ &= x_4 + 2x_5 + 2^2(x_4 \oplus x_5) + 2^3(\bar{x}_5 \bar{x}_4 \oplus x_4) + 2^4.0, \end{aligned}$$

hence

(86) $$g = x_4 + 2x_5 + 2^2(x_4 \oplus x_5) + 2^3(x_4 x_5 \oplus \bar{x}_5).$$

Example 8. Let us solve the pseudo-Boolean equation

(87) $$3\bar{x}_1 + 2x_2 + 4x_3 = 5x_4 + 2\bar{x}_5 + 6.$$

Using the binary developments found in Examples 6 and 7, we see that (87) is equivalent to the following system of Boolean equations (written with the aid of the sum modulo 2):

(88.0) $$\bar{x}_1 = x_4,$$

(88.1) $$\bar{x}_1 \oplus x_2 = x_5,$$

(88.2) $$\bar{x}_1 x_2 \oplus x_3 = x_4 \oplus x_5,$$

(88.3) $$\bar{x}_1 x_2 x_3 = x_4 x_5 \oplus \bar{x}_5.$$

Performing the substitutions (88.0) and (88.1) in (88.2) and (88.3), we find

(88′.2) $$\bar{x}_1 x_2 \oplus x_3 = x_2,$$

(88′.3) $$\bar{x}_1 x_2 x_3 = \bar{x}_1 x_2 \oplus \bar{x}_2.$$

Further, we eliminate x_3 and we get $\bar{x}_1 x_2(\bar{x}_1 x_2 \oplus x_2) = \bar{x}_1 x_2 \oplus \bar{x}_2$, or else $0 = x_1 x_2 \oplus 1$, i.e.

(88″.3) $$x_1 x_2 = 1.$$

It follows from (88″.3), (88′.2), (88.0) and (88.1) that the unique solution is

(89) $$x_1 = 1, \quad x_2 = 1, \quad x_3 = 1, \quad x_4 = 0, \quad x_5 = 1.$$

Example 9. Let us solve the pseudo-Boolean inequality

(90) $$3\bar{x}_1 + 2x_2 + 4x_3 \leq 5x_4 + 2\bar{x}_5 + 6.$$

Using formulas (80) and the results of Examples 6 and 7, we see that

(91.0) $R_0 = x_1 \cup x_4,$

(91.1) $R_1 = (x_1 \oplus x_2)\, x_5 \cup [(x_1 \oplus x_2) \cup x_5]\, R_0,$

(91.2) $R_2 = (\bar{x}_1 x_2 \oplus \bar{x}_3)(x_4 \oplus x_5) \cup [(\bar{x}_1 x_2 \oplus \bar{x}_3) \cup (x_4 \oplus x_5)]\, R_1,$

(91.3) $R_3 = (\bar{x}_1 x_2 x_3 \oplus 1)(x_4 x_5 \oplus \bar{x}_5) \cup [(\bar{x}_1 x_2 x_3 \oplus 1) \cup (x_4 x_5 \oplus \bar{x}_5)]\, R_2.$

Now the problem is to solve the equation $R_3 = 1$; we can do this by applying the idea of the method of bifurcations.

Since R_3 is a disjunction of two terms, we can solve separately the equations

(92.1) $$(\bar{x}_1 x_2 x_3 \oplus 1)(x_4 x_5 \oplus \bar{x}_5) = 1,$$

(92.2) $$[(\bar{x}_1 x_2 x_3 \oplus 1) \cup (x_4 x_5 \oplus \bar{x}_5)]\, R_2 = 1.$$

The equation (92.1) is equivalent to the system of conditions $\bar{x}_1 x_2 x_3 = 0$ and $x_4 x_5 \oplus x_5 = 0$. The latter equality may be also written in the form $\bar{x}_4 x_5 = 0$, so that (92.1) is equivalent to

(92′.1) $$\bar{x}_1 x_2 x_3 \cup \bar{x}_4 x_5 = 0.$$

Since we are interested in obtaining disjoint families of solutions, we may add the condition $\bar{x}_1 x_2 x_3 \cup \bar{x}_4 x_5 = 1$ to the equation (92.2), obtaining thus

$$(\bar{x}_1 x_2 x_3 \cup \bar{x}_4 x_5)\, [(\bar{x}_1 x_2 x_3 \oplus 1) \cup (x_4 x_5 \oplus \bar{x}_5)]\, R_2 = 1,$$

or else

(92′.2) $$[\bar{x}_1 x_2 x_3 (x_4 x_5 \oplus \bar{x}_5) \cup \bar{x}_4 x_5 (\bar{x}_1 x_2 x_3 \oplus 1)]\, R_2 = 1.$$

Now we solve separately the equations

(92.2.1) $$\bar{x}_1 x_2 x_3 (x_4 x_5 \oplus \bar{x}_5)\, R_2 = 1,$$

(92.2.2) $$\bar{x}_4 x_5 (\bar{x}_1 x_2 x_3 \oplus 1)\, R_2 = 1.$$

We solve first (92.2.1), which implies $x_1 = 0,\ x_2 = x_3 = 1$, so that formulas (91) become

$$R_0 = x_4,$$

$$R_1 = x_5 \cup x_5 R_0 = x_5,$$

$$R_2 = (x_4 \oplus x_5) \cup R_1 = x_4 \cup x_5,$$

hence (92.2.1) reduces to

$$\bar{x}_1 x_2 x_3 (x_4 x_5 \oplus \bar{x}_5)(x_4 \cup x_5) = 1,$$

or, equivalently, to

(92′.2.1) $$\bar{x}_1 x_2 x_3 x_4 = 1.$$

The equation (92.2.2) implies $x_4 = 0,\ x_5 = 1$, so that formulas (91) become

$$R_0 = x_1,$$

$$R_1 = (x_1 \oplus x_2) \cup R_0 = x_1 \cup (x_1 \oplus x_2) = x_1 \oplus x_2 \oplus x_1 x_2,$$

$$R_2 = (\bar{x}_1 x_2 \oplus \bar{x}_3) \cup R_1 = (x_1 \oplus x_2 \oplus x_1 x_2) \cup (\bar{x}_1 x_2 \oplus \bar{x}_3)$$

hence (92.2.2) reduces to

$$\bar{x}_4 x_5 (\bar{x}_1 x_2 x_3 \oplus 1)\, [(x_1 \oplus x_2 \oplus x_1 x_2) \cup (\bar{x}_1 x_2 \oplus \bar{x}_3)] = 1,$$

or, equivalently, to

(92.′2.2) $$\bar{x}_4 x_5 (x_1 \cup \bar{x}_3) = 1.$$

In conclusion, the original problem is now reduced to that of solving separately the equations (92′.1), (92′.2.1) and (92′.2.2). But (92′.1) is equivalent to the conditions $x_2 x_3 \leqq x_1$ and $x_5 \leqq x_4$, leading immediately to the families of solutions; the equation (92′.2.1) has two solutions; finally, the solutions of (92′.2.2) are immediately obtained.

We obtain thus the following solutions:

Table 14

x_1	x_2	x_3	x_4	x_5	Corresponding to
—	0	0	—	0	(92′.1)
—	0	0	1	1	
—	0	1	—	0	
—	0	1	1	1	
—	1	0	—	0	
—	1	0	1	1	
1	1	1	—	0	
1	1	1	1	1	
0	1	1	1	—	(92′.2.1)
1	—	—	0	1	(92′.2.2)
0	—	0	0	1	

Chapter IV

Nonlinear Pseudo-Boolean Equations and Inequalities

In Chapter III we have proposed a method for the determination of all the solutions of a system of *linear* pseudo-Boolean equations and/or inequalities. The aim of this chapter is to solve the problem in case of a system of *arbitrary* (i.e. linear and/or nonlinear), equations and/or inequalities, as in P. L. HAMMER and S. RUDEANU [5, 6].

We recall that a Boolean function has bivalent (0, 1) variables and bivalent values, while a pseudo-Boolean function has again bivalent values, but takes real values.

In this chapter we associate to each pseudo-Boolean equation (or inequality, or system of equations and/or inequalities) a "characteristic" Boolean equation which has the same solutions as the original system (§§ 1, 2, 3). This idea allows also the inclusion of logical conditions in the system (§ 4).

The construction of the characteristic equation is based on the reduction of the general case to the linear one; this "linearization" process does not raise computational difficulties.

The problem is now reduced to that of solving the characteristic equation. This task is done using a procedure which gives the solutions grouped into pairwise disjoint "families of solutions" (§ 5).

Another method of linearization, due to R. FORTET [2], will be described in § 7.

§ 1. The Characteristic Function in the Linear Case

Let $\sum(x_1, \ldots, x_n)$ denote a pseudo-Boolean equation, or inequality, or system of pseudo-Boolean equations and/or inequalities.

Definition 1. The *characteristic equation* of $\sum(x_1, \ldots, x_n)$ is a Boolean equation

$$\Phi(x_1, \ldots, x_n) = 1 \tag{1}$$

which has the same solutions as $\sum(x_1, \ldots, x_n)$; the Boolean function $\Phi(x_1, \ldots, x_n)$ will be called the *characteristic function** of $\sum(x_1, \ldots, x_n)$.

In other words, the characteristic function of a pseudo-Boolean system is simply the characteristic function of the set of its solutions.

Now, we recall the well-known interpolation formula for Boolean functions:

$$\Psi(x_1, \ldots, x_n) = \bigcup_{\alpha_1, \ldots, \alpha_n} \Psi(\alpha_1, \ldots, \alpha_n)\, x_1^{\alpha_1} \ldots x_n^{\alpha_n} \tag{2}$$

where $\bigcup\limits_{\alpha_1, \ldots, \alpha_n}$ means that the disjunction is extended over all 2^n possible systems of values $0, 1$ of $\alpha_1, \ldots, \alpha_n$, and the notation x^α means

$$x^\alpha = \begin{cases} x, & \text{if } \alpha = 1, \\ \bar{x}, & \text{if } \alpha = 0. \end{cases} \tag{3}$$

In other words, we have

$$\Psi(x_1, \ldots, x_n) = \bigcup_{\alpha_1, \ldots, \alpha_n}^{1} x_1^{\alpha_1} \ldots x_n^{\alpha_n} \tag{4}$$

where $\bigcup\limits_{\alpha_1, \ldots, \alpha_n}^{1}$ means that the disjunction is extended only over those values of the vector $(\alpha_1, \ldots, \alpha_n)$ for which $\Psi(\alpha_1, \ldots, \alpha_n) = 1$.

Therefore, the characteristic function Φ of $\sum(x_1, \ldots, x_n)$ is given by the following formula:

$$\Phi(x_1, \ldots, x_n) = \bigcup_{\alpha_1, \ldots, \alpha_n}^{\Sigma} x_1^{\alpha_1} \ldots x_n^{\alpha_n}, \tag{5}$$

where $\bigcup\limits_{\alpha_1, \ldots, \alpha_n}^{\Sigma}$ means that the disjunction is extended over all the solutions $(\alpha_1, \ldots, \alpha_n)$ of $\sum(x_1, \ldots, x_n)$.

Now, the results of Chapter III allow the immediate construction of the characteristic function in the linear case. The necessity of this construction will become clear in the next section.

a) Linear Equations

In the case of a single linear pseudo-Boolean equation, the knowledge of all the solutions (obtained, for instance, as in Chapter III), permits the direct determination of the characteristic function, via formula (5).

Example 1. The linear equation (7) in Example 1 of Chapter III, § 1, i.e.

$$4x_1 + \bar{x}_1 - 3x_2 + \bar{x}_2 + 5x_3 - 2x_4 + 5x_5 + 2x_6 - x_7 = 7 \tag{6}$$

* This concept was independently introduced by R. FAURE and Y. MALGRANGE [1], and by P. L. HAMMER [3, 4], who termed it "reduct". See also M. DENIS-PAPIN and Y. MALGRANGE [1].

was shown to have the solutions

Table 1

x_1	x_2	x_3	x_4	x_5	x_6	x_7
0	1	1	0	1	0	1
0	1	1	1	1	1	1
1	0	1	1	0	0	1
0	0	1	0	0	0	0
0	0	1	1	0	1	0
1	1	1	0	0	1	1
1	0	0	1	1	0	1
0	0	0	0	1	0	0
0	0	0	1	1	1	0
1	1	0	0	1	1	1
1	0	0	0	0	1	0

Hence the characteristic function is

$$(7)\qquad \Phi_1 = \bar{x}_1 x_2 x_3 \bar{x}_4 x_5 \bar{x}_6 x_7 \cup \bar{x}_1 x_2 x_3 x_4 x_5 x_6 x_7 \cup$$
$$\cup x_1 \bar{x}_2 x_3 x_4 \bar{x}_5 \bar{x}_6 x_7 \cup \bar{x}_1 \bar{x}_2 x_3 \bar{x}_4 \bar{x}_5 \bar{x}_6 \bar{x}_7 \cup \bar{x}_1 \bar{x}_2 x_3 x_4 \bar{x}_5 x_6 \bar{x}_7 \cup x_1 x_2 x_3 \bar{x}_4 \bar{x}_5 x_6 x_7 \cup$$
$$\cup x_1 \bar{x}_2 \bar{x}_3 x_4 x_5 \bar{x}_6 x_7 \cup \bar{x}_1 \bar{x}_2 \bar{x}_3 \bar{x}_4 x_5 \bar{x}_6 \bar{x}_7 \cup \bar{x}_1 \bar{x}_2 \bar{x}_3 x_4 x_5 x_6 \bar{x}_7 \cup x_1 x_2 \bar{x}_3 \bar{x}_4 x_5 x_6 x_7 \cup$$
$$\cup x_1 \bar{x}_2 \bar{x}_3 \bar{x}_4 \bar{x}_5 x_6 \bar{x}_7 .$$

b) Linear Inequalities

The method given in Chapter III for solving a linear inequality, yielded the solutions grouped into "families of solutions". A family $\mathscr{F}$ of solutions was defined as being a set of solutions characterized by the fact that certain variables have fixed values, while the others remain arbitrary:

$$(8)\quad \mathscr{F}:\ x_{h_1} = \xi_{h_1}, \ldots, x_{h_m}\xi = {}_{h_{m(h)}}; \qquad x_{h_{m(h)+k}} \text{ arbitrary for}$$
$$k = 1, \ldots, n - m(h).$$

In Chapter III we have indicated a procedure for obtaining a system $\mathscr{F}_1, \ldots, \mathscr{F}_p$ of families of solutions with the property that the set $\mathscr{S}$ of *all* the solutions is expressed as the set-theoretical join

$$(9)\qquad \mathscr{S} = \mathscr{F}_1 \cup \cdots \cup \mathscr{F}_p .$$

If we take into account relation (9) and the idempotency law ($z = z \cup z = z \cup z \cup z =$ etc.), formula (5) becomes

$$(10)\qquad \Phi(x_1, \ldots, x_n) = \bigcup_{h=1}^{p} \bigcup_{(\alpha_1, \ldots, \alpha_n) \in \mathscr{F}_h} x_1^{\alpha_1} \ldots x_n^{\alpha_n} .$$

Let us now notice that for a family $\mathscr{F}$, formula (8) implies

$$(11)\qquad \bigcup_{(\alpha_1, \ldots, \alpha_n) \in \mathscr{F}} x_1^{\alpha_1} \ldots x_n^{\alpha_n} = x_{h_1}^{\xi_{h_1}} \ldots x_{h_{m(h)}}^{\xi_{h_{m(h)}}}$$

because the left-hand side of (11) is, in fact, equal to

$$x_{h_1}^{\xi_{h_1}} \dots x_{h_m}^{\xi_{h_m}} \bigcup_{\alpha_{h_{m+1}}, \dots, \alpha_{h_n}} x_{h_{m+1}}^{\alpha_{h_{m+1}}} \dots x_{h_n}^{\alpha_{h_n}}$$

where, $\alpha_{h_{m+1}}, \dots, \alpha_{h_n}$ taking all possible values 0 and 1, make the last disjunction equal to 1 (where h_m stands for $h_{m(h)}$).

Now, applying formula (11), we see that each family $\mathcal{F}_h$ $(h = 1, \dots, p)$ is characterized by a conjunction

$$x_{h_1}^{\xi_{h_1}} \dots x_{h_{m(h)}}^{\xi_{h_{m(h)}}} = C_h. \tag{12}$$

Therefore, from (10), (11) and (12) we have

Theorem 1. *The characteristic function of a linear inequality may be written as*

$$\Phi(x_1, \dots, x_n) = C_1 \cup \dots \cup C_p. \tag{13}$$

Example 2. The linear inequality (21) in Example 2 of Chapter III, § 2, i.e.

$$2\bar{x}_1 - 5x_2 + 3x_3 + 4\bar{x}_4 - 7x_5 + 16x_6 - x_7 \geqq -4, \tag{III.21}$$

was shown to have the following families of solutions:

Table 2

No.	x_1	x_2	x_3	x_4	x_5	x_6	x_7
1	—	—	—	—	—	1	—
2	—	0	—	—	0	0	—
3	—	1	—	0	0	0	—
4	—	1	1	1	0	0	—
5	0	1	0	1	0	0	—
6	—	0	—	0	1	0	—
7	0	0	1	1	1	0	—
8	1	0	1	1	1	0	0
9	0	1	1	0	1	0	—

Formulas (12) and (13) show that the corresponding characteristic function is

$$\Phi_2 = x_6 \cup \bar{x}_2 \bar{x}_5 \bar{x}_6 \cup x_2 \bar{x}_4 \bar{x}_5 \bar{x}_6 \cup x_2 x_3 x_4 \bar{x}_5 \bar{x}_6 \cup \bar{x}_1 x_2 \bar{x}_3 x_4 \bar{x}_5 \bar{x}_6 \cup \\ \cup \bar{x}_2 \bar{x}_4 x_5 \bar{x}_6 \cup \bar{x}_1 \bar{x}_2 x_3 x_4 x_5 \bar{x}_6 \cup x_1 \bar{x}_2 x_3 x_4 x_5 \bar{x}_6 \bar{x}_7 \cup \bar{x}_1 x_2 x_3 \bar{x}_4 x_5 \bar{x}_6. \tag{14}$$

c) Linear Systems

In Chapter III the solutions of a system of linear pseudo-Boolean equations and/or inequalities were also grouped into families of solutions. Therefore the characteristic function of a linear system may be obtained in the same way as in the case of a single linear inequality.

Example 3. The linear system (35) in Example 4 of Chapter III, § 3, i.e.

$$2x_1 - 4x_2 + 8x_3 + 3x_4 - 6x_5 = -2, \tag{15.1}$$

$$5x_1 - 4x_3 + 3x_5 + 2x_6 - x_7 + 9x_8 \leqq 5, \tag{15.2}$$

$$4x_1 + 6x_2 + 4x_4 - 5x_5 - 9x_6 + 8x_7 > -1, \tag{15.3}$$

$$2x_2 - 4x_4 - x_6 + 3x_8 \geqq 1, \tag{15.4}$$

was shown to have the following solutions:

Table 3

x_1	x_2	x_3	x_4	x_5	x_6	x_7	x_8
0	1	1	0	1	0	—	0
0	1	1	0	1	1	1	0
1	1	0	0	0	0	—	0

Hence the characteristic function is

(16) $$\Phi_3 = \bar{x}_1 x_2 x_3 \bar{x}_4 x_5 \bar{x}_6 \bar{x}_8 \cup \bar{x}_1 x_2 x_3 \bar{x}_4 x_5 x_6 x_7 \bar{x}_8 \cup x_1 x_2 \bar{x}_3 \bar{x}_4 \bar{x}_5 \bar{x}_6 \bar{x}_8 .$$

§ 2. The Characteristic Function for a Nonlinear Equation or Inequality

Let us consider a nonlinear pseudo-Boolean equation with the unknowns $x_1, \ldots, x_n$:

(17) $$a_1 P_1 + \cdots + a_m P_m = b,$$

where each P_i $(i = 1, \ldots, m)$ stands for a certain conjunction (i.e. a product of variables with or without bars):

(18) $$P_i = x_{i_1}^{\pi_{i_1}} \ldots x_{i_{k(i)}}^{\pi_{i_{k(i)}}} .$$

Let us replace the product P_i by a single bivalent variable y_i and solve the resulting linear pseudo-Boolean equation

(19) $$a_1 y_1 + \cdots + a_m y_m = b,$$

where $y_1, \ldots, y_m$ are treated as independent variables.

If $\Psi(y_1, \ldots, y_m)$ is the characteristic equation of (19), obtained as in § 1 *a*), then the Boolean function

(20) $$\Phi(x_1, \ldots, x_n) = \Psi\left[x_{1_1}^{\pi_{1_1}} \ldots x_{1_{k(1)}}^{\pi_{1_{k(1)}}}, \ldots, x_{m_1}^{\pi_{m_1}} \ldots x_{m_{k(m)}}^{\pi_{m_{k(m)}}}\right]$$

will be the characteristic function of (17).

In the case of a linear inequality we apply the same procedure.

Example 4. Let us solve the pseudo-Boolean equation

(21) $$-6 x_1 \bar{x}_2 x_3 - 4 x_2 x_4 + 2 x_2 x_4 \bar{x}_5 + 4 \bar{x}_3 \bar{x}_4 = -2 .$$

Putting

(22) $$x_1 \bar{x}_2 x_3 = y_1, \quad x_2 x_4 = y_2, \quad x_2 x_4 \bar{x}_5 = y_3, \quad \bar{x}_3 \bar{x}_4 = y_4,$$

we have the linear equation

(23) $$-6 y_1 - 4 y_2 + 2 y_3 + 4 y_4 = -2,$$

which may be solved as in Chapter III and has the solutions

(24.1) $$y_1 = 0, \quad y_2 = 1, \quad y_3 = 1, \quad y_4 = 0$$

and

(24.2) $$y_1 = 1, \quad y_2 = 0, \quad y_3 = 0, \quad y_4 = 1 .$$

Hence the characteristic function of (23) is

(25) $$\Psi_1 = \bar{y}_1 y_2 y_3 \bar{y}_4 \cup y_1 \bar{y}_2 \bar{y}_3 y_4;$$

from (25) and (22) we derive the characteristic function of (21):

$$\Phi_4 = (\bar{x}_1 \cup x_2 \cup \bar{x}_3) \cdot x_2 x_4 \cdot x_2 x_4 \bar{x}_5 \cdot (x_3 \cup x_4) \cup$$
$$\cup x_1 \bar{x}_2 x_3 (\bar{x}_2 \cup \bar{x}_4)(\bar{x}_2 \cup \bar{x}_4 \cup x_5) \bar{x}_3 \bar{x}_4,$$

or else

(26) $$\Phi_4 = x_2 x_4 \bar{x}_5.$$

The characteristic equation $\Phi_4 = 1$ shows that the solutions of (21) are:

(27) $$x_2 = x_4 = 1, \quad x_5 = 0, \quad x_1 \text{ and } x_3 \text{ arbitrary.}$$

Example 5. In order to solve the pseudo-Boolean inequality

(28) $$7x_1 x_2 x_3 + 5x_2 x_4 x_6 x_7 x_8 - 4x_3 x_8 - 2\bar{x}_1 x_4 x_8 - x_4 \bar{x}_5 x_6 \leqq 3,$$

we set

(29) $$x_1 x_2 x_3 = y_1, \quad x_2 x_4 x_6 x_7 x_8 = y_2, \quad x_3 x_8 = y_3, \quad \bar{x}_1 x_4 x_8 = y_4, \quad x_4 \bar{x}_5 x_6 = y_5.$$

Thus we obtain the inequality

(30) $$7y_1 + 5y_2 - 4y_3 - 2y_4 - y_5 \leqq 3,$$

whose families of solutions, obtained as in Chapter III, are

Table 4

y_1	y_2	y_3	y_4	y_5
0	0	—	—	—
0	1	1	—	—
0	1	0	1	—
1	0	1	—	—

leading to the characteristic function

(31) $$\Psi_2 = \bar{y}_1 \bar{y}_2 \cup \bar{y}_1 y_2 y_3 \cup \bar{y}_1 y_2 \bar{y}_3 y_4 \cup y_1 \bar{y}_2 y_3$$
$$= \bar{y}_1 (\bar{y}_2 \cup y_3 \cup y_4) \cup \bar{y}_2 y_3,$$

hence the characteristic function of (28) is

(32) $$\Phi_5 = (\bar{x}_1 \cup \bar{x}_2 \cup \bar{x}_3)(\bar{x}_2 \cup \bar{x}_4 \cup \bar{x}_6 \cup \bar{x}_7 \cup \bar{x}_8 \cup x_3 \cup \bar{x}_1) \cup$$
$$\cup x_3 x_8 (\bar{x}_2 \cup \bar{x}_4 \cup \bar{x}_6 \cup \bar{x}_7)$$
$$= \bar{x}_1 \cup \bar{x}_2 \cup \bar{x}_3 \bar{x}_4 \cup \bar{x}_3 \bar{x}_6 \cup \bar{x}_3 \bar{x}_7 \cup \bar{x}_3 \bar{x}_8 \cup \bar{x}_4 x_8 \cup \bar{x}_6 x_8 \cup \bar{x}_7 x_8.$$

If we are interested* in simultaneously obtaining the solutions of the equation

$$f(x_1, \ldots, x_n) = 0,$$

of the inequality

$$f(x_1, \ldots, x_n) \geqq 0$$

* As it will be the case in Chapter VI, § 1.

and of the strict inequality

$$f(x_1, \ldots, x_n) > 0,$$

we "linearize" them as above and proceed as in Theorem 5 of Chapter III.

Example 6. Let us consider the equation

$$(33) \quad -7x_1 x_2 x_3 - 5x_2 x_4 x_6 x_7 x_8 + 4x_3 x_8 + 2\bar{x}_1 x_4 x_8 + x_1 \bar{x}_5 x_6 = -3$$

and the inequalities

$$(28) \quad -7x_1 x_2 x_3 - 5x_2 x_4 x_6 x_7 x_8 + 4x_3 x_8 + 2\bar{x}_1 x_4 x_8 + x_1 \bar{x}_5 x_6 \geqq -3,$$

$$(34) \quad -7x_1 x_2 x_3 - 5x_2 x_4 x_6 x_7 x_8 + 4x_3 x_8 + 2\bar{x}_1 x_4 x_8 + x_1 \bar{x}_5 x_6 > -3.$$

With the substitutions (29), we obtain

$$(35) \quad -7y_1 - 5y_2 + 4y_3 + 2y_4 + y_5 = -3$$

instead of (33),

$$(36) \quad -7y_1 - 5y_2 + 4y_3 + 2y_4 + y_5 \geqq -3$$

instead of (28), and

$$(37) \quad -7y_1 - 5y_2 + 4y_3 + 2y_4 + y_5 > -3$$

instead of (34).

As in Chapter III, we seek the basic solutions of the canonical form of (36), i.e. of

$$(38) \quad 7\bar{y}_1 + 5\bar{y}_2 + 4y_3 + 2y_4 + y_5 \geqq 9$$

and obtain

Table 5

$\bar{y}_1$	$\bar{y}_2$	y_3	y_4	y_5	(35)?
1	1	0	0	0	
1	0	1	0	0	
1	0	0	1	0	V
0	1	1	0	0	V

We see that the equation (35) has the solutions

$$(39.1) \quad y_1 = 0, \quad y_2 = 1, \quad y_3 = 0, \quad y_4 = 1, \quad y_5 = 0$$

and

$$(39.2) \quad y_1 = 1, \quad y_2 = 0, \quad y_3 = 1, \quad y_4 = 0, \quad y_5 = 0.$$

Hence, the characteristic function of (35) is

$$(40) \quad \Psi_3 = \bar{y}_1 y_2 \bar{y}_3 y_4 \bar{y}_5 \cup y_1 \bar{y}_2 y_3 \bar{y}_4 \bar{y}_5.$$

As in Chapter III, we obtain now the basic solutions of the strict inequality (37):

Table 6

$\bar{y}_1$	$\bar{y}_2$	y_3	y_4	y_5
1	1	0	0	0
1	0	1	0	0
1	0	0	1	1
0	1	1	1	0
0	1	1	0	1

Hence, the families of solutions of (37) are:

Table 7

y_1	y_2	y_3	y_4	y_5
0	0	—	—	—
0	1	1	—	—
0	1	0	1	1
1	0	1	1	—
1	0	1	0	1

and its characteristic function is

$$(41) \qquad \Psi_4 = \bar{y}_1 \bar{y}_2 \cup \bar{y}_1 y_2 y_3 \cup \bar{y}_1 y_2 \bar{y}_3 y_4 y_5 \cup y_1 \bar{y}_2 y_3 y_4 \cup y_1 \bar{y}_2 y_3 \bar{y}_4 y_5$$

$$= \bar{y}_1 \bar{y}_2 \cup \bar{y}_1 y_3 \cup \bar{y}_1 y_4 y_5 \cup \bar{y}_2 y_3 y_4 \cup \bar{y}_2 y_3 y_5 .$$

Now, the characteristic function of (28) was obtained in Example 5. From (29) and (40) we deduce the characteristic function of (33):

$$(42) \quad \Phi_6 = \bar{x}_1 x_2 \bar{x}_3 x_4 x_5 x_6 x_7 x_8 \cup x_1 x_2 x_3 \bar{x}_4 x_8 \cup x_1 x_2 x_3 \bar{x}_6 x_8 \cup x_1 x_2 x_3 x_5 \bar{x}_7 x_8 ,$$

while from (29) and (41) we obtain the characteristic function of (34):

$$(43) \quad \Phi_7 = \bar{x}_2 \cup (\bar{x}_1 \cup \bar{x}_3)(\bar{x}_4 \cup \bar{x}_6 \cup \bar{x}_7 \cup \bar{x}_8) \cup \bar{x}_1 (x_3 \cup \bar{x}_5) \cup x_3 x_4 \bar{x}_5 x_6 \bar{x}_7 x_8 .$$

Note added in proofs.

There is a variant of the above described procedure enabling us to solve simultaneously the equation $f = 0$ and the inequality $f < 0$ (and this suffices for our purposes in Chap. VI, § 1). As a matter of fact, we can also solve simultaneously the six conditions $f \geqq 0$, $f \leqq 0$, $f > 0$, $f < 0$, $f = 0$ and $f \neq 0$.

The method runs as follows: we solve first the inequality $f \geqq 0$, obtaining as a by-product the solutions of $f = 0$, as it was indicated before; let φ be the characteristic function of $f \geqq 0$ and φ'' that of $f = 0$. Then the characteristic function φ' of the strict inequality $f < 0$ is simply the negation $\varphi' = \bar{\varphi}$ of the function φ. (Furthermore, $\varphi' \cup \varphi''$ is the characteristic function of $f \leqq 0$, $\overline{\varphi'}\, \overline{\varphi''}$ is that of $f > 0$, while $\overline{\varphi''}$ is the characteristic function of $f \neq 0$).

For example, let

$$f = -7 x_1 x_2 x_3 - 5 x_2 x_4 x_6 x_7 x_8 + 4 x_3 x_8 + 2 \bar{x}_1 x_4 x_8 + x_1 \bar{x}_5 x_6 + 3 .$$

The conditions $f \geqq 0$ and $f = 0$ coincide with (28) and (33), respectively. We have seen that their characteristic functions are

$$(32) \quad \varphi = \bar{x}_1 \cup \bar{x}_2 \cup \bar{x}_3 \bar{x}_4 \cup \bar{x}_3 \bar{x}_6 \cup \bar{x}_3 \bar{x}_7 \cup \bar{x}_3 \bar{x}_8 \cup \bar{x}_4 x_8 \cup \bar{x}_6 x_8 \cup \bar{x}_7 x_8$$

and

$$(42) \quad \varphi'' = \bar{x}_1 x_2 \bar{x}_3 x_4 x_5 x_6 x_7 x_8 \cup x_1 x_2 x_3 \bar{x}_4 x_8 \cup x_1 x_2 x_3 \bar{x}_6 x_8 \cup x_1 x_2 x_3 x_5 \bar{x}_7 x_8 ,$$

respectively. Hence the characteristic function of the inequality $f < 0$ is

$$\varphi' = \bar{\varphi} = x_1 x_2 x_3 \bar{x}_8 \cup x_1 x_2 x_3 x_4 x_6 x_7 \cup x_1 x_2 x_4 x_6 x_7 x_8 .$$

We obtain also the characteristic function

$$\varphi' \cup \varphi'' = x_2 \bar{x}_3 x_4 x_5 x_6 x_7 x_8 \cup x_1 x_2 x_3 \bar{x}_4 \cup x_1 x_2 x_3 x_5 \cup x_1 x_2 x_3 \bar{x}_6 \cup$$
$$\cup x_1 x_2 x_3 x_7 \cup x_1 x_2 x_3 \bar{x}_8 \cup x_1 x_2 x_4 x_6 x_7 x_8$$

of the inequality $f \leqq 0$, the characteristic function

$$\bar{\varphi}' \bar{\varphi}'' = \overline{\varphi' \cup \varphi''} = \bar{x}_2 \cup \bar{x}_1 \bar{x}_4 \cup \bar{x}_1 \bar{x}_6 \cup \bar{x}_1 \bar{x}_7 \cup \bar{x}_1 \bar{x}_8 \cup \bar{x}_1 x_3 \cup \bar{x}_1 \bar{x}_5 \cup$$
$$\cup \bar{x}_3 \bar{x}_4 \cup \bar{x}_3 \bar{x}_6 \cup \bar{x}_3 \bar{x}_7 \cup \bar{x}_3 \bar{x}_8 \cup x_4 \bar{x}_5 x_6 \bar{x}_7 x_8 ,$$

of the inequality $f > 0$, and the characteristic function

$$\bar{\varphi}'' = \bar{x}_2 \cup \bar{x}_8 \cup \bar{x}_1 x_3 \cup \bar{x}_1 \bar{x}_4 \cup \bar{x}_1 \bar{x}_5 \cup \bar{x}_1 \bar{x}_6 \cup \bar{x}_1 \bar{x}_7 \cup x_1 \bar{x}_3 \cup \bar{x}_3 \bar{x}_4 \cup$$
$$\cup \bar{x}_3 \bar{x}_5 \cup \bar{x}_3 \bar{x}_6 \cup \bar{x}_3 \bar{x}_7 \cup x_1 x_4 x_6 x_7 \cup x_3 x_4 x_6 x_7 \cup x_4 \bar{x}_5 x_6 x_7$$

of $f \neq 0$.

§ 3. The Characteristic Function for Systems

Let us consider a system of pseudo-Boolean equations and inequalities:

(44. j) $\quad f_j(x_1, \ldots, x_n) = 0 \qquad (j = 1, \ldots, m),$

(44. h) $\quad f_h(x_1, \ldots, x_n) \geqq 0 \qquad (h = m+1, \ldots, m+p),$

and let

(45.1) $\quad \varphi_1(x_1, \ldots, x_n) = 1,$

. . . .

(45.9) $\quad \varphi_q(x_1, \ldots, x_n) = 1$

be the corresponding characteristic equations, determined as in §§ 1, 2. If we denote by Φ the characteristic function of the system (44), we have obviously:

Theorem 2.

$$\Phi(x_1, \ldots, x_n) = \prod_{s-1}^{m+p} \varphi_s(x_1, \ldots, x_n). \tag{46}$$

Example 7. Let us consider the system

(47.1) $\quad 7x_1 x_2 x_3 - 2\bar{x}_1 x_4 x_8 + 5x_2 x_4 x_6 x_7 x_8 - 4x_3 x_8 - x_4 \bar{x}_5 x_6 \leqq 3,$

(47.2) $\quad 3x_1 - 2x_2 \bar{x}_6 + 4x_5 \bar{x}_6 \bar{x}_8 + 2x_1 x_2 x_3 - 7x_8 \geqq -8,$

(47.3) $\quad 8x_4 x_5 \bar{x}_8 - 4\bar{x}_3 \bar{x}_7 x_8 + 3x_1 x_2 + \bar{x}_3 + \bar{x}_4 + \bar{x}_5 < 3,$

(47.4) $\quad 2x_3 + 3x_5 - \bar{x}_5 \bar{x}_6 + 4x_6 \bar{x}_7 x_8 - 2x_5 x_6 x_7 x_8 \geqq 1.$

The characteristic functions of the above inequalities, obtained as in § 2 are:

$$(48.1)\qquad \varphi_1 = \bar{x}_1 \cup \bar{x}_2 \cup \bar{x}_3\,\bar{x}_8 \cup (\bar{x}_3 \cup x_8)\,(\bar{x}_4 \cup \bar{x}_6 \cup \bar{x}_7),$$

$$(48.2)\qquad \varphi_2 = x_1 \cup \bar{x}_2 \cup x_6 \cup \bar{x}_8,$$

$$(48.3)\qquad \varphi_3 = \bar{x}_3\,\bar{x}_7\,x_8 \cup (\bar{x}_1 \cup \bar{x}_2)\,(x_3 \cup x_4 \cup x_5)\,(\bar{x}_4 \cup \bar{x}_5 \cup x_8),$$

$$(48.4)\qquad \varphi_4 = x_3 \cup x_5 \cup x_6\,\bar{x}_7\,x_8,$$

or else

$$(48'.1)\quad \varphi_1 = \bar{x}_1 \cup \bar{x}_2 \cup \bar{x}_3\,\bar{x}_4 \cup \bar{x}_3\,\bar{x}_6 \cup \bar{x}_3\,\bar{x}_7 \cup \bar{x}_3\,\bar{x}_8 \cup \bar{x}_4\,x_8 \cup \bar{x}_6\,x_8 \cup \bar{x}_7\,x_8,$$

$$(48'.2)\quad \varphi_2 = x_1 \cup \bar{x}_2 \cup x_6 \cup \bar{x}_8,$$

$$(48'.3)\quad \varphi_3 = \bar{x}_1\,x_3\,\bar{x}_4 \cup \bar{x}_1\,x_3\,\bar{x}_5 \cup \bar{x}_1\,x_3\,x_8 \cup \bar{x}_1\,x_4\,\bar{x}_5 \cup \bar{x}_1\,x_4\,x_8 \cup$$
$$\cup\, \bar{x}_1\,\bar{x}_4\,x_5 \cup \bar{x}_1\,x_5\,x_8 \cup \bar{x}_2\,x_3\,\bar{x}_4 \cup \bar{x}_2\,x_3\,\bar{x}_5 \cup \bar{x}_2\,x_3\,x_8 \cup$$
$$\cup\, \bar{x}_2\,x_4\,\bar{x}_5 \cup \bar{x}_2\,x_4\,x_8 \cup \bar{x}_2\,\bar{x}_4\,x_5 \cup \bar{x}_2\,x_5\,x_8 \cup \bar{x}_3\,\bar{x}_7\,x_8,$$

$$(48'.4)\quad \varphi_4 = x_3 \cup x_5 \cup x_6\,\bar{x}_7\,x_8.$$

Multiplying these functions, as indicated in Theorem 2, we obtain the characteristic function of the system:

$$(49)\quad \Phi_8 = x_1\,\bar{x}_3\,x_5\,\bar{x}_7\,x_8 \cup \bar{x}_3\,x_6\,\bar{x}_7\,x_8 \cup \bar{x}_2\,x_3\,\bar{x}_4 \cup \bar{x}_2\,x_3\,\bar{x}_5 \cup$$
$$\cup\, \bar{x}_2\,x_3\,x_8 \cup \bar{x}_2\,\bar{x}_4\,x_5 \cup \bar{x}_2\,x_5\,x_8 \cup \bar{x}_2\,x_6\,\bar{x}_7\,x_8 \cup \bar{x}_1\,x_3\,\bar{x}_5\,x_6 \cup$$
$$\cup\, \bar{x}_1\,x_3\,x_6\,x_8 \cup \bar{x}_1\,x_5\,x_6\,x_8 \cup \bar{x}_1\,x_6\,\bar{x}_7\,x_8 \cup \bar{x}_1\,x_3\,\bar{x}_5\,\bar{x}_8 \cup \bar{x}_1\,\bar{x}_4\,x_5\,\bar{x}_8.$$

From Theorem 2 we deduce:

Corollary 1. If the conditions in the original system are grouped into several subsystems $\Sigma_1, \ldots, \Sigma_r$ having the characteristic equations $\chi_1(x_1, \ldots, x_n) = 1, \ldots, \chi_r(x_1, \ldots, x_n) = 1$, then

$$(50)\qquad \Phi(x_1, \ldots, x_n) = \prod_{t=1}^{r} \chi_t(x_1, \ldots, x_n).$$

Remark 1. It is easier to determine the characteristic function of a system of *linear* equations and inequalities as in § 1 *c*), than to compute the product of the different characteristic equations corresponding to its constraints. Therefore, if we have a system consisting of both linear and nonlinear conditions, we compute the characteristic functions of the nonlinear conditions separately, the characteristic functions of the subsystem of linear conditions, and finally their product.

Remark 2. Let us consider a system Σ whose characteristic equation is Φ. If, after obtaining Φ, we are ulteriorly given a further system Σ' which is to be fulfilled, and if we denote by Φ' its characteristic function, then the characteristic function of the completed system $\{\Sigma, \Sigma'\}$ is simply $\Phi\,\Phi'$.

Example 8. Let us solve the system consisting of the nonlinear inequalities (47) from Example 7 and of the linear "sub-system" (15) from Example 3. The characteristic function of (47) is the function Φ_8 in formula (49), while the characteristic function of (15) is the function

$$(16)\qquad \Phi_3 = \bar{x}_1 x_2 x_3 \bar{x}_4 x_5 \bar{x}_6 \bar{x}_8 \cup \bar{x}_1 x_2 x_3 \bar{x}_4 x_5 x_6 x_7 \bar{x}_8 \cup x_1 x_2 \bar{x}_3 \bar{x}_4 \bar{x}_5 \bar{x}_6 \bar{x}_8 .$$

Hence, in view of Remark 1, the characteristic function of the augmented system in (47) and (15) is

$$(51)\qquad \Phi_9 = \Phi_3 \Phi_8 = \bar{x}_1 x_2 x_3 \bar{x}_4 x_5 \bar{x}_6 \bar{x}_8 \cup \bar{x}_1 x_2 x_3 \bar{x}_4 x_5 x_7 \bar{x}_8 .$$

§ 4. The Characteristic Function of Logical Conditions

In several practical problems we are faced with mathematical programs containing logical conditions imposed on the variables (see, for instance, G. B. Dantzig [3], F. Radó [1, 2], L. Németi [1], L. Németi and F. Radó [1]).

In this section we shall briefly examine systems of pseudo-Boolean equations and/or inequalities containing logical conditions.

For this sake, let us consider two pseudo-Boolean systems, $\Sigma'(x_1, \ldots, x_n)$ and $\Sigma''(x_1, \ldots, x_n)$ whose characteristic functions are $\Phi_{\Sigma'}(x_1, \ldots, x_n)$ and $\Phi_{\Sigma''}(x_1, \ldots, x_n)$.

If $\Sigma' \,\&\, \Sigma''$ denotes the problem of finding the values of $(x_1, \ldots, x_n)$ which satisfy both Σ' and Σ'', then Theorem 2 states that the characteristic function $\Phi_{\Sigma' \& \Sigma''}$ of $\Sigma' \,\&\, \Sigma''$ is

$$(52)\qquad \Phi_{\Sigma' \& \Sigma''} = \Phi_{\Sigma'} \Phi_{\Sigma''} .$$

Similar results are obviously valid for other logical problems. For instance:

1. *Disjunction of* Σ' *and* Σ'', briefly $\Sigma' \vee \Sigma''$: finding the values of $(x_1, \ldots, x_n)$ which fulfil at least one of the systems Σ', Σ''. The characteristic function $\Phi_{\Sigma' \vee \Sigma''}$ of $\Sigma' \vee \Sigma''$ (i.e., the Boolean function which has the value 1 if and only if $x_1, \ldots, x_n$ fulfil $\Sigma' \vee \Sigma''$) is

$$(53)\qquad \Phi_{\Sigma' \vee \Sigma''} = \Phi_{\Sigma'} \cup \Phi_{\Sigma''} .$$

An analogous result holds for the logical disjunction of more than two systems.

2. *Negation of* Σ', briefly $\neg \Sigma'$: finding the values of $(x_1, \ldots, x_n)$ which do not satisfy Σ'. The characteristic function of $\neg \Sigma'$ is

$$(54)\qquad \Phi_{\neg \Sigma'} = \bar{\Phi}_{\Sigma'} .$$

3. *Difference of* Σ' *and* Σ'', briefly $\Sigma' \,\&\, \neg \Sigma''$: finding the values of $(x_1, \ldots, x_n)$ which satisfy Σ' but not Σ''. The characteristic function of $\Sigma' \,\&\, \neg \Sigma''$ is

$$(55)\qquad \Phi_{\Sigma' \& \neg \Sigma''} = \Phi_{\Sigma'} \bar{\Phi}_{\Sigma''} .$$

4. *Symmetric Difference of* Σ' *and* Σ'', briefly $\Sigma' \bigtriangledown \Sigma''$: finding the values of $(x_1, \ldots, x_n)$ which fulfil one of the systems Σ', Σ'', but not both. The characteristic function of $\Sigma' \bigtriangledown \Sigma''$ is

$$(56) \qquad \Phi_{\Sigma' \bigtriangledown \Sigma''} = \Phi_{\Sigma'} \bar{\Phi}_{\Sigma''} \cup \bar{\Phi}_{\Sigma'} \Phi_{\Sigma''}.$$

5. *Conditioning of* Σ'' *by* Σ', briefly $\Sigma' \to \Sigma''$: finding those values of $(x_1, \ldots, x_n)$ which either do not satisfy Σ', or satisfy both Σ' and Σ''. The characteristic function of $\Sigma' \to \Sigma''$ is

$$(57) \qquad \Phi_{\Sigma' \to \Sigma''} = \bar{\Phi}_{\Sigma'} \cup \Phi_{\Sigma''}.$$

Similar results can be obtained immediately for other logical conditions ("neither-nor", "if and only if", etc.).

Example 9. If Σ' stands for the single inequality

$$(47.2) \qquad 3x_1 - 2x_2 \bar{x}_6 + 4x_5 \bar{x}_6 \bar{x}_8 + 2x_1 x_2 x_3 - 7x_8 \geqq -8$$

and Σ'' denotes the single inequality

$$(47.4) \qquad 2x_3 + 3x_5 - \bar{x}_5 \bar{x}_6 + 4x_6 \bar{x}_7 x_8 - 2x_5 x_6 x_7 x_8 \geqq 1,$$

then the corresponding characteristic functions, determined in Example 7, are

$$(48'.2) \qquad \varphi_2 = x_1 \cup \bar{x}_2 \cup x_6 \cup \bar{x}_8$$

and

$$(48'.4) \qquad \varphi_4 = x_3 \cup x_5 \cup x_6 \bar{x}_7 x_8,$$

respectively.

Then, taking into account Definition 1 and the above results, we see that the vector $(x_1, \ldots, x_8)$:

1) satisfies at least one of the inequalities (47.2) and (47.4) if and only if

$$(58) \qquad \varphi_2 \cup \varphi_4 = x_1 \cup \bar{x}_2 \cup x_3 \cup x_5 \cup x_6 \cup \bar{x}_8 = 1;$$

2) does not satisfy (47.2) if and only if

$$(59) \qquad \bar{\varphi}_2 = \bar{x}_1 x_2 \bar{x}_6 x_8 = 1;$$

3) satisfies (47.2), but not (47.4), if and only if

$$(60) \qquad \varphi_2 \bar{\varphi}_4 = \bar{x}_3 \bar{x}_5 (x_1 \cup \bar{x}_2 \cup x_6 \cup \bar{x}_8)(\bar{x}_6 \cup x_7 \cup \bar{x}_8) = 1;$$

4) satisfies one of the inequalities (47.2) and (47.4) but not both, if and only if

$$(61) \qquad \varphi_2 \bar{\varphi}_4 \cup \bar{\varphi}_2 \varphi_4 = \bar{x}_3 \bar{x}_5 (x_1 \cup \bar{x}_2 \cup x_6 \cup \bar{x}_8)(\bar{x}_6 \cup x_7 \cup \bar{x}_8) \cup \bar{x}_1 x_2 \bar{x}_6 x_8 (x_3 \cup x_5) = 1;$$

5) either does not satisfy (47.2) or satisfies both (47.2) and (47.4), if and only if

$$(62) \qquad \bar{\varphi}_2 \cup \varphi_4 = \bar{x}_1 x_2 \bar{x}_6 x_8 \cup x_3 \cup x_5 \cup x_6 \bar{x}_7 x_8 = 1.$$

§ 5. Irredundant Solutions of the Characteristic Equation

The knowledge of the characteristic function permits the direct listing of all the families of solutions (as it was shown in § 1).

Example 10. It was shown in Example 5 that the characteristic equation of the pseudo-Boolean inequality

$$(28) \qquad 7x_1 x_2 x_3 + 5x_2 x_4 x_6 x_7 x_8 - 4x_3 x_8 - 2\bar{x}_1 x_4 x_8 - x_4 \bar{x}_5 x_6 \leqq 3$$

is

$$\Phi_5 = \bar{x}_1 \cup \bar{x}_2 \cup \bar{x}_3 \bar{x}_4 \cup \bar{x}_3 \bar{x}_6 \cup \bar{x}_3 \bar{x}_7 \cup \bar{x}_3 \bar{x}_8 \cup \bar{x}_4 x_8 \cup \bar{x}_6 x_8 \cup \bar{x}_7 x_8 = 1 .$$

Hence its families of solutions are

Table 8

No.	x_1	x_2	x_3	x_4	x_5	x_6	x_7	x_8
1	0	—	—	—	—	—	—	—
2	—	0	—	—	—	—	—	—
3	—	—	0	0	—	—	—	—
4	—	—	0	—	—	0	—	—
5	—	—	0	—	—	—	0	—
6	—	—	0	—	—	—	—	0
7	—	—	—	0	—	—	—	1
8	—	—	—	—	—	0	—	1
9	—	—	—	—	—	—	0	1

The same procedure offers immediately the solutions of all the problems discussed in the previous examples.

However, the method does not assure the irredundancy of the obtained list, i.e. those solutions which belong to different families appear several times in our list. For instance, the solution $x_1 = x_2 = x_3 = x_4 = x_6 = x_7 = x_8 = 0$, $x_5 = 1$ in Table 8 belongs to the families $1, 2, 3, 4, 5, 6$ and hence, developing Table 8 into the explicit list of all the solutions, the above solution will appear 6 times.

Therefore it might be desired to have a procedure for transforming the original families in such a way as to obtain a system of families which:

1) contain all the solutions;

2) are pairwise disjoint, i.e. the same solution cannot belong to more than one family.

The technique we shall indicate in order to solve this problem is based on the following

Remark 3. If

$$\Phi = C_1 \cup \cdots \cup C_p \tag{13}$$

is a disjunctive form of the characteristic function corresponding to the families $\mathcal{F}_1, \ldots, \mathcal{F}_p$ (see Theorem 1), then the above property 2 is equivalent to

$$C_i C_j = 0 \quad \text{for} \quad i \neq j . \tag{63}$$

Hence the above problem may be re-formulated as follows: if the original form (13) of the Boolean function Φ does not satisfy (63), find an equivalent disjunctive form

$$\Phi = D_1 \cup \cdots \cup D_q \tag{64}$$

of Φ so that

(65) $$D_h D_k = 0 \quad \text{for} \quad h \neq k;$$

then (64) may be called the *disjointed form* of Φ.

We start the discussion with the "linear" case:

(66) $$\Phi = x_1^{\alpha_1} \cup x_2^{\alpha_2} \cup x_3^{\alpha_3} \cup \cdots \cup x_p^{\alpha_p}.$$

Lemma 1. *The disjointed form of* (66) *is*

(67) $$\Phi = x_1^{\alpha_1} \cup x_1^{\bar{\alpha}_1} x_2^{\alpha_2} \cup x_1^{\bar{\alpha}_1} x_2^{\bar{\alpha}_2} x_3^{\alpha_3} \cup \cdots \cup x_1^{\bar{\alpha}_1} \ldots x_{p-1}^{\bar{\alpha}_{p-1}} x_p^{\alpha_p}.$$

Proof. Obvious from the identity $a \cup b = a \cup \bar{a}\, b$.

Lemma 2. *If*

(68. h) $$\Phi_h = C_{h1} \cup C_{h2} \cup \cdots \cup C_{h\,m(h)} \qquad (h = 1, \ldots, r)$$

are disjointed forms, then

(69) $$\Phi = \prod_{h=1}^{r} \Phi_h = \bigcup_{i_p, \ldots, i_r} C_{1 i_1} C_{2 i_2} \ldots C_{r i_r}$$

is also a disjointed form.

Proof. If $C_{1 i_1} C_{2 i_2} \ldots C_{r i_r}$ and $C_{1 j_1} C_{2 j_2} \ldots C_{r j_r}$ are two distinct terms of (69), then the vectors $(i_1, i_2, \ldots, i_r)$ and $(j_1, j_2, \ldots, j_r)$ are distinct, so that for at least one index s we have $i_s \neq j_s$, implying $C_{s i_s} C_{s j_s} = 0$. Hence the product of the two terms is equal to zero.

Lemma 3. *The procedure indicated in Chapter III, § 3 for obtaining the families of solutions of a linear system leads to a disjointed form of the corresponding characteristic function.*

Proof. See Chapter III, § 3, Remark 2.

Now, a disjointed form of the characteristic function $\varphi(x_1, \ldots, x_n)$ of a single pseudo-Boolean equation or inequality may be obtained as follows:

We find, as in § 1, the characteristic function $\psi(y_1, \ldots, y_m)$ of the associated linear equation (inequality).

Each y_i is a product of variables x_j with or without bars, while $\bar{y}_i$ is a disjunction of variables x_j with or without bars. We replace the y_i and $\bar{y}_i$ by their expressions, we apply Lemma 1 to each disjunction corresponding to a $\bar{y}_i$ appearing in $\psi(y_1, \ldots, y_m)$, after which we perform all the multiplications.

Further, if we want to obtain a disjointed form of the characteristic function Φ of a system, then we simply multiply the disjointed forms of the characteristic functions φ of the different equations and inequalities of the system.

Theorem 3. *The above procedure leads to a disjointed form of the characteristic function $\Phi(x_1, \ldots, x_n)$ of the given system.*

Proof. In view of Lemma 2, it suffices to prove that, constructing as above the characteristic function $\varphi(x_1, \ldots, x_n)$ of a *single* equation or inequality, we obtain a disjointed form.

Let C' and C'' be two distinct conjunctions of $\varphi(x_1, \ldots, x_n)$; we have to prove that $C' C'' = 0$. Two cases are possible:

1. C' and C'' result from the same conjunction $K = y_{i_1} \cdots y_{i_s} \bar{y}_{j_1} \cdots \bar{y}_{j_t}$ of the characteristic function $\psi(y_1, \ldots, y_m)$ of the associated linear equation (inequality). Then $C' = y_{i_1} \ldots y_{i_s} D'$ and $C'' = y_{i_1} \ldots y_{i_s} D''$, where D' and D'' are two distinct conjunctions resulting from $\bar{y}_{j_1} \ldots \bar{y}_{j_t}$. We have $D' D'' = 0$, by Lemma 2, hence $C' C'' = 0$.

2. C' and C'' result from two distinct conjunctions, say K' and K'', of $\psi(y_1, \ldots, y_m)$. Since the function ψ is obtained in a disjointed form, there exists a variable $y_i = x_{h_1}^{\alpha_{h_1}} \ldots x_{h_u}^{\alpha_{h_u}}$ so that $K' = y_i H'$ and $K'' = \bar{y}_i H''$. Therefore C' is of the form $C' = x_{h_1}^{\alpha_{h_1}} \ldots x_{h_u}^{\alpha_{h_u}} E'$, while C'' is of the form $C'' = x_{h_k}^{\bar{\alpha}_{h_k}} E''$, with $1 \leqq k \leqq u$. Hence $C' C'' = 0$.

Example 11. Let us consider the system

$$2x_1 x_2 x_4 - 4x_5 x_6 + 3x_3 \leqq 2, \tag{70.1}$$

$$4x_1 x_3 x_5 + 6x_2 x_4 x_6 \geqq 4. \tag{70.2}$$

Denoting

$$x_5 x_6 = y_1, \quad x_1 x_2 x_4 = y_2, \quad x_2 x_4 x_6 = y_3, \quad x_1 x_3 x_5 = y_4, \tag{71}$$

we can write the inequalities (70) in the form

$$4y_1 + 3\bar{x}_3 + 2\bar{y}_2 \geqq 3, \tag{72.1}$$

$$6y_3 + 4y_4 \geqq 4. \tag{72.2}$$

The characteristic functions of these inequalities, written in disjointed forms, are

$$\psi_1(y_1, x_3, y_2) = y_1 \cup \bar{y}_1 \bar{x}_3, \tag{73.1}$$

$$\psi_2(y_3, y_4) = y_3 \cup \bar{y}_3 y_4, \tag{73.2}$$

hence

$$\varphi_1(x_1, \ldots, x_6) = x_5 x_6 \cup (\bar{x}_5 \cup \bar{x}_6) \bar{x}_3, \tag{74.1}$$

$$\varphi_2(x_1, \ldots, x_6) = x_2 x_4 x_6 \cup (\bar{x}_2 \cup \bar{x}_4 \cup \bar{x}_6) x_1 x_3 x_5. \tag{74.2}$$

Making the $\bar{y}_i$ disjoint and performing the multiplications, we obtain the disjointed forms of φ_1 and φ_2:

$$\varphi_1(x_1, \ldots, x_6) = x_5 x_6 \cup (\bar{x}_5 \cup x_5 \bar{x}_6) \bar{x}_3 = x_5 x_6 \cup \bar{x}_3 \bar{x}_5 \cup \bar{x}_3 x_5 \bar{x}_6, \tag{74'.1}$$

$$\begin{aligned} \varphi_2(x_2, \ldots, x_6) &= x_2 x_4 x_6 \cup (\bar{x}_2 \cup x_2 \bar{x}_4 \cup x_2 x_4 \bar{x}_6) x_1 x_3 x_5 \\ &= x_2 x_4 x_6 \cup x_1 \bar{x}_2 x_3 x_5 \cup x_1 x_2 x_3 \bar{x}_4 x_5 \cup x_1 x_2 x_3 x_4 x_5 \bar{x}_6. \end{aligned} \tag{74'.2}$$

Therefore a disjointed form of the characteristic function of the system (70) is

$$\begin{aligned} \Phi_{10}(x_1, \ldots, x_6) &= (x_5 x_6 \cup \bar{x}_3 \bar{x}_5 \cup \bar{x}_3 x_5 \bar{x}_6)(x_2 x_4 x_6 \cup \\ &\cup x_1 \bar{x}_2 x_3 x_5 \cup x_1 x_2 x_3 \bar{x}_4 x_5 \cup x_1 x_2 x_3 x_4 x_5 \bar{x}_6) \\ &= x_2 x_4 x_5 x_6 \cup x_1 \bar{x}_2 x_3 x_5 x_6 \cup x_1 x_2 x_3 \bar{x}_4 x_5 x_6 \cup x_2 \bar{x}_3 x_4 \bar{x}_5 x_6, \end{aligned} \tag{75}$$

corresponding to the following complete system of disjoint families of solutions:

Table 9

No.	x_1	x_2	x_3	x_4	x_5	x_6
1	—	1	—	1	1	1
2	1	0	1	—	1	1
3	1	1	1	0	1	1
4	—	1	0	1	0	1

§ 6. The Pseudo-Boolean Form of the Characteristic Function

The characteristic function of a pseudo-Boolean system is a Boolean function. However, in Chapter VI and VII it will be necessary to have a pseudo-Boolean expression of the characteristic function, i.e. an expression using only the arithmetical operations "+", "—", and possibly the negation "‾" of single variables.

The following identities are well-known:

$$a_1 \cup a_2 = a_1 + a_2 - a_1 a_2,$$

$$a_1 \cup a_2 \cup a_3 = a_1 + a_2 + a_3 - a_1 a_2 - a_1 a_3 - a_2 a_3 + a_1 a_2 a_3,$$

$$\dots\dots\dots\dots\dots\dots\dots\dots\dots\dots$$

etc., which permit the transformation of every disjunctive form of a Boolean function into a pseudo-Boolean one. An easier way for performing this transformation is offered by the formula

$$\bigcup_{i=1}^{n} a_i = 1 - \prod_{i=1}^{n} \bar{a}_i.$$

If

$$\Phi = C_1 \cup \dots \cup C_m \tag{76}$$

is a disjointed form of the Boolean function Φ, then $C_i C_j = 0$ for all $i \neq j$, and the above identities show that relation (76) may be simply written in the pseudo-Boolean form

$$\Phi = C_1 + \dots + C_m. \tag{76'}$$

Example 12. The characteristic function Φ_{10} given by formula (75) in Example 11 can be written in the pseudo-Boolean form

$$\Phi_{10}(x_1, \dots, x_6) = x_2 x_4 x_5 x_6 + x_1 \bar{x}_2 x_3 x_5 x_6 + x_1 x_2 x_3 \bar{x}_4 x_5 x_6 + x_2 \bar{x}_3 x_4 \bar{x}_5 x_6. \tag{77}$$

§ 7. The Methods of R. Fortet and P. Camion

The procedure for solving linear pseudo-Boolean equations and inequalities, presented in the last section of Chapter III, may be extended to the case of arbitrary pseudo-Boolean conditions. For the sake of this, it suffices to treat every conjunction $x_{i_1}^{\alpha_{i_1}} \dots x_{i_m}^{\alpha_{i_m}}$ as a single bivalent variable.

Example 13. Let us solve the pseudo-Boolean equation

$$4x_1 x_2 + 3x_1 x_3 x_4 + x_3 \bar{x}_4 = 2x_1 x_5 + 3x_2 x_4. \tag{78}$$

Setting

$$x_1 x_2 = y_1, \quad x_1 x_3 x_4 = y_2, \quad x_3 \bar{x}_4 = y_3, \quad x_1 x_5 = y_4, \quad x_2 x_4 = y_5, \tag{79}$$

equation (78) becomes

$$4y_1 + 3y_2 + y_3 = 2y_4 + 3y_5. \tag{80}$$

Using the methods from Chapter III § 4, we find that the equation (80) may be written in the form

$$(y_2 \oplus y_3) + 2(y_2 y_3 \oplus y_2) + 2^2(y_1 \oplus y_2 y_3) + 2^3 y_1 y_2 y_3 = y_5 + 2(y_4 \oplus y_5) + 2^2 y_4 y_5,$$

so that (80) is equivalent to the following system of Boolean equations:

$$y_2 \oplus y_3 = y_5, \tag{81.0}$$

$$y_2 y_3 \oplus y_2 = y_4 \oplus y_5, \tag{81.1}$$

$$y_1 \oplus y_2 y_3 = y_4 y_5, \tag{81.2}$$

$$y_1 y_2 y_3 = 0. \tag{81.3}$$

In view of the transformation formulas (79), the system (81) becomes

$$x_1 x_3 x_4 \oplus x_3 \bar{x}_4 = x_2 x_4, \tag{82.0}$$

$$x_1 x_3 x_4 = x_1 x_5 \oplus x_2 x_4, \tag{82.1}$$

$$x_1 x_2 = x_1 x_2 x_4 x_5, \tag{82.2}$$

$$0 = 0. \tag{82.3}$$

Using, for instance, the method of bifurcations, we find that the system (82) has the following solutions:

Table 10

x_1	x_2	x_3	x_4	x_5
0	—	0	0	—
0	0	—	1	—
1	0	0	—	0

R. FORTET [2] suggests the following procedure for the linearization of a nonlinear pseudo-Boolean equation:

Any conjunction $x_{i_1}^{\alpha_{i_1}} \ldots x_{i_m}^{\alpha_{i_m}}$ is to be replaced by a new variable y_i satisfying

$$x_{i_1}^{\bar{\alpha}_{i_1}} + \cdots + x_{i_m}^{\bar{\alpha}_{i_m}} + y_i \geqq 1, \tag{83}$$

$$-\left(x_{i_1}^{\bar{\alpha}_{i_1}} + \cdots + x_{i_m}^{\bar{\alpha}_{i_m}}\right) + m\,\bar{y}_i \geqq 0. \tag{84}$$

For, we have

Theorem 4. *Relations* (83) *and* (84) *are equivalent to the condition*

$$y_i = x_{i_1}^{\alpha_{i_1}} \ldots x_{i_m}^{\alpha_{i_m}}. \tag{85}$$

Proof. It is obvious that (85) implies (83) and also

$$m\,\bar{y}_i \geqq \sum_{j=1}^{m} x_{i_j}^{\bar{\alpha}_{i_j}}, \tag{86}$$

which is equivalent to (84).

Conversely, the inequality (83) may be written in the form

$$\sum_{j=1}^{m} x_{i_j}^{\bar{\alpha}_{i_j}} \geqq \bar{y}_i \tag{87}$$

showing that if $\bar{y}_i = 1$, then there exists an index j for which $x_{i_j}^{\bar{\alpha}_{i_j}} = 1$, i.e. $x_{i_j}^{\alpha_{i_j}} = 0$. In other words, (87) implies that if $y_i = 0$, then $\prod_{j=1}^{m} x_{i_j}^{\alpha_{i_j}} = 0$, i.e.

$$\prod_{j=1}^{m} x_{i_j}^{\alpha_{i_j}} \leqq y_i. \tag{88}$$

Further, the inequality (84), or, equivalently, (86), shows that if $\bar{y}_i = 0$, then all $x_{i_j}^{\bar{\alpha}_{i_j}} = 0$, i.e. all $x_{i_j}^{\alpha_{i_j}} = 1$. In other words, (84) implies the condition "if $y_i = 1$, then $\prod_{j=1}^{m} x_{i_j}^{\alpha_{i_j}} = 1$", which is equivalent to

$$y_i \leqq \prod_{j=1}^{m} x_{i_j}^{\alpha_{i_j}}. \tag{89}$$

In conclusion, the inequalities (83) and (84) imply (88) and (89), i.e. the equality (85), thus completing the proof.

Example 14. The pseudo-Boolean equation

$$-6x_1\bar{x}_2x_3 - 4x_2x_4 + 2x_2x_4\bar{x}_5 + 4\bar{x}_3\bar{x}_4 = -2 \tag{21}$$

is equivalent to the following linear pseudo-Boolean system

$$-6y_1 - 4y_2 + 2y_3 + 4y_4 = -2, \tag{90.0}$$

$$\bar{x}_1 + x_2 + \bar{x}_3 + y_1 \geqq 1, \tag{90.1'}$$

$$-(\bar{x}_1 + x_2 + \bar{x}_3) + 3\bar{y}_1 \geqq 0, \tag{90.1''}$$

$$\bar{x}_2 + \bar{x}_4 + y_2 \geqq 1, \tag{90.2'}$$

$$-(\bar{x}_2 + \bar{x}_4) + 2\bar{y}_2 \geqq 0, \tag{90.2''}$$

$$\bar{x}_2 + \bar{x}_4 + x_5 + y_3 \geqq 1, \tag{90.3'}$$

$$-(\bar{x}_2 + \bar{x}_4 + x_5) + 3\bar{y}_3 \geqq 0, \tag{90.3''}$$

$$x_3 + x_4 + y_4 \geqq 1, \tag{90.4'}$$

$$-(x_3 + x_4) + 2y_4 \geqq 0. \tag{90.4''}$$

Concluding this section, we shall indicate, in the following example, how the parametric solution of a pseudo-Boolean system, as indicated by R. FORTET [1], may be combined with the method of bifurcations.

Example 15*. Let us solve the system of pseudo-Boolean equations

$$x_{12} + x_{13} + x_{14} + x_{16} = 1, \tag{91.1}$$

$$x_{23} = 1, \tag{91.2}$$

$$x_{34} + x_{36} + x_{38} = 1, \tag{91.3}$$

$$x_{42} + x_{45} = 1, \tag{91.4}$$

$$x_{53} + x_{56} + x_{58} = 1, \tag{91.5}$$

$$x_{64} + x_{67} = 1, \tag{91.6}$$

$$x_{71} + x_{72} + x_{73} + x_{74} + x_{78} = 1, \tag{91.7}$$

$$x_{81} + x_{82} + x_{84} = 1, \tag{91.8}$$

$$x_{71} + x_{81} = 1, \tag{91.9}$$

$$x_{12} + x_{42} + x_{72} + x_{82} = 1, \tag{91.10}$$

$$x_{13} + x_{23} + x_{53} + x_{73} = 1, \tag{91.11}$$

$$x_{14} + x_{34} + x_{64} + x_{74} + x_{84} = 1, \tag{91.12}$$

$$x_{45} = 1, \tag{91.13}$$

$$x_{16} + x_{36} + x_{56} = 1, \tag{91.14}$$

$$x_{67} = 1, \tag{91.15}$$

$$x_{38} + x_{58} + x_{78} = 1, \tag{91.16}$$

$$x_{38}\, x_{82} = 0, \tag{91.17}$$

$$x_{38}\, x_{81}\, x_{12} = 0, \tag{91.18}$$

$$x_{36}\, x_{72} = 0, \tag{91.19}$$

$$x_{58}\, x_{84} = 0, \tag{91.20}$$

$$x_{71}\, x_{16} = 0. \tag{91.21}$$

Notice first that relations (91.4) and (91.13), (91.6) and (91.15), (91.11) and (91.2), imply

$$x_{42} = x_{64} = x_{13} = x_{53} = x_{73} = 0. \tag{92}$$

We begin now the process of splitting.

Case I. We seek the solutions for which

$$x_{71} + x_{72} = 1, \quad x_{81} + x_{84} = 1. \tag{93.1}$$

Taking into account (92) and (93.1) from (91.7), (91.12), (91.8), (91.9), (91.21), (91.20), (91.1), (91.5), (91.14) and (91.3), we deduce successively

$$\begin{aligned} &x_{74} = x_{78} = x_{14} = x_{34} = 0, \quad x_{84} = 1, \quad x_{81} = x_{82} = 0, \\ &x_{71} = 1, \quad x_{72} = x_{16} = x_{58} = 0, \quad x_{12} = x_{56} = 1, \quad x_{36} = 0, \quad x_{38} = 1. \end{aligned} \tag{94}$$

It is easy to verify that the values (92), (93.1) and (94) verify all the equations (91).

---

* See also § 7 in Chapter IX.

Case II. We seek the solutions for which

$$(93.2)\qquad x_{71}+x_{72}=1, \quad x_{81}=x_{84}=0.$$

It follows from (91.7), (91.8) and (91.10) that

$$(95)\qquad x_{72}=x_{74}=x_{78}=0, \quad x_{82}=1, \quad x_{12}=x_{42}=x_{72}=0, \quad x_{71}=1.$$

Equation (91.14) implies $x_{16}\,x_{56}=0$, i.e. $x_{56}\leq\bar{x}_{16}$, hence

$$(96)\qquad x_{16}=s, \quad x_{56}=\bar{s}\,t,$$

where s and t are arbitrary Boolean parameters.

In view of the above discussion, from (91.14), (91.5), (91.16), (91.3), (91.12). (91.1), (91.10), (93.2), (91.17), (91.21) and (91.19), we deduce successively

$$(97)\qquad x_{36}=\overline{x_{16}+x_{56}}=\overline{x_{16}\cup x_{56}}=\overline{s\cup t}=\bar{s}\bar{t}, \quad x_{58}=s\cup\bar{t}, \quad x_{38}=\bar{s}\,t,$$

$$x_{34}=s, \quad x_{14}=\bar{s}, \quad x_{12}=0, \quad x_{71}=\bar{x}_{72}=x_{82},$$

and also the relations

$$\bar{s}\,t\,\bar{x}_{72}=0, \quad s\,\bar{x}_{72}=0, \quad \bar{s}\bar{t}\,x_{72}=0,$$

which is equivalent to the single equation

$$(s\cup t)\,\bar{x}_{72}\cup\bar{s}\bar{t}\,x_{72}=0,$$

whose solution is

$$(98)\qquad x_{72}=s\cup t.$$

It is easy to verify that formulas (92), (93.2), (95), (96), (97) and (98) define a solution of the system (91) for each couple of values given to the parameters s and t.

Case III. We seek the solutions for which

$$(93.3)\qquad x_{71}=x_{72}=0.$$

Taking into account (92) and (93.3), from (91.9), (91.8), (91.10), (91.1), (91.18), (91.3), (91.12), (91.7), (91.16) and (91.5), we deduce successively

$$(99)\qquad x_{81}=1, \quad x_{82}=x_{84}=0, \quad x_{12}=1, \quad x_{14}=x_{16}=0, \quad x_{38}=0,$$

$$x_{34}=u, \quad x_{36}=\bar{u}, \quad x_{74}=\bar{x}_{34}=\bar{u}, \quad x_{78}=x_{74}=u,$$

$$x_{58}=\bar{x}_{78}=\bar{u}, \quad x_{56}=\bar{x}_{58}=u,$$

where u is an arbitrary Boolean parameter.

It is easy to verify that formulas (92), (93.3) and (99) define a solution of the system (91) for both values 0 and 1 given to the parameter u.

Thus, giving to the parameters s, t, u all possible values we find that the system (91) has the following six solutions:

$$(100.1)\qquad x_{12}=x_{23}=x_{38}=x_{84}=x_{45}=x_{56}=x_{67}=x_{71}=1, \text{ the other } x_{ij}=0;$$

$$(100.2)\qquad x_{16}=x_{67}=x_{72}=x_{23}=x_{34}=x_{45}=x_{58}=x_{81}=1, \text{ the other } x_{ij}=0;$$

$$(100.3)\qquad x_{14}=x_{45}=x_{56}=x_{67}=x_{72}=x_{23}=x_{38}=x_{81}=1, \text{ the other } x_{ij}=0;$$

$$(100.4)\qquad x_{14}=x_{45}=x_{58}=x_{82}=x_{23}=x_{36}=x_{67}=x_{71}=1, \text{ the other } x_{ij}=0;$$

$$(100.5)\qquad x_{12}=x_{23}=x_{34}=x_{45}=x_{56}=x_{67}=x_{78}=x_{81}=1, \text{ the other } x_{ij}=0;$$

$$(100.6)\qquad x_{12}=x_{23}=x_{36}=x_{67}=x_{74}=x_{45}=x_{58}=x_{81}=1, \text{ the other } x_{ij}=0.$$

Chapter V

Minimization of Linear Pseudo-Boolean Functions

The results of the preceding two chapters will now be used in order to solve *pseudo-Boolean programs,* that is problems of minimizing (maximizing) pseudo-Boolean functions with variables possibly subject to pseudo-Boolean constraints.

In this chapter we deal with the case of pseudo-Boolean programs with linear objective functions and arbitrary (linear or nonlinear) pseudo-Boolean constraints.

In the first section we describe our method (P. L. HAMMER and S. RUDEANU [4,6]) which is based on the results in Chapter III. More precisely it will be shown that the possession of the families of solutions to the system of constraints offers an immediate way of obtaining the minimizing (maximizing) points. An accelerated method for performing this process is described in § 2.

In the third section we present methods due to P. CAMION [2], which make an intensive use of the Galois field GF (2).

§ 1. Using Partially Minimizing Points

We recall that a pseudo-Boolean function is a real-valued function with bivalent (0, 1) variables:

$$f: B_2^n \to R,$$

where $B_2 = \{0, 1\}$ and $B_2^n = \underbrace{B_2 \times \ldots \times B_2}_{n \text{ times}}$.

Definition 1. A vector $(x_1^*, \ldots, x_n^*) \in B_2^n$ is a *globally minimizing point*, or simply a *minimizing point* of the pseudo-Boolean function $f(x_1, \ldots, x_n)$ if

$$f(x_1^*, \ldots, x_n^*) \leqq f(x_1, \ldots, x_n) \tag{1}$$

for any $(x_1, \ldots, x_n)$ in B_2^n; the value $f(x_1^*, \ldots, x_n^*)$ is the *global minimum**, or simply, the *minimum* of the function f.

* In contrast with the *local minima* which will be studied in Chapter VII.

The minimization of a linear pseudo-Boolean function

$$c_1 x_1 + \cdots + c_n x_n \tag{2}$$

may be performed without any difficulty; indeed, its minimizing points are defined by

$$x_i^* = \begin{cases} 1 & \text{if } c_i < 0, \\ 0 & \text{if } c_i > 0, \\ p_i & \text{if } c_i = 0, \end{cases} \tag{3}$$

where p_i is an arbitrary parameter in B_2.

Example 1. It is obvious that the minimizing points of

$$2 + 3x_1 - 2x_2 - 5x_3 + 2x_6 - x_7 \tag{4}$$

are

$$x_1 = 0, \quad x_2 = 1, \quad x_3 = 1, \quad x_4 = p_4, \quad x_5 = p_5, \quad x_6 = 0, \quad x_7 = 1$$

where p_4 and p_5 are arbitrary parameters in B_2. Hence, the minimum of (4) is -6.

The minimization of a linear pseudo-Boolean function

$$f(x_1, \ldots, x_n) = f(X) = c_1 x_1 + \cdots + c_n x_n \tag{2}$$

subject to certain (linear, or nonlinear) constraints* may be performed in a similar way.

Namely, the methods will comprise three steps:

1. Determination of the solutions of the system of restrictions, grouped into families of solutions (as in Chapters III and IV) $\mathscr{F}_1, \ldots, \mathscr{F}_p$.
2. For each family $\mathscr{F}_k$ of solutions, determination of

$$\min_{X \in \mathscr{F}_k} f(X) \tag{5}$$

and of those points $X^0 \in \mathscr{F}_k$ for which

$$f(X^0) = \min_{X \in \mathscr{F}_k} f(X); \tag{6}$$

these points will be called *partially minimizing points*.

3. Determination (by direct checking) of

$$\min_{k=1,\ldots,p} \; \min_{X \in \mathscr{F}_k} f(X) \tag{7}$$

and of those points X^* in B_2^n for which

$$f(X^*) = \min_{k=1,\ldots,p} \; \min_{X \in \mathscr{F}_k} f(X). \tag{8}$$

It remains now to indicate how to perform step 2.

The vectors $X = (x_1, \ldots, x_n)$ belonging to a family $\mathscr{F}_k$ of solutions are characterized by the fact that the values x_i are fixed for those i

* As it was shown by R. Fortet [1], any pseudo-Boolean program may be reduced to a program of this type, by introduction of certain supplementary variables (see Theorem IV. 4).

which are contained in a certain set I_k of indices:

(9) $$i \in I_k \text{ implies } x_i = x_i^* = \text{fixed (0 or 1)},$$

while x_i remain arbitrary for $i \notin I_k$.

Reasoning as above, it is easy to see that the points X^0 satisfying (6) are given by the following formula:

(10) $$x_i = \begin{cases} x_i^*, & \text{if } i \in I_k, \\ 1, & \text{if } i \notin I_k \text{ and } c_i < 0, \\ 0, & \text{if } i \notin I_k \text{ and } c_i > 0, \\ p_i, & \text{if } i \notin I_k \text{ and } c_i = 0, \end{cases}$$

where p_i are arbitrary parameters in B_2.

Example 2. Let us minimize

(11) $$2 + 3x_1 - 2x_2 - 5x_3 + 2x_4 + 4x_6$$

with the constraints

(12.1) $$2x_1 - 3x_2 + 5x_3 - 4x_4 + 2x_5 - x_6 \leqq 2,$$

(12.2) $$4x_1 + 2x_2 + x_3 + 8x_4 - x_5 - 3x_6 \geqq 4.$$

The families of solutions of (12), determined as in Chapter III, are:

Table 1

No.	x_1	x_2	x_3	x_4	x_5	x_6
1	—	—	0	1	—	—
2	—	1	1	1	—	—
3	0	0	1	1	0	—
4	0	0	1	1	1	1
5	1	0	1	1	0	1
6	1	1	0	0	—	0
7	1	0	0	0	0	0

where the dashes indicate the arbitrary variables.

Putting in Table 1 instead of dashes, the values given by (10) we obtain

Table 2

No.	x_1	x_2	x_3	x_4	x_5	x_6	Value of (11)
1	0	1	0	1	p_1	0	2
2	0	1	1	1	p_2	0	-3
3	0	0	1	1	0	0	-1
4	0	0	1	1	1	1	3
5	1	0	1	1	0	1	6
6	1	1	0	0	p_6	0	3
7	1	0	0	0	0	0	5

Hence the sought minimum is -3 and it is attained in the points $(0, 1, 1, 1, 0, 0)$ and $(0, 1, 1, 1, 1, 0)$.

Example 3. Let us minimize

$$2 + 3x_1 - 2x_2 - 5x_3 + 2x_4 + 4x_5 \tag{13}$$

with the nonlinear constraints

$$x_1 x_2 + 4\bar{x}_1 x_3 - 3x_2 x_3 x_5 + 6\bar{x}_2 x_4 x_6 \geqq -1, \tag{14.1}$$

$$3x_2 x_4 - 5\bar{x}_1 \bar{x}_3 \bar{x}_5 + 4x_4 x_6 \geqq 1. \tag{14.2}$$

The families of solutions of the constraints obtained as in Chapter IV, are:

Table 3

No.	x_1	x_2	x_3	x_4	x_5	x_6
1	0	1	1	1	—	0
2	0	—	1	1	—	1
3	0	1	0	1	1	0
4	0	—	0	1	1	1
5	0	1	0	1	0	1
6	1	0	—	1	—	1
7	1	1	0	1	—	—
8	1	1	1	1	0	—

Putting, instead of dashes, the values indicated by (10) we find the following:

Table 4

No.	x_1	x_2	x_3	x_4	x_5	x_6	Value of (13)
1	0	1	1	1	p_1	0	-3
2	0	1	1	1	p_2	1	1
3	0	1	0	1	1	0	2
4	0	1	0	1	1	1	6
5	0	1	0	1	0	1	6
6	1	0	1	1	p_6	1	6
7	1	1	0	1	p_7	0	5
8	1	1	1	1	0	0	0

Hence the sought minimum is -3 and it is attained in the points $(0, 1, 1, 1, 0, 0)$ and $(0, 1, 1, 1, 1, 0)$.

Let us notice that in case we are interested in the maximization problem instead of the minimization one, then the above procedure remains valid, except formula (10), which is to be replaced by

$$x_i = \begin{cases} x_i^*, & \text{if } i \in I_k, \\ 1, & \text{if } i \notin I_k \text{ and } c_i > 0, \\ 0, & \text{if } i \notin I_k \text{ and } c_i < 0, \\ p_i, & \text{if } i \notin I_k \text{ and } c_i = 0, \end{cases} \tag{15}$$

where p_i are arbitrary parameters in B_2.

§ 2. Accelerated Linear Pseudo-Boolean Programming

The procedure described in the preceding section comprises three steps: the determination of *all* the solutions to the constraints, the determination of the partially minimizing points (corresponding to the various families of solutions), and the choice of the minimizing points (among the partially minimizing ones).

This technique takes no advantage, in the first (and most cumbersome) step, of the informations supplied by the objective function.

In order to utilize more completely the data of the problem, we propose a modified algorithm for the minimization of linear pseudo-Boolean functions under linear constraints.

This accelerated algorithm is based on the following devices.

First, we add a supplementary constraint

$$f(x_1, \ldots, x_n) \leqq M_r, \tag{16. r}$$

where M_r is a parameter defined as follows.

At the beginning of the process ($r = 0$), M_0 is either equal to the value $f(x_1^*, \ldots, x_n^*)$ of the function f at a point $(x_1^*, \ldots, x_n^*)$ satisfying the constraints — if such a point is known a priori — or equal to an upper bound of the function f (for instance the sum of its positive coefficients).

If $(x_1^r, \ldots, x_n^r)$ is the last solution determined at the r-th step of the branching process, then we put $M_{r+1} = f(x_1^r, \ldots, x_n^r)$ and continue the branching process with respect to the system consisting of the old constraints and $(16.\,r + 1)$.

A second accelerating device concerns the order of the branchings. If there are now constraints to which the remarks 1⁰, 3⁰, 4⁰ or 5⁰ in Table III.10.B can be applied, then we proceed as in Chap. III, § 3; in the opposite case, we split the discussion with respect to the first variable still occuring in the canonical form of (16.0), putting it first equal to 1, and afterwards equal to 0.

Accelerating test. A third accelerating device is as follows. Let us write the inequality (16.0) in the form

$$c_{i_1} \tilde{x}_{i_1} + c_{i_2} \tilde{x}_{i_2} + \cdots + c_{i_n} \tilde{x}_{i_n} \geqq m,$$

where

$$c_{i_1} \geqq c_{i_2} \geqq \cdots \geqq c_{i_n} \geqq 0.$$

Suppose that, at a certain step, when the variables $\tilde{x}_{i_h}$ for $h \in H$, are fixed, none of the remarks 1⁰, 3⁰, 4⁰ and 5⁰ can be applied, and we have to split the discussion according to the two possible values

1) $\tilde{x}_{i_j} = 1$,

and

2) $\tilde{x}_{i_j} = 0$,

of a certain variable $\tilde{x}_{i_j}$. We explore the branch 1). If there are no solutions along this branch, then, of course, we turn to the branch 2). But, if there are solutions, let K be the set of those indices $k \notin H$ for which $\tilde{x}_{i_k} = 0$ in the last (and hence, best!) solution of this branch. Now, if

(17) $$c_{i_j} > \sum_{k \in K} c_{i_k},$$

then the exploration of branch 2) will be dropped, because it cannot result in any improvement of the objective function.

Example 4. Let us minimize the pseudo-Boolean function

(18) $$f = 12x_1 + 5x_2 - 9x_3 - 5x_4 + 4x_5 + 8x_6 - 12x_7 - 3x_8 - 10x_9 + x_{10} - 7x_{11} - 7x_{12}$$

under the constraints (III.50):

(III.50.1) $12x_3 + 7\bar{x}_6 + 5x_{11} + 3\bar{x}_2 + 3\bar{x}_{10} + x_1 + x_5 + x_7 + x_{12} \geqq 19,$

(III.50.2) $7x_2 + 6\bar{x}_5 + 3\bar{x}_1 + x_4 \geqq 9,$

(III.50.3) $11\bar{x}_1 + 9\bar{x}_{11} + 7x_4 + 5x_9 + 2\bar{x}_7 + \bar{x}_3 + x_6 + \bar{x}_8 \geqq 28,$

(III.50.4) $12x_5 + 8x_{10} + 7\bar{x}_6 + 6\bar{x}_3 + 5\bar{x}_2 + 5\bar{x}_{12} + 3\bar{x}_8 + \bar{x}_9 \geqq 19,$

(III.50.5) $8x_6 + 7x_1 + 7\bar{x}_9 + 7x_{12} + 5x_3 + 3\bar{x}_4 + 2x_8 + x_2 + \bar{x}_5 + \bar{x}_{10} \geqq 19,$

(III.50.6) $5x_8 + 4x_4 + 3x_7 + 2x_1 + x_9 + \bar{x}_{11} + \bar{x}_{12} \geqq 6.$

We add the supplementary constraint

(18.0) $$f \leqq 30,$$

which can be brought to the canonical form

(18′.0) $$12\bar{x}_1 + 12x_7 + 10x_9 + 9x_3 + 8\bar{x}_6 + 7x_{11} + 7x_{12} + 5\bar{x}_2 + 5x_4 + 4\bar{x}_5 + 3x_8 + \bar{x}_{10} \geqq 0.$$

We deduce from (III.50.3), in view of remark 5^0 in Table III.10.B, that $\bar{x}_1 = 1$. After the introduction of this value into the system, none of the remarks 1^0, 3^0, 4^0 and 5^0 in Table III.10.B can be applied, so that we begin the splitting process. The order of the splits is induced by the ordering of the variables in (18′): x_7, x_9, x_3, $\bar{x}_6$, etc.

Case 1. $x_7 = 1$. Applying remark 5^0 to (III.50.3), (III.50.1) and (III.50.3), we deduce, in turn, that $\bar{x}_{11} = 1$, $x_3 = 1$, $x_4 = 1$, and that the inequality (III.50.6) becomes redundant. Since none of the remarks 1^0, 3^0, 4^0 and 5^0 applies to the remaining system, we continue the splitting process.

Case 1.1. $x_7 = 1$, $x_9 = 1$. Applying remark 5^0 to (III.50.3), (III.50.5), (III.50.1), (III.50.1), (III.50.2) and (III.50.4), we deduce, in turn, that $x_6 = 1$, $\bar{x}_2 = 1$, $\bar{x}_{10} = 1$, $\bar{x}_5 = 1$, and $x_5 = 1$, which is a contradiction.

Case 1.2. $x_7 = 1$, $x_9 = 0$. We have to introduce a new split.

Case 1.2.1. $x_7 = 1$, $x_9 = 0$, $\bar{x}_6 = 1$. The inequality (III.50.1) is solved, while remark 5⁰, applied to (III.50.3) and (III.50.5), shows that $\bar{x}_8 = 1$ and $x_{12} = 1$, respectively. Hence (III.50.3) and (III.50.5) become redundant. The next split is

Case 1.2.1.1. $x_7 = 1$, $x_9 = 0$, $\bar{x}_6 = 1$, $\bar{x}_2 = 1$. We apply, in turn, remark 5⁰ to (III.50.2) and (III.50.4), obtaining $\bar{x}_5 = 1$ and $x_{10} = 1$. The whole system (III.50) is now solved.

We have thus obtained the solution

$$(87)\quad x_1 = 0,\quad x_2 = 0,\quad x_3 = 1,\quad x_4 = 1,\quad x_5 = 0,\quad x_6 = 0,\quad x_7 = 1,$$
$$x_8 = 0,\quad x_9 = 0,\quad x_{10} = 1,\quad x_{11} = 0,\quad x_{12} = 1,$$

for which the objective function takes the value

$$f(0, 0, 1, 1, 0, 0, 1, 0, 0, 1, 0, 1) = -32.$$

Hence we replace the condition (18.0) by

$$(18.1)\qquad f \leqq -32,$$

which can be brought to the canonical form

$$(18'.1)\qquad 12\bar{x}_1 + 12x_7 + 10x_9 + 9x_3 + 8\bar{x}_6 + 7x_{11} + 7x_{12} + 5\bar{x}_2 + 5x_4 + $$
$$+ 4\bar{x}_5 + 3x_8 + \bar{x}_{10} \geqq 62.$$

We should now come back to the case 1.2.1.2.: $x_7 = 1$, $x_9 = 0$, $\bar{x}_6 = 1$, $\bar{x}_2 = 0$, but we apply first the accelerating test. Here the variables corresponding to the set H are $\bar{x}_1$, x_7, x_{11}, x_3, x_4, x_9, $\bar{x}_6$, x_8 and x_{12}, while $\bar{x}_{10}$ is the single variable corresponding to the set K. The coefficients of $\bar{x}_2$ and $\bar{x}_{10}$ are 5 and 1, respectively. Since $5 > 1$, i.e. (17) is satisfied, the case 1.2.1.2 is to be dropped.

The next case is 1.2.2: $x_7 = 1$, $x_9 = 0$, $\bar{x}_6 = 0$. Here the variables corresponding to H are $\bar{x}_1$, x_7, x_{11}, x_3, x_4 and x_9, while those corresponding to K are x_8 and $\bar{x}_{10}$. The coefficient of $\bar{x}_6$ in (18′) is 8, while those of x_8 and $\bar{x}_{10}$ are 3 and 1. Since $8 > 3 + 1$, the case 1.2.2 is to be dropped.

For the case 2, i.e. $x_7 = 0$, the single variable corresponding to the set H is $\bar{x}_1$, which has the coefficient 12, while the variables corresponding to the set K, i.e. x_{11}, x_9, x_8 and $\bar{x}_{10}$, have the coefficients 7, 10, 3 and 1, respectively. Since $12 < 7 + 10 + 3 + 1$, the case 2 is to be actually examined.

Case 2. $x_7 = 0$. It follows from (18′.1) and remark 5⁰, that $x_9 = 1$. The next split is

Case 2.1. $x_7 = 0$, $x_3 = 1$. Applying remark 5⁰ to (III.50.3) and then to (18′.1), we obtain $\bar{x}_{11} = 1$ and $\bar{x}_6 = 1$, $x_{12} = 1$, $\bar{x}_2 = 1$, $\bar{x}_5 = 1$, $x_8 = 1$, respectively. Now the inequality (III.50.5) has no solutions.

Case 2.2. $x_7 = 0$, $x_3 = 0$. Remark 4⁰ shows that the single solution to (18′.1) is $\bar{x}_6 = x_{11} = x_{12} = \bar{x}_2 = x_4 = \bar{x}_5 = x_8 = \bar{x}_{10} = 1$, but these values do not satisfy (III.50.3).

Thus the branching process is finished; we have reached a single solution of the system (III.50), namely the vector (0, 0, 1, 1, 0, 0, 1, 0, 0, 1, 0, 1), which is the minimizing point, the corresponding minimum being -32.

The process is illustrated in Fig. 1; the notation $\varnothing$ at the end of a branch means that there are no solutions corresponding to that branch, (S) indicates a solution, while [T] means that the accelerating test has shown that the corresponding branch was dropped.

We remark that only 31 vertices from the total amount of $12^{12} = 4096$ were actually explored.

The accelerated methods for linear pseudo Boolean programming was programmed, for an ELLIOTT 803-B computer, by Cs. Fábián and Gh. Weisz.

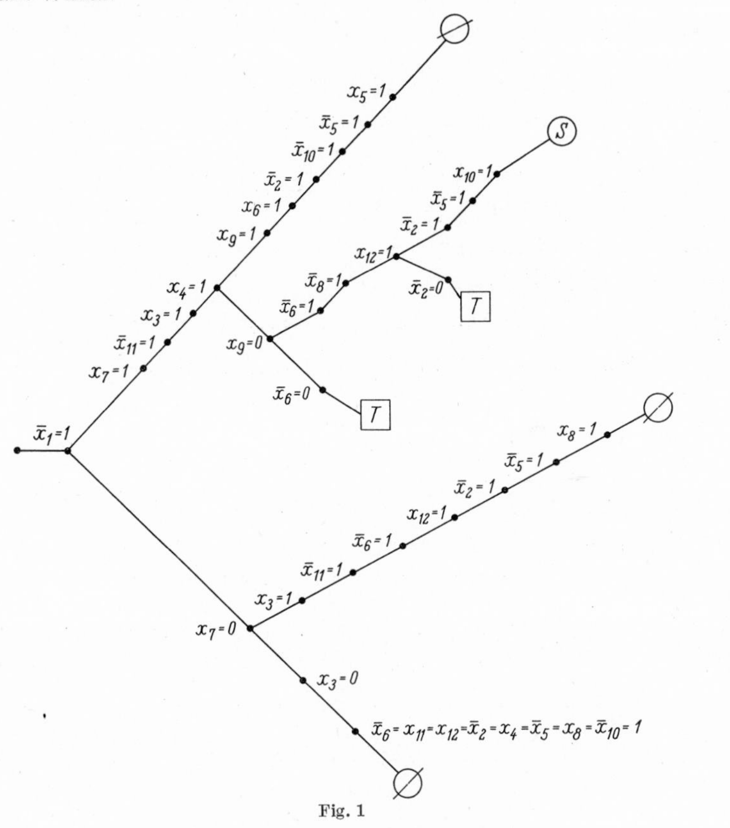

Fig. 1

§ 3. The Method of P. Camion

P. Camion [2] proposes the following methods for solving a pseudo-Boolean program (linear or not):

Consider the problem of minimizing the pseudo-Boolean function

$$f(x_1, \ldots, x_n),$$

under pseudo-Boolean constraints having the characteristic equation*

$$\Phi(x_1, \ldots, x_n) = 1. \tag{19}$$

* These methods, as well as that proposed in § 1, need the knowledge of the characteristic equation of the constraints. Once in the possession of this equation, the procedures utilize different ideas for obtaining the minimizing points.

Method I (see also M. CARVALLO [2]). Find, by any method, the characteristic function $\Psi(x_1, \ldots, x_n, y_1, \ldots, y_n)$ of the inequality

$$f(y_1, \ldots, y_n) < f(x_1, \ldots, x_n).$$

Let

$$\chi(x_1, \ldots, x_n) = 0$$

be the necessary and sufficient condition for the equation

$$\Phi(y_1, \ldots, y_n)\,\Psi(x_1, \ldots, x_n, y_1, \ldots, y_n) = 1 \tag{20}$$

to have a solution in $y_1, \ldots, y_n$.

Then the solutions of the problem are obviously characterized by the Boolean equation

$$\Phi(x_1, \ldots, x_n)\,\chi(x_1, \ldots, x_n) = 1. \tag{21}$$

Example 5. Let us minimize the pseudo-Boolean function

$$f = 2 - 3x_1 + x_2 + 5x_3 \tag{22}$$

under the constraints

$$2x_1 + 3x_2 + 5x_3 \geqq 3, \tag{23.1}$$

$$4x_1 + 5x_2 - 3x_3 \geqq 1. \tag{23.2}$$

The inequality $f(y_1, y_2, y_3) < f(x_1, x_1, x_3)$ becomes

$$-3y_1 + y_2 + 5y_3 < -3x_1 + x_2 + 5x_3$$

and its characteristic function is

$$\Psi(x_1, x_2, x_3, y_1, y_2, y_3) = x_3\,\bar{y}_3 \cup (\bar{x}_3\,\bar{y}_3 \cup x_3\,y_3)\,[\bar{x}_1\,y_1 \cup (x_1\,y_1 \cup \bar{x}_1\,\bar{y}_1)\,x_2\,\bar{y}_2]. \tag{24}$$

Also, the characteristic function of the system (23) of constraints is

$$\Phi(x_1, x_2, x_3) = x_2 \cup x_1\,x_3. \tag{25}$$

Hence the equation (20) becomes

$$\Phi(y_1, y_2, y_3)\,\Psi(x_1, x_2, x_3, y_1, y_2, y_3)$$
$$= y_2\,\bar{y}_3(x_3 \cup \bar{x}_1\,y_1) \cup x_3\,y_1\,y_3(\bar{x}_1 \cup x_2\,\bar{y}_2) = 1. \tag{26}$$

By eliminating y_1, y_2, y_3, we obtain the equation $\bar{x}_1 \cup x_3 = 1$, or

$$\chi(x_1, x_2, x_3) = x_1\,\bar{x}_3 = 0. \tag{27}$$

It follows from (25) and (27) that equation (21) becomes

$$x_1\,x_2\,\bar{x}_3 = 1. \tag{28}$$

Hence the single minimizing point is $(1, 1, 0)$ and the sought minimum is

$$f_{\min} = f(1, 1, 0) = 0.$$

Method I'. Find the parametric solutions of the constraints; introduce them into the objective function and apply the previous method in absence of constraints (Φ is now identically equal to 1).

Example 6. Consider the same pseudo-Boolean program as in the previous example.

The parametric solution of the characteristic equation $x_2 \cup x_1 x_3 = 1$ of the constraints is

(29.1) $$x_1 = z_1,$$

(29.2) $$x_2 = \bar{z}_1 \cup \bar{z}_3 \cup z_2,$$

(29.3) $$x_3 = z_3,$$

hence the objective function becomes

(30) $$f = -2 - 3z_1 + \bar{z}_1 + z_1 \bar{z}_3 + z_1 z_2 z_3 + 5z_3.$$

We now form the inequality

(31) $$-4t_1 + 5t_3 + t_1 \bar{t}_3 + t_1 t_2 t_3 < -4z_1 + 5z_3 + z_1 \bar{z}_3 + z_1 z_2 z_3$$

and we determine its characteristic equation:

$$\Psi(z_1, z_2, z_3, t_1, t_2, t_3) = z_3 (z_1 z_2 t_1 \bar{t}_2 \cup \bar{t}_3) \cup \bar{z}_1 t_1 (z_3 \cup \bar{t}_3) = 1.$$

Eliminating t_1, t_2 and t_3 from (31), we obtain the equation $z_3 \cup \bar{z}_1 = 1$, or

(32) $$\chi(z_1, z_2, z_3) = z_1 \bar{z}_3 = 0,$$

so that the equation (21), which becomes

(33) $$\chi(z_1, z_2, z_3) = 1,$$

reduces to

(34) $$z_1 \bar{z}_3 = 1$$

and has the solution

(35) $$z_1 = 1, \quad z_2 \text{ arbitrary}, \quad z_3 = 0,$$

which, introduced into formulas (29), give

(36) $$x_1 = 1, \quad x_2 = 1, \quad x_3 = 0,$$

i.e. the same result as in the preceding example.

Method II. Let

(37) $$x_i = \varphi_i(z_1, \ldots, z_n)$$

be the parametric solution of the characteristic equation (19) of the constraints. Then the objective function may be written in the form

(38) $$f(x_1, \ldots, x_n) = f(\varphi_1(z_1, \ldots, z_n), \ldots, \varphi_n(z_1, \ldots, z_n))$$
$$= \sum_{i=1}^{p} a_i C_i - \sum_{j=p+1}^{q} b_j C_j + c$$

where C_i and C_j are conjunctions in the variables $z_1, \ldots, z_n$, the coefficients a_i, b_j being positive. Hence

(39) $$f(x_1, \ldots, x_n) = g(z_1, \ldots, z_n) - \sum_{j=p+1}^{q} b_j + c$$

where

(40) $$g(z_1, \ldots, z_n) = \sum_{i=1}^{p} a_i C_i + \sum_{j=p+1}^{q} b_j \bar{C}_j.$$

The function $g(z_1, \ldots, z_n)$ is non-negative and has a binary development

(41) $$g = \sum_{k=0}^{r} 2^k \psi_k(C_1, \ldots, C_q) = \sum_{k=0}^{r} 2^k \Theta_k(z_1, \ldots, z_n).$$

Now let Θ_s be the last Θ_k which is not identically equal to 1; solve the equation $\Theta_s = 0$ and introduce its parametric solution into the other Θ_k's. Let Θ_t be the last Θ_i which is not identically equal to 1 after the preceding substitution; solve the equation $\Theta_t = 0$ and introduce its parametric solution into the Θ_i with $i \leq t$, etc.

At the end of the above process, introduce the values found for $z_1, \ldots, z_n$ into the relations (37), thus obviously obtaining the minimizing points.

Example 7. Let us resume the pseudo-Boolean program treated in Examples 5 and 6.

It was shown in Example 6 that the parametric solution of the characteristic equation of the system of constraints is

(29.1) $$x_1 = z_1,$$

(29.2) $$x_2 = \bar{z}_1 \cup \bar{z}_3 \cup z_2,$$

(29.3) $$x_3 = z_3,$$

and that the function (22) becomes

(30) $$f = -2 - 3z_1 + \bar{z}_1 + z_1 \bar{z}_3 + z_1 z_3 z_2 + 5z_3.$$

Hence

(42) $$g = (\bar{z}_1 + z_1 \bar{z}_3 + z_1 z_2 z_3 + 5z_3) + 3\bar{z}_1 = 4\bar{z}_1 + 5z_3 + z_1 \bar{z}_3 + z_1 z_2 z_3.$$

Using the methods described in the last section of Chapter III, we find that the binary development of the function g is

(43) $$g = (z_3 \oplus z_1 \bar{z}_3 \oplus z_1 z_2 z_3) + 2z_1 z_2 z_3 + 2^2(\bar{z}_1 \oplus z_3) + 2^3 z_3.$$

The first equation is $z_3 = 0$, reducing g to the function

$$g' = z_1 + 2^2 \bar{z}_1,$$

hence $\bar{z}_1 = 0$. We have thus found the values (35), leading to (36), as in Example 6.

Chapter VI

Minimization of Nonlinear Pseudo-Boolean Functions

Any nonlinear pseudo-Boolean program may be reduced to a linear one, if the linearization process of R. FORTET (see § 7 of Chapter IV) is applied to the objective function as well as to the constraints.

Other methods for solving any pseudo-Boolean program (linear or nonlinear), due to P. CAMION, were described in § 3 of Chapter V.

In this chapter, we describe our algorithm (given first in P. L. HAMMER, I. ROSENBERG and S. RUDEANU [1, 2] and in the present modified form in P. L. HAMMER and S. RUDEANU [5,6]) for finding the minimum of a pseudo-Boolean function as well as the minimizing points. This procedure is, in fact, a combination of BELLMAN's dynamic programming approach with Boolean techniques (§ 1). In §§ 2—5 the method is extended to the case when the variables have to fulfil certain pseudo-Boolean conditions (equations, inequalities, systems, logical conditions); the method proposed in § 3 seems to be the most efficient.

The method requires the solution of a certain system of pseudo-Boolean equations and inequalities. This can be done with the procedures given in Chapters III and IV.

A. Minima without Constraints

§ 1. The Basic Algorithm

We recall that a pseudo-Boolean function is a real-valued function with bivalent (0, 1) variables:

$$f : B_2^n \to R. \tag{1}$$

We recall also that a vector $(x_1^*, \ldots, x_n^*) \in B_2^n$ is a *minimizing point* of the pseudo-Boolean function $f(x_1, \ldots, x_n)$, if

$$f(x_1^*, \ldots, x_n^*) \leqq f(x_1, \ldots, x_n) \tag{2}$$

for any $(x_1, \ldots, x_n)$ in B_2^n; the value $f(x_1^*, \ldots, x_n^*)$ is the *minimum* of f.

We see from the definition that

$$f(x_1^*, \ldots, x_n^*) \leqq f(x_1^*, \ldots, x_{i-1}^*, \bar{x}_i^*, x_{i+1}^*, \ldots, x_n^*) \tag{3. i}$$

for all $i = 1, 2, \ldots, n$.

The conditions (3. i) are hence *necessary* for $(x_1^*, \ldots, x_n^*)$ to be a minimizing point, but not sufficient: they characterize the *local* minima of the function f. However — and this is the basic idea of the algorithm we are going to describe here for obtaining the minimizing points — if we successively solve the inequalities (3.1), (3.2), ..., etc., introducing in each inequality the parametric solution of the preceding ones, then we obtain all the minimizing points and only them. This method was first given by I. ROSENBERG and the present authors in P. L. HAMMER, I. ROSENBERG and S. RUDEANU [1], starting from an idea of R. FORTET [1, 2].

We describe below the proposed algorithm, followed by an example and by some computational remarks. The section will conclude with the proof of the fact that the algorithm does obtain *all* the minimizing points and *only* them.

Description of the algorithm

Our algorithm is made up of two main stages: in the first one, the minimum of the given pseudo-Boolean function $f(x_1, \ldots, x_n)$ is found, while in the second one all the minimizing points are determined.

In fact, the recursive procedure we are going to propose here, may be viewed as a dynamic approach to bivalent programming (as it was shown in P. L. HAMMER [9]).

First stage. Let us denote, for the sake of recurrency,

$$f_1(x_1, \ldots, x_n) = f(x_1, \ldots, x_n). \tag{4.1}$$

It was seen in Chapter I that any pseudo-Boolean function is a polynomial, linear in each of its variables. Hence the function f_1 may be written in the form

$$f_1(x_1, \ldots, x_n) = x_1 g_1(x_2, \ldots, x_n) + h_1(x_2, \ldots, x_n) \tag{5.1}$$

and the inequality (3.1) becomes

$$(x_1 - \bar{x}_1) g_1(x_2, \ldots, x_n) \leqq 0. \tag{6.1}$$

For satisfying (6.1) it is necessary and sufficient that x_1 should satisfy to the following conditions: if $g_1(x_2, \ldots, x_n) < 0$, then $x_1 = 1$; if $g_1(x_2, \ldots, x_n) > 0$, then $x_1 = 0$; if $g_1(x_2, \ldots, x_n) = 0$, then $x_1 = p_1 =$ arbitrary bivalent parameter (for, in this case, the inequality is satisfied for each value of x_1).

Let us now introduce a Boolean function φ_1 depending on the variables $x_2, \ldots, x_n$ and on a new variable p_1:

$$\varphi_1 : B_2^n \to B_2, \tag{7.1}$$

defined by

$$\varphi_1(p_1, x_2, \ldots, x_n) = \begin{cases} 1, & \text{if } g_1(x_2, \ldots, x_n) < 0, \\ 0, & \text{if } g_1(x_2, \ldots, x_n) > 0, \\ p_1, & \text{if } g_1(x_2, \ldots, x_n) = 0. \end{cases} \tag{8.1}$$

In view of the above discussions, we have:

Lemma 1. *For every* $(p_1, x_2, \ldots, x_n) \in B_2^n$, *if*

$$x_1 = \varphi_1(p_1, x_2, \ldots, x_n), \tag{9.1}$$

then the vector $(x_1, x_2, \ldots, x_n) \in B_2^n$ *satisfies* (6.1) *and, conversely, if* $(x_1, x_2, \ldots, x_n) \in B_2^n$ *is a solution of* (6.1), *then there exists an element* $p_1 \in B_2$ *so that relation* (9.1) *holds.*

Let us now notice that the function φ_1 may also be written in the form

$$\varphi_1(p_1, x_2, \ldots, x_n) = \varphi_1'(x_2, \ldots, x_n) \cup p_1 \varphi_1''(x_2, \ldots, x_n), \tag{10.1}$$

where the functions φ_1' and φ'' are defined as follows:

$$\varphi_1'(x_2, \ldots, x_n) = \begin{cases} 1, & \text{if } g_1(x_2, \ldots, x_n) < 0, \\ 0, & \text{if } g_1(x_2, \ldots, x_n) \geqq 0, \end{cases} \tag{11.1}$$

and

$$\varphi_1''(x_2, \ldots, x_n) = \begin{cases} 1, & \text{if } g_1(x_2, \ldots, x_n) = 0, \\ 0, & \text{if } g_1(x_2, \ldots, x_n) \neq 0. \end{cases} \tag{12.1}$$

We see that φ_1' and φ_1'' are the *characteristic functions* (as defined in Chapter IV) of the inequality $g_1 < 0$ and of the equation $g_1 = 0$, respectively*. Hence a Boolean expression of φ_1 may be constructed and this will be used in the second stage of the algorithm.

Let us now construct the function

$$f_2(x_2, \ldots, x_n) = f_1(\varphi_1'(x_2, \ldots, x_n), x_2, \ldots, x_n). \tag{4.2}$$

To do this, we express the function φ_1' in pseudo-Boolean form (i.e., using the arithmetical operations $+$, $-$, and possibly the negation of single variables), as it was done in § 6 of Chapter IV, and introduce it into (5.1):

$$f_2(x_2, \ldots, x_n) = \varphi_1'(x_2, \ldots, x_n)\, g_1(x_2, \ldots, x_n) + h_1(x_2, \ldots, x_n).$$

We proceed now with the function f_2 in the same way as with f_1, etc. At the step i of the first stage we are faced with a pseudo-Boolean function f_i of $n - i + 1$ variables $x_i, x_{i+1}, \ldots, x_n$:

$$f_i : B_2^{n-i+1} \to R, \tag{1. i}$$

which is written in the form

$$f_i(x_i, x_{i+1}, \ldots, x_n) = x_i\, g_i(x_{i+1}, \ldots, x_n) + h_i(x_{i+1}, \ldots, x_n), \tag{5. i}$$

and we have to solve the inequality

$$(x_i - \bar{x}_i)\, g_i(x_{i+1}, \ldots, x_n) \leqq 0. \tag{6. i}$$

---

* Therefore, φ_i' and φ_i'' can be determined with the method suggested in Chap. IV, § 2, Example 6, or with the variant of it described in the note added after that example.

To do this, we introduce a Boolean function φ_i of $n - i + 1$ variables $p_i, x_{i+1}, \ldots, x_n$:

(7. i) $$\varphi_i : B_2^{n-i+1} \to B_2,$$

defined by

(8. i) $$\varphi_i(p_i, x_{i+1}, \ldots, x_n) = \begin{cases} 1, & \text{if } g_i(x_{i+1}, \ldots, x_n) < 0, \\ 0, & \text{if } g_i(x_{i+1}, \ldots, x_n) > 0, \\ p_i, & \text{if } g_i(x_{i+1}, \ldots, x_n) = 0, \end{cases}$$

or, equivalently, by

(10. i) $$\varphi_i(p_i, x_{i+1}, \ldots, x_n) = \varphi_i'(x_{i+1}, \ldots, x_n) \cup p_i \varphi_i''(x_{i+1}, \ldots, x_n)$$

where

(11. i) $$\varphi_i'(x_{i+1}, \ldots, x_n) = \begin{cases} 1, & \text{if } g_i(x_{i+1}, \ldots, x_n) < 0, \\ 0, & \text{if } g_i(x_{i+1}, \ldots, x_n) \geqq 0 \end{cases}$$

and

(12. i) $$\varphi_i''(x_{i+1}, \ldots, x_n) = \begin{cases} 1, & \text{if } g_i(x_{i+1}, \ldots, x_n) = 0, \\ 0, & \text{if } g_i(x_{i+1}, \ldots, x_n) \neq 0. \end{cases}$$

We write down the formula

(9. i) $$x_i = \varphi_i(p_i, x_{i+1}, \ldots, x_n)$$

and pass to the $(i + 1)$-th step, constructing the function

(4. $i + 1$) $$f_{i+1}(x_{i+1}, \ldots, x_n) = f_i(\varphi_i'(x_{i+1}, \ldots, x_n), x_{i+1}, \ldots, x_n).$$

Continuing in this way, we obtain at step n a function $f_n(x_n)$ which is written in the form

(5. n) $$f_n(x_n) = x_n g_n + h_n$$

where g_n and h_n are constants.

We have to solve the inequality

(6. n) $$(x_n - \bar{x}_n) g_n \leqq 0;$$

using the function

(8. n) $$\varphi_n(p_n) = \begin{cases} 1, & \text{if } g_n < 0, \\ 0, & \text{if } g_n > 0, \\ p_n, & \text{if } g_n = 0, \end{cases}$$

the solution is

(9. n) $$x_n = \varphi_n(p_n).$$

As it will be proved below, the minimum of the original function f is

(4. $n + 1$) $$f_{\min} = f_{n+1} = f_n(\varphi_n'),$$

where the constant φ_n' is given by

(11. n) $$\varphi_n' = \begin{cases} 1, & \text{if } g_n < 0, \\ 0, & \text{if } g_n \geqq 0. \end{cases}$$

The first stage of the algorithm has come to an end.

Second stage. Let us consider formula

(9. n) $$x_n = \varphi_n(p_n)$$

and introduce it into

(9. $n-1$) $$x_{n-1} = \varphi_{n-1}(p_{n-1}, x_n),$$

obtaining thus

(9'. $n-1$) $$x_{n-1} = \varphi_{n-1}(p_{n-1}, \varphi_n(p_n));$$

then we introduce the expressions (9. n) and (9'. $n-1$) into

(9. $n-2$) $$x_{n-2} = \varphi_{n-2}(p_{n-2}, x_{n-1}, x_n),$$

obtaining the new relation

(9'. $n-2$) $$x_{n-2} = \varphi_{n-2}(p_{n-2}, \varphi_{n-1}(p_{n-1}, \varphi_n(p_n)), \varphi_n(p_n)),$$

etc.

We obtain in this way a sequence of relations of the form

(13. n) $$x_n = \psi_n(p_n),$$

(13. $n-1$) $$x_{n-1} = \psi_{n-1}(p_{n-1}, p_n),$$

$$\cdots\cdots\cdots\cdots\cdots\cdots$$

(13.1) $$x_1 = \psi_1(p_1, \ldots, p_{n-1}, p_n).$$

The following remark is very important for the effectiveness of the algorithm.

Remark 1. Practice has shown that the functions ψ_i do actually depend only on a small number of parameters $p_{i_1}, \ldots, p_{i_m}$.

It will be proved below (Theorem 1) that, giving to the parameters $p_{i_1}, \ldots, p_{i_m}$ which actually appear in formulas (13), all possible values, we obtain all the minimizing points of the function f and only them.

Example 1. Let us minimize the pseudo-Boolean function

(14.1) $$f = f_1 = 2x_1 + 3x_2 - 7x_3 - 5x_1x_2x_3 + 3x_2x_4 + 9x_4x_5 - 2x_1x_5,$$

which may also be written in the form

$$f_1 = x_1(2 - 5x_2x_3 - 2x_5) + 3x_2 - 7x_3 + 3x_2x_4 + 9x_4x_5.$$

Hence

(15.1) $$g_1 = 2 - 5x_2x_3 - 2x_5$$

so that, using the methods given in Chapter IV, we obtain

(16.1) $$\varphi_1' = x_2x_3, \quad \varphi_1'' = (\bar{x}_2 \cup \bar{x}_3)x_5,$$

therefore formula (9.1) becomes

(17.1) $$x_1 = \varphi_1' \cup p_1\varphi_1'' = x_2x_3 \cup p_1(\bar{x}_2 \cup \bar{x}_3)x_5.$$

We replace x_1 by $\varphi_1' = x_2x_3$ in f_1 and obtain the function

(14.2) $$f_2 = -3x_2x_3 + 3x_2 - 7x_3 + 3x_2x_4 + 9x_4x_5 - 2x_2x_3x_5.$$

Hence

(15.2) $$g_2 = 3 - 3x_3 + 3x_4 - 2x_3x_5$$

so that

(16.2) $$\varphi_2' = x_3\bar{x}_4x_5, \quad \varphi_2'' = x_3\bar{x}_4\bar{x}_5;$$

therefore relation (9.2) becomes

(17.2) $$x_2 = \varphi_2' \cup p_2 \varphi_2'' = x_3 \bar{x}_4 x_5 \cup p_2 x_3 \bar{x}_4 \bar{x}_5 .$$

We replace x_2 by $\varphi_2' = x_3 \bar{x}_4 x_5$ in f_2 and obtain the function

(14.3) $$f_3 = -7x_3 + 9x_4 x_5 - 2x_3 \bar{x}_4 x_5 .$$

Hence

(15.3) $$g_3 = -7 - 2\bar{x}_4 x_5 ,$$

so that

(16.3) $$\varphi_3' = 1, \quad \varphi_3'' = 0$$

and relation (9.3) becomes

(17.3) $$x_3 = \varphi_3' \cup p_3 \varphi_3'' = 1 .$$

We replace x_3 by $\varphi_3' = 1$ in f_3 and get the function

(14.4) $$f_4 = -7 + 9x_4 x_5 - 2\bar{x}_4 x_5 = -7 - 2x_5 + 11x_4 x_5 ,$$

for, $\bar{x}_4 = 1 - x_4$.

Hence

(15.4) $$g_4 = 11x_5 ,$$

(16.4) $$\varphi_4' = 0, \quad \varphi_4'' = \bar{x}_5$$

and relation (9.4) is reduced to

(17.4) $$x_4 = \varphi_4' \cup p_4 \varphi_4'' = p_4 \bar{x}_5 .$$

We replace x_4 by $\varphi_4' = 0$ in f_4 and obtain

(14.5) $$f_5 = -7 - 2x_5 ,$$

hence

(15.5) $$g_5 = -2 ,$$

(16.5) $$\varphi_5' = 1, \quad \varphi_5'' = 0$$

and relation (9.5) becomes

(17.5) $$x_5 = \varphi_5' \cup p_5 \varphi_5'' = 1 .$$

The minimum value of the function f is obtained by putting $x_5 = \varphi_5' = 1$ in f_5:

(14.6) $$f_{\min} = f_6 = -9 .$$

It remains to determine the minimizing points. Relation (17.5) is simply

(18.5) $$x_5 = 1 ;$$

introducing this value into (17.4), we get

(18.4) $$x_4 = 0 ;$$

further, relation (17.3) is simply

(18.3) $$x_3 = 1 ;$$

introducing the above values into (17.2), we have

(18.2) $$x_2 = 1$$

and finally, from (17.1) we get

(18.1) $$x_1 = 1 .$$

Therefore the single minimizing point is $(1, 1, 1, 0, 1)$.

Example 2. The minimization of the pseudo-Boolean function

(19) $$2x_1 + 3x_2 - 7x_3 - 5x_1 x_2 x_3 + 3x_2 x_4 + 9x_4 x_5$$

may be carried out as in Table 1.

Hence $f_{\min} = -7$ and the minimizing points are

$$x_5 = p_5,$$
$$x_4 = p_4 \bar{p}_5,$$
$$x_3 = 1,$$
$$x_2 = p_2(\bar{p}_4 \cup p_5),$$
$$x_1 = p_2(\bar{p}_4 \cup p_5).$$

Table 1

No.	f_i	g_i	φ_i'	φ_i''	x_i
1	$2x_1 + 3x_2 - 7x_3 - 5x_1 x_2 x_3 + 3x_2 x_4 + 9x_4 x_5$	$2 - 5x_2 x_3$	$x_2 x_3$	0	$x_2 x_3$
2	$-3x_2 x_3 + 3x_2 - 7x_3 + 3x_2 x_4 + 9x_4 x_5$	$3 - 3x_3 + 3x_4$	0	$x_3 \bar{x}_4$	$p_2 x_3 \bar{x}_4$
3	$-7x_3 + 9x_4 x_5$	-7	1	0	1
4	$-7 + 9x_4 x_5$	$9x_5$	0	$\bar{x}_5$	$p_4 \bar{x}_5$
5	-7	0	0	1	p_5
6	$f_6 = f_{\min} = -7$				

For the various values of the parameters p_2, p_4, p_5, we obtain *all* the minimizing points $(x_1, \ldots, x_5)$ of f_1:

(20.1) $x_1 = 0, \quad x_2 = 0, \quad x_3 = 1, \quad x_4 = 0, \quad x_5 = 0;$

(20.2) $x_1 = 0, \quad x_2 = 0, \quad x_3 = 1, \quad x_4 = 0, \quad x_5 = 1;$

(20.3) $x_1 = 0, \quad x_2 = 0, \quad x_3 = 1, \quad x_4 = 1, \quad x_5 = 0;$

(20.4) $x_1 = 1, \quad x_2 = 1, \quad x_3 = 1, \quad x_4 = 0, \quad x_5 = 0;$

(20.5) $x_1 = 1, \quad x_2 = 1, \quad x_3 = 1, \quad x_4 = 0, \quad x_5 = 1.$

We are now going to give below a series of remarks concerning the above described algorithm. Remarks 2, 3, 9 and 10 are aimed to accelerate the computations.

Remark 2. In the preceding discussion, we have started by eliminating the variable x_1, followed by x_2, x_3, etc. In fact, there is no special reason for preferring this order, and we may adopt any other one which seems to be more appropriate. For instance, if we denote by n_i the number of terms in g_i, where $f = x_i g_i + h_i$, then it seems convenient to start with the elimination of that variable x_{i_0} for which $n_{i_0} = \min_i n_i$.

Remark 3. If, after eliminating the variables $x_{i_1}, x_{i_2}, \ldots, x_{i_m}$, where $m < n$, we obtain $f_{m+1} =$ constant, then the variables $x_{i_{m+1}}, \ldots, x_{i_n}$ are arbitrary (because, for any $q = 1, 2, \ldots, n-m$ we have $g_{i_{m+q}} = 0$, hence $\varphi'_{i_{m+q}} = 0$, $\varphi''_{i_{m+q}} = 1$; therefore $x_{i_{m+q}} = \varphi'_{i_{m+q}} + p_{i_{m+q}} \varphi''_{i_{m+q}} = p_{i_{m+q}}$), and $f_{m+1} = f_{\min}$.

Remark 4. The function φ_i may also be written in the form

$$(21.\,i)\qquad \varphi_i(p_i, x_{i+1}, \ldots, x_n) = \varphi_i'(x_{i+1}, \ldots, x_n) \cup p_i\, \varphi_i'''(x_{i+1}, \ldots, x_n)$$

where φ_i''' is the function defined by

$$(22.\,i)\qquad \varphi_i'''(x_{i+1}, \ldots, x_n) = \begin{cases} 1, & \text{if } g_i(x_{i+1}, \ldots, x_n) \leqq 0, \\ 0, & \text{if } g_i(x_{i+1}, \ldots, x_n) > 0 \end{cases}$$

(for it is obvious that $\varphi_i''' = \varphi_i' \cup \varphi_i''$, hence (10. i) implies $\varphi_i = \varphi_i' \cup p_i\, \varphi_i'' = \varphi_i' \cup p_i\, \varphi_i' \cup p_i\, \varphi_i'' = \varphi_i' \cup p_i\, \varphi_i'''$).

Remark 5. We have $\varphi_i'(x_{i+1}, \ldots, x_n) = \varphi_i(0, x_{i+1}, \ldots, x_n)$, so that relation (4. $i+1$) becomes

$$(23.\,i+1)\qquad f_{i+1}(x_{i+1}, \ldots, x_n) = f_i(\varphi_i(0, x_{i+1}, \ldots, x_n), x_{i+1}, \ldots, x_n).$$

Remark 6. The function f_{i+1} may also be written in the form

$$(24.\,i+1)\qquad f_{i+1}(x_{i+1}, \ldots, x_n) = f_i(\varphi_i(1, x_{i+1}, \ldots, x_n), x_{i+1}, \ldots, x_n),$$

for, relations (8. i) and (11. i) imply that the function

$$\varphi_i\, g_i = \varphi_i'\, g_i = \begin{cases} g_i, & \text{if } g_i < 0, \\ 0, & \text{if } g_i \geqq 0, \end{cases}$$

does not depend on p_i, hence formula $f_i = x_i\, g_i + h_i$ implies, in turn,

$$\begin{aligned} f_{i+1} &= f_i(\varphi_i', x_{i+1}, \ldots, x_n) = \varphi_i'\, g_i + h_i = \varphi_i\, g_i + h_i \\ &= \varphi_i(0, x_{i+1}, \ldots, x_n)\, g_i(x_{i+1}, \ldots, x_n) + h_i(x_{i+1}, \ldots, x_n) \\ &= \varphi_i(1, x_{i+1}, \ldots, x_n)\, g_i(x_{i+1}, \ldots, x_n) + h_i(x_{n+1}, \ldots, x_n) \\ &= f_i(\varphi_i(1, x_{i+1}, \ldots, x_n), x_{i+1}, \ldots, x_n). \end{aligned}$$

Remark 7. The above Remarks 5 and 6 show that the function f_{i+1} may be written in the form

$$(25.\,i+1)\qquad f_{i+1}(x_{i+1}, \ldots, x_n) = f_i(\varphi_i(p_i, x_{i+1}, \ldots, x_n), x_{i+1}, \ldots, x_n)$$

whatever the element p_i may be (0 or 1).

Remark 8. The function f_{i+1} may also be written in the form

$$(26.\,i+1)\qquad f_{i+1}(x_{i+1}, \ldots, x_n) = f_i(\varphi_i'''(x_{i+1}, \ldots, x_n), x_{i+1}, \ldots, x_n)$$

(for, as it was shown in Remark 4, $\varphi_i''' = \varphi_i' \cup \varphi_i'' = \varphi_i(1, x_{i+1}, \ldots, x_n)$ so that relation (24. $i+1$) may be written in the form (26. $i+1$)).

Remark 9. When the problem is to find only the minimum of the function (and not the minimizing points), then the algorithm reduces to its first stage, which, in its turn, becomes more simple. Namely, at the i-th stage it suffices to determine either φ_i' or φ_i''' and to obtain f_{i+1} using either formula (4. $i+1$) or formula (26. $i+1$) respectively.

Remark 10. When the problem is to find not all, but a single minimizing point, then the first stage of the algorithm is performed as in Remark 9, while the second stage is also simplified in an obvious way, the rôle of the functions φ_i being played by the simpler functions φ_i' or φ_i'''.

We conclude now this section with the proof of the validity of the proposed algorithm.

Lemma 2. *The vector* $(x_1^*, \ldots, x_n^*)$ *is a minimizing point of the function* $f_1(x_1, x_2, \ldots, x_n)$ *if and only if the following two conditions hold:*

(α) *there exists an element* $p_1^* \in B_2$ *such that* $x_1^* = \varphi_1(p_1^*, x_2^*, \ldots, x_n^*)$;

(β) *the vector* $(x_2^*, \ldots, x_n^*)$ *is a minimizing point of the function* $f_2(x_2, \ldots, x_n)$.

Proof. Assume first that $(x_1^*, x_2^*, \ldots, x_n^*)$ is a minimizing point of f_1. Then $(x_1^*, x_2^*, \ldots, x_n^*)$ satisfies (6.1), so that property (α) holds in view of Lemma 1.

Further, property (β) is also valid. For, let $(y_2^*, \ldots, y_n^*)$ be an arbitrary vector in B_2^{n-1} and put $y_1^* = \varphi_1'(y_2^*, \ldots, y_n^*)$. Then

$$(27)\quad f_2(y_2^*, \ldots, y_n^*) = f_1(y_1^*, y_2^*, \ldots, y_n^*) \geqq f_1(x_1^*, x_2^*, \ldots, x_n^*) = f_1(\varphi_1(p_1^*, x_2^*, \ldots, x_n^*), x_2^*, \ldots, x_n^*).$$

On the other hand, Remark 7 shows that $f_2(x_2^*, \ldots, x_n^*) = f_1(\varphi_1(p_1^*, x_2^*, \ldots, x_n^*), x_2^*, \ldots, x_n^*)$. Hence the inequality (27) becomes $f_2(y_2^*, \ldots, y_n^*) \geqq f_2(x_2^*, \ldots, x_n^*)$, proving thus ($\beta$).

Conversely, assume now that $(x_1^*, \ldots, x_n^*)$ is a vector satisfying (α) and (β); we have to prove that it is a minimizing point of f_1.

For, if not, let $(y_1^*, \ldots, y_n^*)$ be a minimizing point of f_1. Then

$$(28)\qquad f_1(y_1^*, \ldots, y_n^*) < f_1(x_1^*, \ldots, x_n^*).$$

On the other hand, $(y_1^*, \ldots, y_n^*)$ is a solution of the inequality (6.1), i.e. we have $(y_1^* - \overline{y_1^*})\, g_1(y_2^*, \ldots, y_n^*) \leqq 0$. Hence, by Lemma 1, there exists an element q_1^* such that $y_1^* = \varphi_1(q_1^*, y_2^*, \ldots, y_n^*)$. Since by ($\alpha$), we have $x_1^* = \varphi_1(p_1^*, x_2^*, \ldots, x_n^*)$, the relation (28) becomes

$$(29)\quad f_1(\varphi_1(q_1^*, y_2^*, \ldots, y_n^*), y_2^*, \ldots, y_n^*) < f_1(\varphi_1(p_1^*, x_2^*, \ldots, x_n^*), x_2^*, \ldots, x_n^*).$$

In view of Remark 7, the inequality (29) may be written in the form

$$(30)\qquad f_2(y_2^*, \ldots, y_n^*) < f_2(x_2^*, \ldots, x_n^*),$$

contradicting thus (β).

Theorem 1. *The vector* $(x_1^*, \ldots, x_n^*)$ *is a minimizing point of the pseudo-Boolean function* $f(x_1, \ldots, x_n)$, *if and only if there exist* values* $p_1^*, \ldots, p_n^*$ *in* B_2, *so that*

$$(31.1)\qquad x_1^* = \varphi_1(p_1^*, x_2^*, \ldots, x_n^*),$$

$$(31.2)\qquad x_2^* = \varphi_2(p_2^*, x_3^*, \ldots, x_n^*),$$

$$\cdots\cdots\cdots\cdots$$

$$(31.n)\qquad x_n^* = \varphi_n(p_n^*).$$

* Notice that Theorem 1 does not contradict Remark 1. For, if the function φ_i does not actually depend on p_i, then, of course, there exists a value p_i^* so that relation (31.i) should take place.

The minimum of $f(x_1, \ldots, x_n)$ *is* f_{n+1} *defined as in* (4. $n+1$).

Proof. Both the necessity and the sufficiency of the conditions (31) result from a repeated application of Lemma 2, written for vectors of the form $(x_i^*, \ldots, x_n^*)$ $(i = 1, 2, \ldots, n)$. (For $i = n$, Lemma 2 states that the element x_n^* is a minimizing point of $f_n(x_n)$ if and only if there exists an element p_n^* such that $x_n^* = \varphi_n(p_n^*)$.)

Now, Remark 7 and the first part of the theorem imply that the minimum of the function f is

$$f_{\min} = f_1(x_1^*, \ldots, x_n^*) = f_2(x_2^*, \ldots, x_n^*) = \cdots = f_n(x_n^*) = f_{n+1},$$

completing the proof.

B. Minima with Constraints

In this division, we shall examine several alternative methods for solving the problem of the determination of all the minimizing points of a pseudo-Boolean function, whose variables are subject to pseudo-Boolean constraints.

The main tools will be the solution of pseudo-Boolean equations, inequalities and logical conditions, as it was done in Chapters III and IV, and the algorithm for minimizing pseudo-Boolean functions, described in § 1.

In § 2 we deal with the case when the conditions are of the form $f_j = 0$, where each f_j is either non-negative, or non-positive, and integer-valued.

The general problem of the determination of the minimizing points of an arbitrary pseudo-Boolean function with arbitrary constraints, will be dealt with in §§ 3, 4, 5 in three alternative ways. The best of these methods seems to be that given in § 3.

§ 2. Lagrangean Multipliers

Let us consider the problem of minimizing an *integer-valued* pseudo-Boolean function

(32) $$f(x_1, \ldots, x_n)$$

subject to the pseudo-Boolean equations

(33) $$f_j(x_1, \ldots, x_n) = 0 \qquad (j = 1, \ldots, m)$$

where each f_j is an *integer-valued* pseudo-Boolean function, either non-negative, or non-positive. Without loss of generality, we may suppose that all $f_j \geqq 0$, i.e.

(34. j) $$f_j(x_1, \ldots, x_n) \geqq 0 \text{ for all } (x_1, \ldots, x_n) \in B_2^n \qquad (j = 1, \ldots, m).$$

Problems of this type arise frenquently in applications; many a problem of the theory of graphs and of networks or of switching theory, may be formulated in this form.

The method we describe below was first given in P. L. HAMMER, I. ROSENBERG and S. RUDEANU [2].

Let us now consider an expression of the function f as a polynomial in $x_1, \ldots, x_n$ and possibly in $\bar{x}_1, \ldots, \bar{x}_n$, and let S^+ and S^- be the sum of the positive and of the negative coefficients of f, respectively.

We have

Theorem 2. *Assume the pseudo-Boolean functions f_j $(j=1, \ldots, m)$ are integer-valued and fulfil* (34).

(α) *If the vector $X^* = (x_1^*, \ldots, x_n^*)$ minimizes f, being subject to the constraints* (33), *then X^* minimizes also the pseudo-Boolean function*

$$F(x_1, \ldots, x_n) = f(x_1, \ldots, x_n) + (S^+ - S^- + 1) \sum_{j=1}^{m} f_j(x_1, \ldots, x_n). \tag{35}$$

(β) *If X^* minimizes F and*

$$F(x_1^*, \ldots, x_n^*) \leqq S^+, \tag{36}$$

then X^* minimizes f under the conditions (33).

(γ) *If X^* minimizes F and*

$$F(x_1^*, \ldots, x_n^*) > S^+, \tag{37}$$

then the conditions (33) *are inconsistent.*

Proof. It is easy to notice that for any X in B_2^n, we have

$$S^- \leqq f(X) \leqq S^+. \tag{38}$$

(α) Now let X^* be a vector minimizing f under the restrictions (33) and suppose there exists a $Y^* \in B_2^n$ such that

$$F(Y^*) < F(X^*). \tag{39}$$

It follows that Y^* also fulfils (33), for if it would not, then, according to (34), at least for a j_0, relation $f_{j_0}(Y^*) > 0$ should hold, i.e. $f_{j_0}(Y^*) \geqq 1$ (for, each f_j is integer-valued), hence $\sum_{j=1}^{m} f_j(Y^*) \geqq 1$, implying

$$F(Y^*) \geqq f(Y^*) + S^+ - S^- + 1 \geqq S^+ + 1. \tag{40}$$

On the other hand

$$F(X^*) = f(X^*) \leqq S^+, \tag{41}$$

and from (40) and (41) we obtain

$$F(X^*) < F(Y^*), \tag{42}$$

contradicting (39). Hence Y^* does fulfil the conditions (33).

We now see that $F(X^*) = f(X^*)$ and $F(Y^*) = f(Y^*)$, so that relation (39) becomes $f(Y^*) < f(X^*)$, contradicting the fact that X^* minimizes f under the restrictions (33).

(β) Conversely, let X^* be a minimum of F satisfying (36). Then X^* satisfies (33), for if not, we could reason as above, deducing

$$F(X^*) \geqq f(X^*) + S^+ - S^- + 1 > S^+.$$

Now it is obvious that X^* minimizes f under the conditions (33).

(γ) Let X^* be a minimizing point of F, satisfying (37). Assume that relations (33) would be consistent and let Y^* be a vector satisfying them. Hence $F(Y^*) = f(Y^*) \leqq S^+$, so that $\min F \leqq S^+$, contradicting (37).

Remark 11. The above theorem remains valid if we replace S^+ and S^- by any upper bound and lower bound of f, respectively.

Corollary 1. For finding all the points in B_2^n which minimize the function f under the conditions $f_j = 0$, we apply the algorithm in § 1 to the function F. If the value $F_{n+1} = F_{\min} > S^+$, then the problem has no solutions. If $F_{n+1} = F_{\min} \leqq S^+$, then this value is the minimum of the function f under the restrictions $f_j = 0$, and the solutions of the considered problem coincide with the minimizing points of F (which may be obtained applying the second stage of the algorithm).

Example 3. Let us minimize the function

$$f = 2x_1 x_3 - x_2 \bar{x}_3 + 4x_3 x_4 x_5 - x_1 x_5, \tag{43}$$

with the condition

$$x_1 x_2 + 2x_2 x_3 + \bar{x}_4 x_5 = 0. \tag{44}$$

Here $S^+ = 6$, $S^- = -2$, hence we have to minimize the function

$$F = 2x_1 x_3 - x_2 \bar{x}_3 + 4x_3 x_4 x_5 - x_1 x_5 + 9(x_1 x_2 + 2x_2 x_3 + \bar{x}_4 x_5). \tag{45}$$

We find, with the basic algorithm, that $F_{\min} = F_6 = -1 < S^+$, hence the problem is consistent. The minimizing points are

$$x_1 = \bar{p}_2 p_5, \quad x_2 = p_2 \cup \bar{p}_5, \quad x_3 = 0, \quad x_4 = p_5 \cup p_4, \quad x_5 = p_5. \tag{46}$$

Remark 12. In many a problem the conditions are expressed by Boolean equations

$$\varphi_j = c_{j_1} \cup \cdots \cup c_{j_{k(j)}} = 0 \qquad (j = 1, \ldots, m) \tag{47. j}$$

where $c_{j_1}, c_{j_2}, \ldots, c_{j_{k(j)}}$ are elementary conjunctions. Since the equations (47) may also be written in the pseudo-Boolean form

$$f_j = c_{j_1} + c_{j_2} + \cdots + c_{j_{k(j)}} = 0 \qquad (j = 1, \ldots, m); \tag{48. j}$$

the functions f_j being non-negative, we can apply the above described method.

Example 4. The maximization of the pseudo-Boolean function

$$f = x_1 + x_2 + x_3 + x_4 + x_5 + x_6 \tag{49}$$

under the condition

$$x_1 x_4 \cup x_1 x_5 \cup x_1 x_6 \cup x_2 x_4 \cup x_3 x_6 \cup x_5 x_6 = 0 \tag{50}$$

is equivalent to the minimization of the function $-f$ under the restriction (50), or the minimization of the unrestricted pseudo-Boolean function

$$\begin{aligned} F = & -x_1 - x_2 - x_3 - x_4 - x_5 - x_6 + \\ & + 7(x_1 x_4 + x_1 x_5 + x_1 x_6 + x_2 x_4 + x_3 x_6 + x_5 x_6). \end{aligned} \tag{51}$$

Applying the basic algorithm, we find $F_{\min} = F_7 = -3$ (hence the sought maximum is $+3$) and the minimizing points

$$x_1 = \bar{p}_5, \quad x_2 = \bar{p}_4 \cup \bar{p}_5, \quad x_3 = 1, \quad x_4 = p_4 p_5, \quad x_5 = p_5, \quad x_6 = 0. \tag{52}$$

Remark 13. Consider the case when only part of the constraints (33) satisfy the conditions (34). Reasoning as in Theorem 2, we can transform the problem into another one, having those constraints which fulfil (34) introduced into the function to be minimized, and maintaining the other constraints in their original form.

The method of Lagrangean multipliers can be extended to the general case when the constraints need not be as in Theorem 2. For, in this case, let $\Phi(x_1, \ldots, x_n)$ be the characteristic function (written in a pseudo-Boolean form) of the system of constraints. As $\bar{\Phi}$ takes only the values 0 and 1, it fulfils condition (34). Therefore the problem of minimizing f under the given constraints, reduces to that of minimizing the restriction-free pseudo-Boolean function $f + \lambda \bar{\Phi}$, where $\lambda = S^+ - S^- + 1$, as above.

Example 5. Let us consider the problem of minimizing

$$f = 3x_1 \bar{x}_2 - 8\bar{x}_1 x_3 x_6 + 4x_2 x_5 \bar{x}_6 - 7\bar{x}_5 x_6 + 3x_4 - 5x_4 x_5 x_6 \tag{53}$$

under the conditions

$$x_1 x_2 + 4\bar{x}_1 x_3 - 3x_2 x_3 x_5 + 6\bar{x}_2 x_4 x_6 \geqq -1, \tag{54.1}$$

$$3x_2 x_4 - 5\bar{x}_1 \bar{x}_3 \bar{x}_5 + 4x_4 x_6 \geqq 1. \tag{54.2}$$

A pseudo-Boolean form of the characteristic function of the system (54) (found as in Chapter IV) is

$$\begin{aligned} \Phi = & \bar{x}_1 x_2 x_3 x_4 \bar{x}_6 + \bar{x}_1 x_3 x_4 x_6 + \bar{x}_1 x_2 \bar{x}_3 x_4 x_5 \bar{x}_6 + \\ & + \bar{x}_1 \bar{x}_3 x_4 x_5 x_6 + \bar{x}_1 x_2 \bar{x}_3 x_4 \bar{x}_5 x_6 + x_1 \bar{x}_2 x_4 x_6 + x_1 x_2 \bar{x}_3 x_4 + \\ & + x_1 x_2 x_3 x_4 \bar{x}_5. \end{aligned} \tag{55}$$

Hence the problem is now to minimize the unconstrained pseudo-Boolean function*

$$f + 31\bar{\Phi}, \tag{56}$$

the minimum of which is -12, while the minimizing points are

$$x_1 = 0, \quad x_2 = p, \quad x_3 = 1, \quad x_4 = 1, \quad x_5 = 0, \quad x_6 = 1. \tag{57}$$

---

* $\bar{\Phi}$ can be computed either as $1 - \Phi$, or using the "negated" form $\bar{\Phi} = \bigcup \bar{\varphi}_s$ of formula (IV.46).

§ 3. Minimization Using Families of Solutions

The method described in § 1 of Chapter V may be extended to the case of arbitrary (linear or nonlinear) objective functions. Namely, the knowledge of the p families of solutions to the constraints allows us to transform the original problem into (at most!) p minimization problems for unrestricted pseudo-Boolean functions, each of which has less variables than the original function.

This procedure seems to be the best of the different approaches proposed in this chapter for solving the general problem.

Let us consider again the problem of minimizing a pseudo-Boolean function

$$f(x_1, \ldots, x_n) \tag{57}$$

subject to the constraints

$$f_j(x_1, \ldots, x_n) \lesseqqgtr 0 \qquad (j = 1, \ldots, m), \tag{58}$$

and let $\mathscr{F}_1, \ldots, \mathscr{F}_p$ be the families of solutions of (58).

We may proceed now as follows.

(1. k) Introduce the fixed variables of the family $\mathscr{F}_k$ into (57);

(2. k) Minimize the unrestricted pseudo-Boolean function obtained at (1. k); let v^k be the corresponding minimum.

(3). Choose those points for which $v^{k_0} = \min\limits_{k=1,\ldots,p} v^k$.

Example 6. The minimization of the pseudo-Boolean function

$$3x_1\bar{x}_2 - 8\bar{x}_1 x_3 x_6 + 4x_2 x_5 \bar{x}_6 - 7\bar{x}_5 x_6 + 3x_4 - 5x_4 x_5 x_6 \tag{53}$$

under the constraints

$$2x_1 - 3x_2 + 5x_3 - 4x_4 + 2x_5 - x_6 \leqq 2, \tag{59.1}$$

$$4x_1 + 2x_2 + x_3 + 8x_4 - x_5 - 3x_6 \geqq 4 \tag{59.2}$$

leads to the minimization of 7 unrestricted pseudo-Boolean functions corresponding to the 7 families of solutions.

Proceeding as indicated above, we obtain Table 2, showing that the minimum is -12 and it is reached in the points $(0,1,1,1,0,1)$ and $(0,0,1,1,0,1)$.

Table 2

No.	Families of solutions						Functions to be minimized	Partially minimizing points						Partial minimum
	x_1	x_2	x_3	x_4	x_5	x_6		x_1	x_2	x_3	x_4	x_5	x_6	
1	—	—	0	1	—	—	$3x_1\bar{x}_2 + 4x_2 x_5 \bar{x}_6 - 7\bar{x}_5 x_6 + 3 - 5x_5 x_6$	$p_1 p_2$	p_2	0	1	0	1	-4
2	—	1	1	1	—	—	$-8\bar{x}_1 x_6 + 4x_5\bar{x}_6 - 7\bar{x}_5 x_6 + 3 - 5x_5 x_6$	0	1	1	1	0	1	-12
3	0	0	1	1	0	—	$3 - 15x_6$	0	0	1	1	0	1	-12
4	0	0	1	1	1	1	-10	0	0	1	1	1	1	-10
5	1	0	1	1	0	1	-1	1	0	1	1	0	1	-1
6	1	1	0	0	—	0	$4x_5$	1	1	0	0	0	0	0
7	1	0	0	0	0	0	3	1	0	0	0	0	0	3

Example 7. The minimization of the same pseudo-Boolean function (53) under the constraints

$$x_1 x_2 + 4\bar{x}_1 x_3 - 3x_2 x_3 x_5 + 6\bar{x}_2 x_4 x_6 \geqq -1, \tag{54.1}$$

$$3x_2 x_4 - 5\bar{x}_1 \bar{x}_3 \bar{x}_5 + 4x_4 x_6 \geqq 1 \tag{54.2}$$

eads to the minimization of 8 unrestricted pseudo-Boolean functions corresponding to the 8 families of solutions.

Proceeding as indicated above, we obtain Table 3, showing that the minimum is -12 and it is reached in the same points as in Example 5: $(0, 0, 1, 1, 0, 1)$ and $(0, 1, 1, 1, 0, 1)$.

Table 3

No.	Families of solutions						Functions to be minimized	Partially minimizing points						Partial minimum
	x_1	x_2	x_3	x_4	x_5	x_6		x_1	x_2	x_3	x_4	x_5	x_6	
1	0	1	1	1	—	0	$4x_5 + 3$	0	1	1	1	0	0	3
2	0	—	1	1	—	1	$-10 - 2\bar{x}_5$	0	p_2^2	1	1	0	1	-12
3	0	1	0	1	1	0	7	0	1	0	1	1	0	7
4	0	—	0	1	1	1	-2	0	p_2^4	0	1	1	1	-2
5	0	1	0	1	0	1	-4	0	1	0	1	0	1	-4
6	1	0	—	1	—	1	$1 - \bar{x}_5$	1	0	p_3^6	1	0	1	0
7	1	1	0	1	—	—	$4x_5 \bar{x}_6 - 7\bar{x}_5 x_6 + 3 - 5x_5 x_6$	1	1	0	1	0	1	-4
8	1	1	1	1	0	—	$-7x_6 + 3$	1	1	1	1	0	1	-4

§ 4. Minimization Using Parametric Forms of the Solutions

R. Fortet [1, 2] has proposed the following approach to the general problem of minimizing a pseudo-Boolean function under constraints: find parametric arithmetical expressions ("codage") of the solutions to the constraints, introduce them into the objective function and minimize the pseudo-Boolean function with independent variables obtained in this way. In the same paper, the parametrization of certain important types of constraints is given.

The concrete method we suggest for achieving this purpose, consists in the following four steps:

1) Determine the characteristic equation of the constraints, as in Chapters III and IV.

2) Find a parametric solution of the characteristic equation, using one of the well-known methods for solving Boolean equations.

3) Write the solution obtained at the step 2 in a pseudo-Boolean form.

4) Introduce this solution into the objective function and minimize the new function obtained in this way, by means of the basic algorithm.

In view of the results obtained in Chapter III and IV and in § 1 of this Chapter, the above algorithm may be applied to any bivalent program.

Example 8. Let us resume the problem studied in Example 7: minimize the pseudo-Boolean function (53) under the constraints (54).

The characteristic function of the conditions (54), as shown in Example 5, is

$$(60)\quad \Phi = \bar{x}_1 x_2 x_3 x_4 \bar{x}_6 \cup \bar{x}_1 x_3 x_4 x_6 \cup \bar{x}_1 x_2 \bar{x}_3 x_4 x_5 \bar{x}_6 \cup \bar{x}_1 \bar{x}_3 x_4 x_5 x_6 \cup \cup \bar{x}_1 x_2 \bar{x}_3 x_4 \bar{x}_5 x_6 \cup x_1 \bar{x}_2 x_4 x_6 \cup x_1 x_2 \bar{x}_3 x_4 \cup x_1 x_2 x_3 x_4 \bar{x}_5 .$$

We have to solve the Boolean equation $\Phi = 1$.

Since Φ is of the form $\Phi = x_4 \Phi'$, we obtain

$$(61.1)\quad x_4 = 1$$

and the equation $\Phi' = 1$, which may be written in the form

$$(62.1)\quad x_3(\bar{x}_1 x_2 \cup \bar{x}_1 x_6 \cup \bar{x}_2 x_6 \cup x_2 \bar{x}_5) \cup \cup \bar{x}_3(x_2 x_5 \cup x_5 x_6 \cup x_2 x_6 \cup x_1 x_6 \cup x_1 x_2) = 1 .$$

In Chapter II we have studied Boolean equations written in the form $\varphi = 0$. Taking into account the results presented there and the principle of duality, we see that an equation of the form $A x \cup B \bar{x} = 1$ is consistent if and only if $A \cup B = 1$; if this condition is satisfied, the parametric solution is $x = A p \cup \bar{B} \bar{p}$. For the above equation (62.1), the result of the elimination is

$$(62.2)\quad x_2 \cup x_6 = 1,$$

which has the parametric solution

$$(61.2)\quad x_2 = p \cup \bar{x}_6 .$$

In view of (61.2), the equation (62.1) becomes

$$x_3(\bar{x}_1 \cup \bar{x}_5 \cup x_6 \bar{p}) \cup \bar{x}_3(x_1 \cup x_5 \cup x_6 p) = 1$$

and has the parametric solution

$$(61.3)\quad x_3 = (\bar{x}_1 \cup \bar{x}_5 \cup x_6 \bar{p}) q \cup \bar{x}_1 \bar{x}_5 (\bar{x}_6 \cup \bar{p}) \bar{q} .$$

Thus the parametric solution of the equation $\Phi = 1$ is given by formulas (61), the variables x_1, x_5 and x_6 being arbitrary.

Now we write formulas (61) in a pseudo-Boolean form:

$$(61'.1)\quad x_4 = 1,$$

$$(61'.2)\quad x_2 = p_6 x_6 + \bar{x}_6,$$

$$(61'.3)\quad x_3 = (\bar{x}_1 + x_1 \bar{x}_5 + x_1 x_5 x_6 \bar{p}) q + \bar{x}_1 \bar{x}_5 (\bar{x}_6 + x_6 \bar{p}) \bar{q},$$

we introduce these expressions into the objective function (53) and obtain the new function

$$(63)\quad F = 3 x_1 x_6 \bar{p} - 8 \bar{x}_1 x_6 q - 8 \bar{x}_1 \bar{x}_5 x_6 \bar{p} \bar{q} + 4 x_5 - 7 x_6 - 2 x_5 x_6 + 3 .$$

Using the algorithm in § 1, we find $F_{\min} = -12$ and the minimizing points $x_1 = 0$, $x_5 = 0$, $x_6 = 1$, $q = p \cup r$, p and r arbitrary; hence, from (61'), $x_4 = 1$, $x_2 = p$ and $x_3 = 1$.

§ 5. Extension of the Basic Algorithm

The procedure given in § 1 for minimizing pseudo-Boolean functions without restrictions can be extended to the general case when the variables are subject to pseudo-Boolean constraints. This generalized algorithm runs as follows.

Let

(64.1) $$f = f_1(x_1, x_2, \ldots, x_n)$$

be the objective function and let

(65.1) $$\psi_1(x_1, \ldots, x_n) = 1$$

be the characteristic equation associated to the constraints; we suppose that the equation (65.1) is consistent.

We express the functions f and ψ_1 in the forms

$$f(x_1, \ldots, x_n) = x_1\, g_1(x_2, \ldots, x_n) + h_1(x_2, \ldots, x_n)$$

and

(66.1) $$\psi_1(x_1, \ldots, x_n) = x_1\, \chi_1(x_2, \ldots, x_n) \cup \bar{x}_1\, \omega_1(x_2, \ldots, x_n),$$

respectively. Let $\Phi_1(p_1, x_2, \ldots, x_n)$ be the Boolean function defined by the following table:

Table 4

$\chi_1(x_2, \ldots, x_n)$	$\omega_1(x_2, \ldots, x_n)$	$g_1(x_2, \ldots, x_n)$	$\Phi_1(p_1, x_2, \ldots, x_n)$
0			0
1	0		1
1	1	<0	1
1	1	>0	0
1	1	$=0$	p_1

where p_1 is an arbitrary parameter.

If $\varphi_1(p_1, x_2, \ldots, x_n)$ denotes the Boolean function defined by formula (8.1) in § 1, then the function Φ_1 may also be written in the form

$$\Phi_1 = \chi_1\, \bar{\omega}_1 \cup \chi_1\, \omega_1\, \varphi_1,$$

as shown by Table 4, or, equivalently, in the form

(67.1) $$\Phi_1 = \chi_1(\bar{\omega}_1 \cup \varphi_1).$$

We write down the formula

(68.1) $$x_1 = \Phi_1(p_1, x_2, \ldots, x_n);$$

on the other hand, as in § 1, we consider the Boolean function $\Phi_1'(x_2, \ldots, x_n) = \Phi_1(0, x_2, \ldots, x_n)$, which may be defined by

(69.1) $$\Phi_1' = \chi_1(\bar{\omega}_1 \cup \varphi_1')$$

where $\varphi_1'(x_2, \ldots, x_n) = \varphi_1(0, x_2, \ldots, x_n)$. We write the function Φ_1' in a pseudo-Boolean form and introduce it instead of x_1 into the objective function $f(x_1, x_2, \ldots, x_n)$. We obtain thus a new function

$$f_2(x_2, \ldots, x_n) = f(\Phi_1'(x_2, \ldots, x_n), x_2, \ldots, x_n); \tag{64.2}$$

and we consider also the new equation

$$\psi_2(x_2, \ldots, x_n) = \chi_1(x_2, \ldots, x_n) \cup \omega_1(x_2, \ldots, x_n) = 1, \tag{65.2}$$

to which we apply the same operations, etc.

After n steps, we obtain a sequence of relations of the form

$$x_1 = \Phi_1(p_1, x_2, \ldots, x_n), \tag{68.1}$$

$$x_2 = \Phi_2(p_2, x_3, \ldots, x_n), \tag{68.2}$$

$$\cdots\cdots\cdots\cdots\cdots\cdots$$

$$x_n = \Phi_n(p_n), \tag{68. n}$$

and a constant $f_{n+1} = f_n(\Phi_n')$ which is the minimum of the function f under the given constraints.

As in § 1, the second stage of the algorithm uses formulas (68) in order to detect all the minimizing points under the given constraints.

We shall now prove the validity of this algorithm. To do this, let us introduce the set G_1 of all the vectors $(x_1, x_2, \ldots, x_n) \in B_2^n$ satisfying the conditions

$$\psi_1(x_1, x_2, \ldots, x_n) = 1 \tag{65.1}$$

and

$$\text{if } \psi_1(\bar{x}_1, x_2, \ldots, x_n) = 1, \text{ then } (x_1 - \bar{x}_1)\, g_1(x_2, \ldots, x_n) \leqq 0. \tag{70.1}$$

Lemma 3. *The vector* $(x_1^*, x_2^*, \ldots, x_n^*)$ *minimizes the function* $f(x_1, x_2, \ldots, x_n)$ *under the constraint* (65.1) *if and only if the following two conditions hold:*

$$(x_1^*, x_2^*, \ldots, x_n^*) \in G_1, \tag{71.1}$$

$$f(x_1^*, x_2^*, \ldots, x_n^*) \leqq f(x_1, x_2, \ldots, x_n) \text{ for all } (x_1, x_2, \ldots, x_n) \in G_1. \tag{72.1}$$

Proof. The necessity of the conditions (71.1) and (72.1) is obvious.

Let us prove the sufficiency. For this sake, we notice first that all the vectors in G_1 satisfy the constraint (65.1). Hence it will suffice to show that for any vector $(y_1, y_2, \ldots, y_n)$ not in G_1, but satisfying (65.1), there exists a vector $(x_1, x_2, \ldots, x_n) \in G_1$ such that $f(x_1, x_2, \ldots, x_n) \leqq f(y_1, y_2, \ldots, y_n)$.

Let us therefore assume that the vector $(y_1, y_2, \ldots, y_n)$ satisfies the above conditions. This means that

$$\psi_1(y_1, y_2, \ldots, y_n) = 1, \tag{73}$$

while the condition (70.1) does not hold. But the falsity of (70.1) for $(y_1, y_2, \ldots, y_n)$ means that

(74) $$\psi_1(\bar{y}_1, y_2, \ldots, y_n) = 1$$

and

(75) $$(y_1 - \bar{y}_1)\, g_1(y_2, \ldots, y_n) > 0.$$

But the inequality (75) is equivalent to

(76) $$(\bar{y}_1 - y_1)\, g_1(y_2, \ldots, y_n) < 0.$$

Now relations (74), (73), and (76) show that $(\bar{y}_1, y_2, \ldots, y_n) \in G_1$; further, (76) implies also that

$$f(\bar{y}_1, y_2, \ldots, y_n) < f(y_1, y_2, \ldots, y_n),$$

thus completing the proof.

Lemma 4. *A vector $(x_1, x_2, \ldots, x_n)$ belongs to G_1 if and only if the following two conditions hold:*

(65.2) $$\chi_1(x_2, \ldots, x_n) \cup \omega_1(x_2, \ldots, x_n) = 1,$$

(77.1) *there exists $p_1 \in B_2$, so that $x_1 = \Phi_1(p_1, x_2, \ldots, x_n)$, Φ_1 being the function defined by Table 4.*

Proof. Assume first that $(x_1, x_2, \ldots, x_n) \in G_1$.

Then (65.1) holds, implying also (65.2), which is known to be the consistency condition for (65.1).

Therefore, the following situations may arise concerning $\chi_1(x_2, \ldots, x_n)$, $\omega_1(x_2, \ldots, x_n)$, $g_1(x_2, \ldots, x_n)$ and x_1:

$$\chi_1 = 0, \quad \omega_1 = 1, \quad \text{hence} \quad x_1 = 0, \quad \text{by (65.1) and (66.1)};$$

$$\chi_1 = 1, \quad \omega_1 = 0, \quad \text{hence} \quad x_1 = 1, \quad \text{by (65.1) and (66.1)};$$

$$\chi_1 = \omega_1 = 1, \quad \text{i.e.} \quad \psi_1(0, x_2, \ldots, x_n) = \psi_1(1, x_2, \ldots, x_n) = 1,$$

showing that $\psi_1(\bar{x}_1, x_2, \ldots, x_n) = 1$; in view of the condition (70.1), this implies

$$(x_1 - \bar{x}_1)\, g_1(x_2, \ldots, x_n) \leqq 0,$$

so that $x_1 = 1$, for $g_1 < 0$, $x_1 = 0$ for $g_1 > 0$, and $x_1 = p_1$ (arbitrary) for $g_1 = 0$.

We have thus shown that x_1 is determined as in Table 4, i.e. the property (77.1) is verified.

Conversely, let $(x_1, x_2, \ldots, x_n)$ be a vector satisfying (65.2) and (77.1); we shall prove that $(x_1, x_2, \ldots, x_n) \in G_1$.

Property (77.1) shows that x_1 is determined as in Table 4, while (65.2) implies that the situation $\chi_1(x_2, \ldots, x_n) = \omega_1(x_2, \ldots, x_n) = 0$ is impossible. Therefore, the following cases may occur concerning $\chi_1(x_2, \ldots, x_n)$ and $\omega_1(x_2, \ldots, x_n)$: ($\alpha$) $\chi_1 = 0$, $\omega_1 = 1$; (β) $\chi_1 = 1$, $\omega_1 = 0$; (γ) $\chi_1 = \omega_1 = 1$.

In case (α), we have $x_1 = 0$ by Table 4; in case (β), Table 4 implies $x_1 = 1$. Hence in all possible cases $[(\alpha), (\beta)$ and $(\gamma)]$, the equation (65.1) is satisfied. If, moreover, we have $\psi_1(\bar{x}_1, x_2, \ldots, x_n) = 1$, then we are in case (γ), so that Table 4 shows that the inequality $(x_1 - \bar{x}_1)\, g_1 \leqq 0$ is fulfilled. Thus $(x_1, x_2, \ldots, x_n) \in G_1$, completing the proof.

Lemma 5. *A vector $(x_1^*, x_2^*, \ldots, x_n^*)$ minimizes the function $f(x_1, x_2, \ldots, x_n)$ under the constraint* (65.1) *if and only if the following conditions hold:*

(78.1) $$\chi_1(x_2^*, \ldots, x_n^*) \cup \omega_1(x_2^*, \ldots, x_n^*) = 1;$$

(79.1) *there exists a* $p_1^* \in B_2$, *so that* $x_1^* = \Phi_1(p_1^*, x_2^*, \ldots, x_n^*)$;

(80.1) $f(x_1^*, x_2^*, \ldots, x_n^*) \leqq f(x_1, x_2, \ldots, x_n)$ *for all the vectors* $(x_1, x_2, \ldots, x_n)$ *satisfying* (65.2) *and* (77.1).

Proof. Obvious from Lemmas 3 and 4.

Lemma 6. *A vector $(x_1^*, x_2^*, \ldots, x_n^*)$ minimizes the function $f(x_1, x_2, \ldots, x_n)$ under the constraint* (65.1) *if and only if it satisfies* (78.1) *and*

(81.1) $f_2(x_2^*, \ldots, x_n^*) \leqq f_2(x_2, \ldots, x_n)$ *for all the vectors* $(x_2, \ldots, x_n)$ *satisfying* (65.2),

where f_2 is defined by (64.2).

Proof. The following statement is an immediate consequence of Lemma 5: a vector $(x_1^*, x_2^*, \ldots, x_n^*)$ minimizes $f(x_1, x_2, \ldots, x_n)$ under the constraint (65.1) if and only if it satisfies (78.1) and

(82.1) $f(\Phi_1(p_1^*, x_2^*, \ldots, x_n^*), x_2^*, \ldots, x_n^*) \leqq f(\Phi_1(p_1, x_2, \ldots, x_n), x_2, \ldots, x_n)$, for all the vectors $(x_1, x_2, \ldots, x_n)$ satisfying (65.2) and (77.1).

But Table 4 shows that the product $\Phi_1 g_1$ does not depend on p_1, so that we deduce, as in Remarks 5 and 6 from § 1, that

$$f(\Phi_1(p_1, x_2, \ldots, x_n), x_2, \ldots, x_n) = f(\Phi_1(0, x_2, \ldots, x_n), x_2, \ldots, x_n)$$
$$= f(\Phi_1'(x_2, \ldots, x_n), x_2, \ldots, x_n) = f_2(x_2, \ldots, x_n),$$

so that the property (82.1) coincides with (80.1). This completes the proof.

Theorem 3. *Assume that the Boolean equation* (65.1) *is consistent. Then the vector $(x_1^*, \ldots, x_n^*)$ is a minimizing point of the pseudo-Boolean function $f(x_1, \ldots, x_n)$ under the constraints* (65.1), *if and only if there exist values $p_1^*, \ldots, p_n^*$ in B_2, so that*

(83.1) $$x_1^* = \Phi_1(p_1^*, x_2^*, \ldots, x_n^*),$$

(83.2) $$x_2^* = \Phi_2(p_2^*, x_2^*, \ldots, x_n^*),$$

. . . .

(83. n) $$x_n^* = \Phi_n(p_n^*).$$

The minimum is f_{n+1} defined in (64. $n+1$) (*i.e.* $f_{n+1} = f_n(\Phi_n')$).

Proof. Immediate from Lemma 6.

Example 9. Let us solve the problem in Example 8 with the method suggested in the present section: minimize the pseudo-Boolean function (53), i.e.

(84.1) $f_1 = 3x_1\bar{x}_2 - 8\bar{x}_1x_3x_6 + 4x_2x_5\bar{x}_6 - 7\bar{x}_5x_6 + 3x_4 - 5x_4x_5x_6,$

under the conditions (54).

The characteristic function of the constraints was found in Example 8:

(85.1) $\psi_1 = \bar{x}_1x_2x_3x_4\bar{x}_6 \cup \bar{x}_1x_3x_4x_6 \cup \bar{x}_1x_2\bar{x}_3x_4x_5\bar{x}_6 \cup \bar{x}_1\bar{x}_3x_4x_5x_6 \cup$
$\cup \bar{x}_1x_2\bar{x}_3x_4\bar{x}_5x_6 \cup x_1\bar{x}_2x_4x_6 \cup x_1x_2\bar{x}_3x_4 \cup x_1x_2x_3x_4\bar{x}_5.$

We begin* the computations with x_5 and writing $\psi_1 = x_5\chi_5 \cup \bar{x}_5\omega_5$, we find** $\chi_5 = \psi_1|_{x_5=1} = \bar{x}_1x_2x_4 \cup \bar{x}_1x_4x_6 \cup \bar{x}_2x_4x_6 \cup x_2\bar{x}_3x_4$ and $\omega_5 = \psi_1|_{x_5=0} = x_1x_2x_4 \cup$ $\cup x_1x_4x_6 \cup x_2x_4x_6 \cup x_3x_4x_6 \cup x_2x_3x_4$. Further, the coefficient of x_5 in f_1 is $g_5 = 4x_2\bar{x}_6 + 7x_6 - 5x_4x_6$, so that $\varphi_5 = \bar{x}_2\bar{x}_6p_5$; hence formula (67) gives

(86.1) $\Phi_5 = \bar{x}_1\bar{x}_2\bar{x}_3x_4x_6 \cup \bar{x}_1x_2\bar{x}_3x_4\bar{x}_6.$

Replacing x_5 by $\Phi_5 = \Phi_5' = \bar{x}_1\bar{x}_2\bar{x}_3x_4x_6 + \bar{x}_1x_2\bar{x}_3x_4\bar{x}_6$ in (84.1), we obtain

(84.2) $f_2 = 3x_1\bar{x}_2 - 8\bar{x}_1x_3x_6 - 7x_6 + 3x_4 + 4\bar{x}_1x_2\bar{x}_3x_4\bar{x}_6 + 2\bar{x}_1\bar{x}_2\bar{x}_3x_4x_6;$

the equation $\psi_1 = x_5\chi_5 \cup \bar{x}_5\omega_5 = 1$, is to be replaced by $\chi_5 \cup \omega_5 = 1$, which in our case reduces to

(85.2) $\psi_2 = x_2x_4 \cup x_4x_6 = 1.$

We eliminate now x_4, which gives $\chi_4 = x_2 \cup x_6, \omega_4 = 0, g_4 = 3 + 4\bar{x}_1x_2\bar{x}_3\bar{x}_6 +$ $+ 2\bar{x}_1\bar{x}_2\bar{x}_3x_6$, hence $\varphi_4 = 0$ and

(86.2) $\Phi_4 = x_2 \cup x_6.$

Replacing x_4 by $\Phi_4 = \Phi_4' = x_2 + \bar{x}_2x_6$ in (84.2), we get

(84.3) $f_3 = 3x_1\bar{x}_2 - 8\bar{x}_1x_3x_6 - 7x_6 + 3x_2 + 3\bar{x}_2x_6 + 4\bar{x}_1x_2\bar{x}_3\bar{x}_6 + 2\bar{x}_1\bar{x}_2\bar{x}_3x_6;$

the equation $\psi_2 = x_4\chi_4 \cup \bar{x}_4\omega_4 = 1$ is to be replaced by $\chi_4 \cup \omega_4 = 1$, i.e. by

(85.3) $\psi_3 = x_2 \cup x_6 = 1.$

Now we eliminate x_6, which gives $\chi_6 = 1, \omega_6 = x_2, g_6 = -8\bar{x}_1x_3 - 7 + 3\bar{x}_2 -$ $- 4\bar{x}_1x_2\bar{x}_3 + 2\bar{x}_1\bar{x}_2\bar{x}_3$, hence $\varphi_6 = 1$ and

(86.3) $\Phi_6 = 1.$

We introduce $x_6 = \Phi_6 = 1$ into (84.3) and obtain

(84.4) $f_4 = 3x_1\bar{x}_2 - 8\bar{x}_1x_3 - 4 + 2\bar{x}_1\bar{x}_2\bar{x}_3;$

the equation (86.3) is to be replaced by $\chi_6 \cup \omega_6 = 1$, which in our case reduces to the identity $1 = 1$.

Now the original problem has been reduced to that of minimizing f_4 without constraints. As in § 1, we find

(86.4) $\varphi_3 = \bar{x}_1,$

---

* As a matter of fact, the equation $\psi_1 = 1$ obviously implies $x_4 = 1$; however, we abstain from any remark not resulting from the algorithm.

** In this example, the indices of the functions f_i and ψ_i refer to the corresponding step of the algorithm, while the index of χ_j, ω_j, g_j, φ_j and Φ_i shows which variable x_j is eliminated.

hence

(84.4) $$f_4 = 3x_1\bar{x}_2 - 8\bar{x}_1 - 4;$$

further, we get

(86.5) $$\varphi_1 = 0,$$

hence

(84.5) $$f_5 = -12.$$

Applying Remark 3 in § 1, we deduce that -12 is the sought minimum and that the value of x_2 remains arbitrary:

(86.6) $$\varphi_2 = p_2.$$

In the second stage of the algorithm we find from (86):

$$x_2 = p_2, \quad x_1 = 0, \quad x_3 = 1, \quad x_6 = 1, \quad x_4 = 1, \quad x_5 = 0;$$

thus the minimizing points under the given constraints are $(0, p_2, 1, 1, 0, 1)$.

Chapter VII

Extensions of Pseudo-Boolean Programming

In this chapter, various generalizations of the problem of minimizing a pseudo-Boolean function will be studied.

§ 1. Local Minima of Pseudo-Boolean Functions

We recall that a pseudo-Boolean function was defined as a real-valued function with bivalent (0, 1) variables:

$$f: B_2^n \to R. \tag{1}$$

In Chapters V and VI various methods were proposed for obtaining the minimum value and the minimizing points of a pseudo-Boolean function. In this section we shall be concerned with "local minima", as defined below.

Definition 1. A vector $(x_1^*, \ldots, x_n^*)$ satisfying

$$(2.\,i)\quad f(x_1^*, \ldots, x_n^*) \leqq f(x_1^*, \ldots, x_{i-1}^*, \overline{x_i^*}, x_{i+1}^*, \ldots, x_n^*) \text{ for } i = 1, \ldots, n,$$

is called a *locally minimizing point*, while $f(x_1^*, \ldots, x_n^*)$ is said to be a *local minimum* of the function f.

Putting

$$(3.\,i)\quad f(x_1, \ldots, x_n) = x_i\, g_i(x_1, \ldots, x_{i-1}, x_{i+1}, \ldots, x_n) + h_i(x_1, \ldots, x_{i-1}, x_{i+1}, \ldots, x_n) \qquad (i = 1, \ldots, n)$$

we have:

Theorem 1. *The locally minimizing points of the pseudo-Boolean function f coincide with the solutions of the system of pseudo-Boolean inequalities*

$$(4.\,i)\quad (2x_i - 1)\, g_i(x_1, \ldots, x_{i-1}, x_{i+1}, \ldots, x_n) \leqq 0 \qquad (i = 1, \ldots, n).$$

Proof. Taking into account (3), we can write (2. i) in the form $x_i\, g_i \leqq \bar{x}_i\, g_i$, which is equivalent to (4. i).

Theorem 2. *Let* $\psi_i(x_1, \ldots, x_{i-1}, x_{i+1}, \ldots, x_n)$ *and* $\chi_i(x_1, \ldots, x_{i-1}, x_{i+1}, \ldots, x_n)$ *be the characteristic functions* of the inequality* $g_i(x_1, \ldots, x_{i-1}, x_{i+1}, \ldots, x_n) < 0$ *and of the equation* $g_i(x_1, \ldots, x_{i-1}, x_{i+1}, \ldots, x_n) = 0$, *respectively* $(i = 1, \ldots, n)$. *Then the characteristic function* $\Phi(x_1, \ldots, x_n)$ *of the system* (4) *is*

$$\Phi = \prod_{i=1}^{n} (x_i \psi_i \cup \bar{x}_i \bar{\psi}_i \cup \chi_i). \tag{5}$$

Remark 1. The determination of the functions ψ_i and χ_i may be reduced, as in § 2 of Chapter IV, to the determination of a single characteristic function, namely that of the inequality $g_i \leqq 0$.

Proof. Let $\varphi_i(x_1, \ldots, x_n)$ be the characteristic function of the inequality (4. i). According to Theorem 2 in Chapter IV, we have $\Phi = \prod_{i=1}^{n} \varphi_i$, so that it suffices to prove

$$\varphi_i = x_i \psi_i \cup \bar{x}_i \bar{\psi}_i \cup \chi_i \qquad (i = 1, \ldots, n). \tag{6. i}$$

But (4. i) holds if and only if

$$x_i = \begin{cases} 1, & \text{if } g_i < 0, \\ 0, & \text{if } g_i > 0, \\ \text{arbitrary}, & \text{if } g_i = 0, \end{cases}$$

which coincides with (6. i).

Example 1. Let us determine the locally minimizing points of the pseudo-Boolean function

$$f = 3x_1 x_2 + 9x_2 x_3 x_4 - 7\bar{x}_1 x_5 x_6 + 2x_3 x_4 \bar{x}_6 + 4x_1 \bar{x}_2 x_3 \bar{x}_4 \bar{x}_5 x_6 - 5x_7 + 5x_8 + 2x_7 \bar{x}_8. \tag{7}$$

The inequalities (4.i) become:

$$(2x_1 - 1)(3x_2 + 7x_5 x_6 + 4\bar{x}_2 x_3 \bar{x}_4 \bar{x}_5 x_6) \leqq 0, \tag{8.1}$$

$$(2x_2 - 1)(3x_1 + 9x_3 x_4 - 4x_1 x_3 \bar{x}_4 \bar{x}_5 x_6) \leqq 0, \tag{8.2}$$

$$(2x_3 - 1)(9x_2 x_4 + 2x_4 \bar{x}_6 + 4x_1 \bar{x}_2 \bar{x}_4 \bar{x}_5 x_6) \leqq 0, \tag{8.3}$$

$$(2x_4 - 1)(9x_2 x_3 + 2x_3 \bar{x}_6 - 4x_1 \bar{x}_2 x_3 \bar{x}_5 x_6) \leqq 0, \tag{8.4}$$

$$(2x_5 - 1)(-7\bar{x}_1 x_6 - 4x_1 \bar{x}_2 x_3 \bar{x}_4 x_6) \leqq 0, \tag{8.5}$$

$$(2x_6 - 1)(-7\bar{x}_1 x_5 - 2x_3 x_4 + 4x_1 \bar{x}_2 x_3 \bar{x}_4 \bar{x}_5) \leqq 0, \tag{8.6}$$

$$(2x_7 - 1)(-5 + 2\bar{x}_8) \leqq 0, \tag{8.7}$$

$$(2x_8 - 1)(5 - 2x_7) \leqq 0. \tag{8.8}$$

---

* We recall that the knowledge of the characteristic function is equivalent to the knowledge of the solutions; see Chapter IV.

Proceeding as in Chapter IV, we find

$$\psi_1 = 0, \quad \chi_1 = \bar{x}_2[\bar{x}_6 \cup \bar{x}_5(\bar{x}_3 \cup x_4)],$$

$$\psi_2 = x_1 x_3 \bar{x}_4 \bar{x}_5 x_6, \quad \chi_2 = \bar{x}_1(\bar{x}_3 \cup \bar{x}_4),$$

$$\psi_3 = 0, \quad \chi_3 = (\bar{x}_4 \cup \bar{x}_2 x_6)(\bar{x}_1 \cup x_2 \cup x_4 \cup x_5 \cup \bar{x}_6),$$

$$\psi_4 = x_1 \bar{x}_2 x_3 \bar{x}_5 x_6, \quad \chi_4 = \bar{x}_3 \cup \bar{x}_2 x_6(\bar{x}_1 \cup x_5),$$

$$\psi_5 = x_6(\bar{x}_1 \cup \bar{x}_2 x_3 \bar{x}_4), \quad \chi_5 = \bar{x}_6 \cup x_1(x_2 \cup \bar{x}_3 \cup x_4),$$

$$\psi_6 = \bar{x}_1 x_5 \cup x_3 x_4, \quad \chi_6 = (x_1 \cup \bar{x}_5)[\bar{x}_3 \cup \bar{x}_4(\bar{x}_1 \cup x_2 \cup x_5)],$$

$$\psi_7 = 1, \quad \chi_7 = 0,$$

$$\psi_8 = 0, \quad \chi_8 = 0,$$

hence, by (6.i), we get

$$\varphi_1 = \bar{x}_1 \cup \bar{x}_2[\bar{x}_6 \cup \bar{x}_5(\bar{x}_3 \cup x_4)],$$

$$\varphi_2 = x_1 x_2 x_3 \bar{x}_4 \bar{x}_5 x_6 \cup \bar{x}_2(\bar{x}_1 \cup \bar{x}_3 \cup x_4 \cup x_5 \cup \bar{x}_6) \cup \bar{x}_1(\bar{x}_3 \cup \bar{x}_4),$$

$$\varphi_3 = \bar{x}_3 \cup (\bar{x}_4 \cup \bar{x}_2 x_6)(\bar{x}_1 \cup x_2 \cup x_4 \cup x_5 \cup \bar{x}_6),$$

$$\varphi_4 = x_1 \bar{x}_2 x_3 x_4 \bar{x}_5 x_6 \cup \bar{x}_4(\bar{x}_1 \cup x_2 \cup \bar{x}_3 \cup x_5 \cup \bar{x}_6) \cup \bar{x}_3 \cup \bar{x}_2 x_6(\bar{x}_1 \cup x_5),$$

$$\varphi_5 = x_5(\bar{x}_1 \cup \bar{x}_2 x_3 \bar{x}_4) \cup \bar{x}_5 x_1(x_2 \cup \bar{x}_3 \cup x_4) \cup \bar{x}_6 \cup x_1(x_2 \cup \bar{x}_3 \cup x_4),$$

$$\varphi_6 = x_6(\bar{x}_1 x_5 \cup x_3 x_4) \cup \bar{x}_6(x_1 \cup \bar{x}_5)(\bar{x}_3 \cup \bar{x}_4) \cup (x_1 \cup \bar{x}_5)[\bar{x}_3 \cup \bar{x}_4(\bar{x}_1 \cup x_2 \cup x_5)],$$

$$\varphi_7 = x_7,$$

$$\varphi_8 = \bar{x}_8.$$

Applying formula (5), we obtain the characteristic function of the system (8):

$$\Phi = \{x_1 \bar{x}_2[(\bar{x}_3 \cup x_4)\bar{x}_5 x_6 \cup (\bar{x}_3 \cup \bar{x}_4)\bar{x}_6] \cup \cup \bar{x}_1[(\bar{x}_3 \cup \bar{x}_4)\bar{x}_5 \bar{x}_6 \cup (\bar{x}_2 \cup \bar{x}_3 \cup \bar{x}_4) x_5 x_6]\} x_7 \bar{x}_8. \tag{9}$$

Hence the locally minimizing points are the following:

Table 1

x_1	x_2	x_3	x_4	x_5	x_6	x_7	x_8
1	0	0	—	0	1	1	0
1	0	1	1	0	1	1	0
1	0	0	—	—	0	1	0
1	0	1	0	—	0	1	0
0	—	0	—	0	0	1	0
0	—	1	0	0	0	1	0
0	0	—	—	1	1	1	0
0	1	0	—	1	1	1	0
0	1	1	0	1	1	1	0

where the dashes mean that the corresponding variables take both values 0 and 1.

If we are interested in finding only one locally minimizing point, instead of all, we can apply the following *iterative procedure*:

1) Select an initial vector $(x_1^{(1)}, \ldots, x_n^{(1)})$, for instance by putting $x_i^{(1)}$ equal to 1 if the free term of g_i is negative, and equal to 0 in the other case.

2) Verify whether $(x_i^{(1)}, \ldots, x_n^{(1)})$ fulfils the conditions (4. i) $(i = 1, \ldots, n)$. If yes, $(x_i^{(1)}, \ldots, x_n^{(1)})$ is a local optimum. If not, let

$$(2x_{i_0}^{(1)} - 1)\, g_{i_0} = \max_i (2x_i^{(1)} - 1)\, g_i,$$

and let

$$x_i^{(2)} = \begin{cases} \overline{x_{i_0}^{(1)}}, & \text{if } i = i_0, \\ x_i^{(1)}, & \text{if } i \neq i_0. \end{cases}$$

Repeat step 2 until obtaining a locally minimizing point.

Example 2. Let us consider the function

$$2x_1 x_2 - 3x_1 x_4 - 5x_2 - 8x_2 x_3 x_4 - 3x_2 x_3 + 2x_3 x_6 - 5x_4 x_6 + 7x_5 x_6 x_7 - \\ - 4x_7 x_8 + 2x_4 x_6 x_8$$

and determine its locally minimizing points with the aid of the iterative procedure.

Denoting $(2x_i - 1)\, g_i(x_1, \ldots, x_{i-1}, x_{i+1}, \ldots, x_n)$ by G_i, we have

$$G_1 = (2x_1 - 1)(2x_2 - 3x_4),$$

$$G_2 = (2x_2 - 1)(-5 + 2x_1 - 8x_3 x_4 - 3x_3),$$

$$G_3 = (2x_3 - 1)(-8x_2 x_4 - 3x_2 + 2x_6),$$

$$G_4 = (2x_4 - 1)(-3x_1 - 8x_2 x_3 - 5x_6 + 2x_6 x_8),$$

$$G_5 = (2x_5 - 1)(7x_6 x_7),$$

$$G_6 = (2x_6 - 1)(2x_3 - 5x_4 + 7x_5 x_7 + 2x_4 x_8),$$

$$G_7 = (2x_7 - 1)(7x_5 x_6 - 4x_8),$$

$$G_8 = (2x_8 - 1)(-4x_7 + 2x_4 x_6)$$

and

Table 2

No.	x_1	x_2	x_3	x_4	x_5	x_6	x_7	x_8	G_1	G_2	G_3	G_4	G_5	G_6	G_7	G_8	i_0
1	0	1	0	0	0	0	0	0	+	−	+	0	0	0	0	0	3
2	0	1	1	0	0	0	0	0	+	−	−	+	0	−	0	0	4
3	0	1	1	1	0	0	0	0	+	−	−	−	0	+	0	0	1
4	1	1	1	1	0	0	0	0	−	−	−	−	0	+	0	0	6
5	1	1	1	1	0	1	0	0	−	−	−	−	0	−	0	−	∨

Hence, we have found the locally minimizing point $(1, 1, 1, 1, 0, 1, 0, 0)$.

Remark 2. There are certain classes of pseudo-Boolean functions (see for instance § 5) for which the locally minimizing points coincide with the globally minimizing points. For these functions, *all* the minimizing points may be determined as follows:

1. Determine one (locally) minimizing point $(x_1^*, \ldots, x_n^*)$ using the iterative procedure given above.

2. Solve the equation

$$f(x_1, \ldots, x_n) = f(x_1^*, \ldots, x_n^*)$$

with the method described in Chapters III and IV.

Definition 2. Let $f(x_1, \ldots, x_n)$ be a pseudo-Boolean function and $\sum$ a system of pseudo-Boolean equations and/or inequalities. A solution $(x_1^*, \ldots, x_n^*)$ of $\sum$ is called a *locally minimizing point of f under the conditions* $\sum$, if, for any $i = 1, \ldots, n$, the point $(x_1^*, \ldots, x_{i-1}^*, \overline{x_i^*}, x_{i+1}^*, \ldots, x_n^*)$ either is not a solution of $\sum$, or it satisfies $\sum$ and relation

$$f(x_1^*, \ldots, x_n^*) \leqq f(x_1^*, \ldots, x_{i-1}^*, \overline{x_i^*}, x_{i+1}^*, \ldots, x_n^*) \tag{2. i}$$

holds. The value $f(x_1^*, \ldots, x_n^*)$ is said to be a *local minimum* of the function f under the conditions $\sum$.

Theorem 3. *For any $i = 1, \ldots, n$ let $\varphi_i(x_1, \ldots, x_n)$ be the characteristic function of the inequality*

$$(2x_i - 1)\, g_i(x_1, \ldots, x_{i-1}, x_{i+1}, \ldots, x_n) \leqq 0 \tag{4. i}$$

[given by formula (6. *i*)], *and let $\Theta(x_1, \ldots, x_n)$ be the characteristic function of the system $\sum$. Then the characteristic equation of the locally minimizing points of the function f under the conditions $\sum$ is*

$$\Theta(x_1, \ldots, x_n) \prod_{i=1}^{n} [\bar{\Theta}(x_1, \ldots, x_{i-1}, \bar{x}_i, x_{i+1}, \ldots, x_n) \cup \cup\, \varphi_i(x_1, \ldots, x_n)] = 1. \tag{10}$$

Proof. A locally minimizing point was defined by the following two conditions:

1⁰. $(x_1, \ldots, x_n)$ is a solution of $\sum$;

2⁰. if $(x_1, \ldots, x_{i-1}, \bar{x}_i, x_{i+1}, \ldots, x_n)$ is a solution of $\sum$, then relation (2. *i*) [or, equivalently, (4. *i*)] holds.

But condition 1⁰. is simply expressed by

$$\Theta(x_1, \ldots, x_n) = 1, \tag{11}$$

while the property 2⁰. is translated by the fact that relation

$$\text{if} \quad \Theta(x_1, \ldots, x_{i-1}, \bar{x}_i, x_{i+1}, \ldots, x_n) = 1, \quad \text{then} \quad \varphi_i(x_1, \ldots, x_n) = 1 \tag{12. i}$$

holds for all $i = 1, \ldots, n$.

Each condition (12. i) is equivalent to

(13. i) $\overline{\Theta}(x_1, \ldots, x_{i-1}, \bar{x}_i, x_{i+1}, \ldots, x_n) \cup \varphi_i(x_1, \ldots, x_n) = 1,$

so that the sought points are characterized by the system (11), (13.1), ..., (13. n), which is equivalent to the single Boolean equation (10).

Example 3. Determine the locally minimizing points of the function

(7) $$f = 3x_1 x_2 + 9x_2 x_3 x_4 - 7\bar{x}_1 x_5 x_6 + 2x_3 x_4 \bar{x}_6 + 4x_1 \bar{x}_2 x_3 \bar{x}_4 \bar{x}_5 x_6 - 5x_7 + 5x_8 + 2x_7 \bar{x}_8$$

under the conditions

(14.1) $7x_1 x_2 x_3 - 2\bar{x}_1 x_4 x_8 + 5x_2 x_4 x_6 x_7 x_8 - 4x_3 x_8 - x_4 \bar{x}_5 x_6 \leqq 3,$

(14.2) $3x_1 - 2x_2 \bar{x}_6 + 14x_5 \bar{x}_6 x_8 + 2x_1 x_2 x_3 - 7x_8 \geqq -8,$

(14.3) $8x_4 x_5 \bar{x}_8 - 4\bar{x}_3 \bar{x}_7 x_8 + 3x_1 x_2 + \bar{x}_3 + \bar{x}_4 + \bar{x}_5 \leqq 2,$

(14.4) $2x_3 + 3x_5 - \bar{x}_5 x_6 + 4x_6 \bar{x}_7 x_8 - 2x_5 x_6 x_7 x_8 \geqq 1,$

(14.5) $3\bar{x}_1 - 5x_2 + 6x_3 - 7x_7 - 8\bar{x}_8 \geqq 1,$

(14.6) $-2x_1 + 4x_3 + 7\bar{x}_4 + x_6 - 3x_8 \geqq 6.$

Proceeding as in Chapter IV, we find that the characteristic function of the system (14) is

$$\Theta(x_1, \ldots, x_8) = \bar{x}_2 x_3 \bar{x}_4 \bar{x}_7 x_8 \cup \bar{x}_1 x_2 x_3 \bar{x}_4 x_6 \bar{x}_7 x_8 \cup \bar{x}_1 x_2 x_3 \bar{x}_4 x_5 \bar{x}_7 x_8 \cup \cup \bar{x}_1 \bar{x}_2 x_3 \bar{x}_4 x_7 x_8 \cup \bar{x}_1 \bar{x}_2 x_3 \bar{x}_4 \bar{x}_7 \bar{x}_8.$$

Applying formula (10), we obtain the characteristic equation of the problem:

(15) $$x_3 \bar{x}_4 (x_1 \bar{x}_2 \bar{x}_6 \bar{x}_7 x_8 \cup \bar{x}_1 x_2 x_5 x_6 \bar{x}_7 x_8 \cup \bar{x}_1 \bar{x}_2 x_5 x_6 x_7 x_8 \cup \cup \bar{x}_1 \bar{x}_2 \bar{x}_5 \bar{x}_6 x_7 x_8 \cup \bar{x}_1 \bar{x}_2 x_5 x_6 \bar{x}_7 \bar{x}_8 \cup \bar{x}_1 \bar{x}_2 \bar{x}_5 \bar{x}_6 \bar{x}_7 \bar{x}_8) = 1,$$

hence the locally minimizing points of the function (7) under the conditions (14) are

Table 3

x_1	x_2	x_3	x_4	x_5	x_6	x_7	x_8
1	0	1	0	—	0	0	1
0	1	1	0	1	1	0	1
0	0	1	0	1	1	1	1
0	0	1	0	0	0	1	1
0	0	1	0	1	1	0	0
0	0	1	0	0	0	0	0

Computational remarks. 1) It seems convenient to compute the product $\Theta\, \Theta_1 \ldots \Theta_n$ in (10) [where $\Theta_i = \overline{\Theta}(x_1, \ldots, x_{i-1}, \bar{x}_i, x_{i+1}, \ldots, x_n) \cup \cup \varphi_i(x_1, \ldots, x_n)$] in the following order:

$$(\ldots(((\Theta\, \Theta_1)\, \Theta_2)\, \Theta_3) \ldots \Theta_n).$$

2) If Θ is brought to the disjunctive canonical form, then the multiplications in (10) reduce to successive deletions of certain conjunctions.

Note added in proofs:

The method of Lagrangeian multipliers, described in Chap. VI, § 2, can be extended to the case when we are interested in finding the local minima of a pseudo-Boolean function

$$f(x_1, \ldots, x_n) \tag{VI.32}$$

under a system of constraints of the form

$$f_j(x_1, \ldots, x_n) = 0 \qquad (j = 1, \ldots, m), \tag{VI.33}$$

where the functions f and f_j are integer-valued and

$$f_j(x_1, \ldots, x_n) \geqslant 0 \quad \text{for all} \quad (x_1, \ldots, x_n) \in B_2^n \qquad (j = 1, \ldots, m). \tag{VI.34}$$

Let S^+ and S^- be the sum of the positive and of the negative coefficients of f, respectively.

Theorem. *Under the above conditions*: (α) *if the vector* $(x_1^*, \ldots, x_n^*)$ *is a locally minimizing point of the function* f *under the constraints* (VI.33), *then* $(x_1^*, \ldots, x_n^*)$ *is also a locally minimizing point of the pseudo-Boolean function*

$$F(x_1, \ldots, x_n) = f(x_1, \ldots, x_n) + (S^+ - S^- + 1) \sum_{j=1}^{m} f_j(x_1, \ldots, x_n); \tag{VI.35}$$

(β) *if* $(x_1^*, \ldots, x_n^*)$ *is a locally minimizing point of* F *and*

$$F(x_1^*, \ldots, x_n^*) \leqslant S^+, \tag{VI.36}$$

then $(x_1^*, \ldots, x_n^*)$ *is also a locally minimizing point of* f *under the conditions* (VI. 33);

(γ) *if, for any locally minimizing point* $(x_1^*, \ldots, x_n^*)$,

$$F(x_1^*, \ldots, x_n^*) > S^+ \tag{VI.37}$$

then the conditions (VI.33) *are inconsistent.*

The proof is analogous to that of Theorem VI. 2.

Corollary. Theorems X. 4 and X. 9 are immediate consequences of the above theorem.

*

Applications. The following type of problems appears frequently in the theory of graphs and in other fields of combinatorial mathematics:

Let S be a finite set and P a certain property defined for the subsets of S. If the set S has n elements $1, 2, \ldots, n$, then every subset $X \subseteqq S$ may be characterized by a vector $(x_1, \ldots, x_n)$, where the component x_h takes the value 1 if $h \in X$, and 0 if $h \notin X$. Assume that property P satisfies either the conditions α, γ, or β, γ below:

α. P is hereditary, i.e. if X satisfies P, then every subset $Y \subseteqq X$ satisfies also P;

β. P is dually — hereditary, i.e. if X satisfies P, then every superset $Y \supseteqq X$ satisfies also P;

γ. P may be characterized by pseudo-Boolean (or Boolean) conditions on the variables $x_1, \ldots, x_n$.

If P satisfies α and γ (β and γ) then the problem of finding all the *maximal* (*minimal*) *sets with property* P is equivalent to that of determining the locally maximizing (minimizing) points of the function $\sum_{h=1}^{n} x_h$ under the constraints characterizing P.

At the beginning of this section, we have described an *iterative procedure* for finding one locally minimizing point without constraints. The procedure may be extended as follows to the case when constraints appear.

1. Find an initial vector $X^1 = (x_1^{(1)}, \ldots, x_n^{(1)})$ in the following way: choose one familly of solutions and let I be the corresponding set of indices for which x_i has fixed values x_i^*; put

$$x_i^{(1)} = \begin{cases} x_i^* & \text{if} \quad i \in I, \\ 1 & \text{if} \quad i \notin I, \quad \gamma_i < 0, \\ 0 & \text{if} \quad i \notin I, \quad \gamma_i \geqq 0, \end{cases}$$

where γ_i is the free term of $g_i(x_1, \ldots, x_{i-1}, x_{i+1}, \ldots, x_n)$.

2. If all $G_i(x_1^{(1)}, \ldots, x_n^{(1)}) \leqq 0$ (see Example 2), then X^1 is the sought locally minimizing point. If there are positive G_i, then check whether there is an i_0 so that

(i) $G_{i_0} > 0$,

(ii) $X_2 = (x_1^{(1)}, \ldots, x_{i_0-1}^{(1)}, \overline{x_{i_0}^{(1)}}, x_{i_0+1}^{(1)}, \ldots, x_n^{(1)})$ is a solution to the constraints.

If there is no i_0 satisfying both (i) and (ii), then X_1 is a locally minimizing point. In the opposite case, repeat point 2 for X_2, etc., until arriving to a point X_k for which (i) and (ii) are not both fulfilled. X_k is a locally minimizing point.

Example 4. A locally minimizing point of (7) under the constraints (14) may be determined as follows:

Table 4

No.	x_1	x_2	x_3	x_4	x_5	x_6	x_7	x_8	G_1	G_2	G_3	G_4	G_5	G_6	G_7	G_8	i_0
1	0	0	1	0	0	0	0	1	0	0	0	−	0	0	+	+	7
2	0	0	1	0	0	0	1	1	0	0	0	−	0	0	−	+	none

Hence, $(0, 0, 1, 0, 0, 0, 1, 1)$ is a locally minimizing point, because its single neighbouring point giving a smaller value for f, i.e. $(0, 0, 1, 0, 0, 0, 1, 0)$ does not belong to the solutions of the constraints.

As it was mentioned in Remark 2, there are classes of pseudo-Boolean functions for which the locally minimizing points coincide with the globally minimizing ones. However, the locally minimizing points of such a function under certain constraints need not coincide with the globally minimizing points under those constraints. More precisely, the following theorem is valid.

Theorem 4. *Let $f(x_1, \ldots, x_n)$ be a non-constant pseudo-Boolean function with $n \geq 3$, for which the locally minimizing points coincide with the globally minimizing ones. Then there exists pseudo-Boolean constraints C such that the locally minimizing points of the function f under the constraints C do not coincide with the globally minimizing points of f under these conditions.*

In fact, we shall prove that there is even a constraint C consisting of a single linear pseudo-Boolean equation and having the above indicated effect.

Proof. Since the function f is non-constant, there are two points $P = (\xi_1, \ldots, \xi_n)$ and $Q = (\eta_1, \ldots, \eta_n)$ such that

$$f(\eta_1, \ldots, \eta_n) > f(\xi_1, \ldots, \xi_n) = \min f(x_1, \ldots, x_n). \tag{16}$$

(α) We shall prove that it is always possible to choose P and Q so as to satisfy (16) and

$$\sum_{i=1}^{n} |\xi_i - \eta_i| \geq 2. \tag{17}$$

For, assume that for all the points $(\eta_1, \ldots, \eta_n)$ verifying (17), relation $f(\eta_1, \ldots, \eta_n) = f(\xi_1, \ldots, \xi_n)$ holds. Then there exists a point $R = (\zeta_1, \ldots, \zeta_n)$ satisfying $f(\zeta_1, \ldots, \zeta_n) > f(\xi_1, \ldots, \xi_n)$ and $\sum_{i=1}^{n} |\xi_i - \zeta_i| < 2$. Since $P \neq R$, the latter inequality shows that in fact $\sum_{i=1}^{n} |\xi_i - \zeta_i| = 1$. Hence there exists an index i_0 such that $R = (\xi_1, \ldots, \xi_{i_0-1}, \bar{\xi}_{i_0}, \xi_{i_0+1}, \ldots, \xi_n)$.

Now put $S = (\bar{\xi}_1, \ldots, \bar{\xi}_n)$; we deduce $\sum_{i=1}^{n} |\xi_i - \bar{\xi}_i| = n > 2$, so that $f(S) = f(P)$, according to our assumption.

Summarizing, we have: $f(R) > f(P) = f(S) = \min f(x_1, \ldots, x_n)$ and $\sum_{i=1}^{n} |\bar{\xi}_i - \zeta_i| = \sum_{\substack{i=1 \\ i \neq i_0}}^{n} |\bar{\xi}_i - \xi_i| = n - 1 \geq 2$, so that we can take S and R in the rôles of P and Q, respectively.

(β) We have just proved that it is always possible to determine two points $P = (\xi_1, \ldots, \xi_n)$ and $Q = (\eta_1, \ldots, \eta_n)$, satisfying (16) and (17). Without loss of generality, we may assume that $P = (0, \ldots, 0)$.

The point Q is not a locally minimizing point of the function f [for, otherwise it would be a globally minimizing point, contradicting (16)]. Hence there exists a non-empty set K of indices such that $f(Q_k) = f(\eta_1, \ldots, \eta_{k-1}, \bar{\eta}_k, \eta_{k+1}, \ldots, \eta_n) < f(\eta_1, \ldots, \eta_n)$ for all $k \in K$, while $f(Q_k) \geqq f(Q)$ for all $k \notin K$. Now, put

$$I = \{i \mid \eta_i = 1\}, \qquad I_0 = I \cap K,$$

$$J = \{j \mid \eta_j = 0\}, \qquad J_0 = J \cap K,$$

and consider a linear pseudo-Boolean equation of the form

$$\sum_{h=1}^{n} a_h x_h = 0, \tag{18.0}$$

where

$$\sum_{i \in I} a_i = 0, \tag{18.1}$$

$$a_i \neq 0 \quad \text{for all} \quad i \in I_0, \tag{18.2}$$

$$a_j \neq 0 \quad \text{for all} \quad j \in J_0. \tag{18.3}$$

Then, obviously, the points P and Q are solutions of the equation (18.0), while for $k \in K$, the points Q_k are not. Therefore P is a globally minimizing point of the function f under the condition (18.0), while Q is a locally minimizing point under this condition, but not a globally minimizing one.

§ 2. Applications to the Theory of Graphs*

In this section we shall prove two results on local optima of linear pseudo-Boolean functions, in view of their applications to graph theory.

Theorem 5. *Let*

$$f(x_1, \ldots, x_n) = \sum_{i=1}^{n} c_i x_i \tag{19}$$

be a pseudo-Boolean function with

$$c_i > 0 \qquad (i = 1, \ldots, n) \tag{20}$$

and let $\Phi(x_1, \ldots, x_n)$ be an increasing Boolean function which is not identically equal to 0.

Then a vector $(x_1, \ldots, x_n)$ is a locally minimizing point of the function (19) *under the constraint*

$$\Phi(x_1, \ldots, x_n) = 1 \tag{21}$$

if and only if it is a solution of the system

$$x_i = \bar{\Phi}(x_1, \ldots, x_{i-1}, 0, x_{i+1}, \ldots, x_n) \qquad (i = 1, \ldots, n). \tag{22}$$

* See also Chapter X.

Proof. We shall apply Theorem 3 in § 1. In case of the function (19), the inequality (4. i) reduces to $(x_i - \bar{x}_i)\, c_i \leqq 0$; since $c_i > 0$, it follows that $\varphi_i(x_1, \ldots, x_n) = \bar{x}_i$. Hence the characteristic equation (10) of the locally minimizing points becomes

$$(23) \qquad \Phi(x_1, \ldots, x_n) \prod_{i=1}^{n} [\bar{\Phi}(x_1, \ldots, x_{i-1}, \bar{x}_i, x_{i+1}, \ldots, x_n) \cup \bar{x}_i] = 1.$$

This equation is equivalent to the system

$$(24) \qquad (A_i\, x_i \cup B_i\, \bar{x}_i)\,(\bar{A}_i\, \bar{x}_i \cup \bar{B}_i\, x_i \cup \bar{x}_i) = 1 \qquad (i = 1, \ldots, n),$$

where we have set

$$(25.1) \qquad A_i = \Phi(x_1, \ldots, x_{i-1}, 1, x_{i+1}, \ldots, x_n) \qquad (i = 1, \ldots, n),$$

$$(25.0) \qquad B_i = \Phi(x_1, \ldots, x_{i-1}, 0, x_{i+1}, \ldots, x_n) \qquad (i = 1, \ldots, n).$$

Performing all the absorptions and multiplications in (24), we obtain

$$(26) \qquad A_i\, \bar{B}_i\, x_i \cup B_i\, \bar{x}_i = 1 \qquad (i = 1, \ldots, n)$$

or else, taking into account that $B_i \leqq A_i$

$$(27) \qquad A_i(\bar{B}_i\, x_i \cup B_i\, \bar{x}_i) = 1 \qquad (i = 1, \ldots, n),$$

which is equivalent to the system (28) and (29) below:

$$(28) \qquad A_i = 1 \qquad (i = 1, \ldots, n),$$

$$(29) \qquad \bar{B}_i\, x_i \cup B_i\, \bar{x}_i = 1 \qquad (i = 1, \ldots, n).$$

But the system (29) means that $x_i = \bar{B}_i$, i.e. (29) coincides with (22), so that it remains to show that the condition (28), which means

$$(30) \qquad \Phi(x_1, \ldots, x_{i-1}, 1, x_{i+1}, \ldots, x_n) = 1 \qquad (i = 1, \ldots, n)$$

is implied by (22).

Indeed, in the opposite case there exists a solution $(x_1^*, \ldots, x_n^*)$ of (22) and an index i such that

$$(31) \qquad \Phi(x_1^*, \ldots, x_{i-1}^*, 1, x_{i+1}^*, \ldots, x_n^*) = 0.$$

Since Φ is monotone increasing, relation (31) implies $\Phi(x_1^*, \ldots, x_{i-1}^*, 0, x_{i+1}^*, \ldots, x_n^*) = 0$, so that we have necessarily

$$(32) \qquad \Phi(x_1^*, \ldots, x_n^*) = 0.$$

But relation (32), together with the condition that Φ is increasing and does not reduce to the constant 0, imply the existence of an index $j \neq i$ such that $x_j^* = 0$.

Hence the condition (32) may be written in the form

$$\Phi(x_1^*, \ldots, x_{j-1}^*, 0, x_{j+1}^*, \ldots, x_n^*) = 0, \tag{33}$$

implying thus $x_j^* = 1$, by (22). This contradiction completes the proof.

Corollary 1. Taking $c_1 = \cdots = c_n = 1$, we obtain Theorem 3 in S. RUDEANU [9]. As it was shown there, from that theorem it results

Corollary 2*. The minimal externally stable sets of a graph** are characterized by the system of Boolean equations

$$x_i = \bigcup_{h=1}^{n} \prod_{\substack{k=1 \\ k \neq i}}^{n} (\bar{a}_{hk}\, \bar{\delta}_{hk} \cup \bar{x}_k) \qquad (i = 1, \ldots, n), \tag{34}$$

where δ_{hk} is the Kronecker symbol.

[We recall that a graph with n vertices $1, \ldots, n$ is described by a Boolean $n \times n$ matrix $A = (a_{ij})$, where $a_{ij} = 1$ if there exists an arc from i to j, and $a_{ij} = 0$ in the opposite case. Further, a set X of vertices is characterized by a vector $(x_1, \ldots, x_n)$, where $x_i = 1$ if $i \in X$ and $x_i = 0$ if $i \notin X$.]

Theorem 6. *Let* (19) *be a pseudo-Boolean function satisfying* (20) *and let* $\Psi(x_1, \ldots, x_n)$ *be an increasing Boolean function which is not identically equal to* 1.

Then a vector $(x_1, \ldots, x_n)$ *is a locally maximizing point of the function* (19) *under the constraint*

$$\Psi(x_1, \ldots, x_n) = 0 \tag{35}$$

if and only if it is a solution of the system

$$x_i = \overline{\Psi}(x_1, \ldots, x_{i-1}, 1, x_{i+1}, \ldots, x_n) \qquad (i = 1, \ldots, n). \tag{36}$$

Proof. Similar to that of Theorem 5.

Corollary 3. Taking $c_1 = \cdots = c_n = 1$, we obtain Theorem 1 in S. RUDEANU [9]. As it was shown there, from that theorem it results

Corollary 4. The maximal internally stable sets of a graph are characterized by the system of Boolean equations

$$x_i = \bar{a}_{ii} \prod_{\substack{j=1 \\ j \neq i}}^{n} (\bar{a}_{ij}\, \bar{a}_{ji} \cup \bar{x}_j) \qquad (i = 1, \ldots, n). \tag{37}$$

The *maximal cliques* of a graph may also be determined using the above ideas. A clique is a set of vertices having the property that any pair of them are related by arcs in both directions, i.e. $x_i\, x_j = 1$ implies

* The reading of Corollaries 2, 4 and 5 can be postponed until that of Chapter X, §§ 1—3.

** We adopt here the terminology of C. BERGE [1].

$a_{ij} a_{ji} = 1$. This property is translated by the system of equations

$$(\bar{a}_{ij} \cup \bar{a}_{ji}) x_i x_j = 0 \qquad (i, j = 1, \ldots, n), \tag{38}$$

showing that *a set of vertices is a clique if and only if it is an internally stable set of the graph having the same vertices but the matrix* $(\bar{a}_{ij} \cup \bar{a}_{ji})$.

Corollary 5. The maximal cliques of a graph are characterized by the system of Boolean equations

$$x_i = a_{ii} \prod_{\substack{j=1 \\ j \neq i}}^{n} (a_{ij} a_{ji} \cup \bar{x}_j) \qquad (i = 1, \ldots, n). \tag{39}$$

§ 3. Near-Minima of Pseudo-Boolean Functions

Let us consider now the following two generalizations of the minimization problem for pseudo-Boolean functions.

Let $f_{\min}$ be the minimum of a pseudo-Boolean function $f(x_1, \ldots, x_n)$.

(i) Find the points $(x_1, \ldots, x_n)$ satisfying

$$f(x_1, \ldots, x_n) \geqq f_{\min} + r, \tag{40}$$

where r is an integer constant.

(ii) Find the points $(x_1, \ldots, x_n)$ so that

$$q f(x_1, \ldots, x_n) \geqq p f_{\min}, \tag{41}$$

where p and q are integer constants satisfying

$$\frac{p}{q} > 1. \tag{42}$$

Now we can proceed as follows:

1) We apply the first part of the basic minimization algorithm given in § 1 of Chapter VI, finding thus $f_{\min}$. As a matter of fact, the procedure is more simple than that used in the general case, according to Remark 9 in Chapter VI.

2) Using the methods given in Chapter IV, we solve either the pseudo-Boolean inequality (40), in case of problem (i), or (41), in case of problem (ii).

In the more general case when the variables are subject to certain constraints, the above procedure is to be modified as follows.

In step 1) we find the characteristic function $\Phi(x_1, \ldots, x_n)$ of the system of constraints and we determine then the minimum $f_{\min}$ under the constraint $\Phi = 1$.

In step 2) we find the characteristic function $\Psi(x_1, \ldots, x_n)$ of the inequality (40) [or, of (41) in case of problem (ii)]; the sought points coincide with the solutions of the Boolean equation

$$\Phi(x_1, \ldots, x_n) \Psi(x_1, \ldots, x_n) = 1. \tag{43}$$

Example 5. Let

(44) $$f(x_1, \ldots, x_6) = 3x_1 \bar{x}_2 - 8\bar{x}_1 x_3 x_6 + 4x_2 x_5 \bar{x}_6 - 7\bar{x}_5 x_6 + 3x_4 - 5x_4 x_5 x_6$$

and let $f_{\min}$ be the minimum of the function f under the constraints

(45.1) $$2x_1 - 3x_2 + 5x_3 - 4x_4 + 2x_5 - x_6 \leqq 2,$$

(45.2) $$4x_1 + 2x_2 + x_3 + 8x_4 - x_5 - 3x_6 \geqq 4;$$

determine the points $(x_1, \ldots, x_6)$ for which

(46) $$f(x_1, \ldots, x_6) \leqq f_{\min} + 2.$$

Proceeding as in Example 6 of Chapter VI, we find that the characteristic function of the system (45) of constraints is

(47) $$\Phi = \bar{x}_3 x_4 \cup x_2 x_3 x_4 \cup \bar{x}_1 \bar{x}_2 x_3 x_4 \bar{x}_5 \cup \bar{x}_1 \bar{x}_2 x_3 x_4 x_5 x_6 \cup$$
$$\cup x_1 \bar{x}_2 x_3 x_4 \bar{x}_5 x_6 \cup x_1 x_2 \bar{x}_3 \bar{x}_4 \bar{x}_6 \cup x_1 \bar{x}_2 \bar{x}_3 \bar{x}_4 \bar{x}_5 \bar{x}_6.$$

while

(48) $$f_{\min} = -12.$$

Hence the inequality (46) becomes

(49) $$3x_1 \bar{x}_2 - 8\bar{x}_1 x_3 x_6 + 4x_2 x_5 \bar{x}_6 - 7\bar{x}_5 x_6 + 3x_4 - 5x_4 x_5 x_6 \leqq -10.$$

Using the methods given in Chapter IV, we find the characteristic function of (49):

(50) $$\Psi = \bar{x}_1 x_3 x_6 (x_4 \cup \bar{x}_5).$$

Applying formula (43), we obtain the equation

(51) $$\Phi \Psi = \bar{x}_1 x_3 x_4 x_6 = 1,$$

hence the sought points are

(52) $x_1 = 0$, x_2 arbitrary, $x_3 = 1$, $x_4 = 1$, x_5 arbitrary, $x_6 = 1$.

§ 4. Minimax Problems in Pseudo-Boolean Programming

Several problems occuring in practice admit the following type of mathematical model: given a pseudo-Boolean function $f(x_1, \ldots, x_m, y_1, \ldots, y_n)$, it is required to find

(53) $$\min_{x_i} \max_{y_j} f(x_1, \ldots, x_m, y_1, \ldots, y_n),$$

the variables being possibly subject to certain constraints.

In other problems we are faced with the determination of

(54) $$\max_{y_j} \min_{x_i} f(x_1, \ldots, x_m, y_1, \ldots, y_n).$$

Remark 3. Notice that (53) need not coincide with (54)*.

* It may be of interest to characterize those pseudo-Boolean functions for which the corresponding values (53) and (54) coincide.

For instance, taking $f(x, y) = -3xy + 2x + y$, we find $\min_x \max_y f = f(0, 1) = 1$, while $\max_y \min_x f = f(0, 0) = f(1, 1) = 0$.

Consider first the problem of determining (53) when the variables are not subject to constraints. We suggest the following procedure:

Notice first that an obvious modification of the minimization algorithm given in § 1 of Chapter VI yields a maximization algorithm (it suffices to take the functions $\bar{\varphi}_i(p_i, x_{i+1}, \ldots, x_n)$ instead of the functions $\varphi_i(p_i, x_{i+1}, \ldots, x_n)$). Apply the first n steps of this maximization algorithm to the variables $y_1, \ldots, y_n$ (taken in an arbitrary order), obtaining thus a function $f_{n+1}(x_1, \ldots, x_m)$. Apply the minimization algorithm to f_{n+1}, obtaining thus a value f_{n+m+1}, which is the required minimax. The "minimaxing" points are then obtained as in the common case.

For the determination of (54) we proceed in a dual way.

Example 6. Find

$$\min_{x_3, x_4, x_5} \max_{x_1, x_2} (2x_1 + 3x_2 - 7x_3 - 5x_1 x_2 x_3 + 3x_2 x_4 + 9x_4 x_5 - 2x_1 x_5). \tag{55}$$

We have

$$f_1 = x_1(2 - 5x_2 x_3 - 2x_5) + 3x_2 - 7x_3 + 3x_2 x_4 + 9x_4 x_5 \tag{56.1}$$

hence, with the notations of § 1 in Chapter VI, we get

$$\bar{\varphi}_1 = (\bar{x}_2 \cup \bar{x}_3)\, \bar{x}_5 \cup (\bar{x}_2 \cup \bar{x}_3)\, x_5\, p_1. \tag{57.1}$$

We obtain now

$$\begin{aligned} f_2 &= (\bar{x}_2 \cup \bar{x}_3)\, \bar{x}_5 (2 - 5x_2 x_3 - 2x_5) + 3x_2 - 7x_3 + 3x_2 x_4 + 9x_4 x_5 \\ &= x_2(3 - 2\bar{x}_5 + 2\bar{x}_3 \bar{x}_5 + 3x_4) + 2\bar{x}_5 - 7x_3 + 9x_4 x_5, \end{aligned} \tag{56.2}$$

hence

$$\bar{\varphi}_2 = 1. \tag{57.2}$$

It follows that

$$\begin{aligned} f_3 &= 3 - 7x_3 + 3x_4 + 2\bar{x}_3 \bar{x}_5 + 9x_4 x_5 \\ &= x_3(-7 - 2\bar{x}_5) + 3 + 3x_4 + 2\bar{x}_5 + 9x_4 x_5, \end{aligned} \tag{56.3}$$

hence

$$\varphi_3 = 1. \tag{57.3}$$

We get now

$$f_4 = -4 + 3x_4 + 9x_4 x_5 = x_4(3 + 9x_5) - 4, \tag{56.4}$$

therefore

$$\varphi_4 = 0. \tag{57.4}$$

It follows that

$$f_5 = -4, \tag{56.5}$$

hence

$$\varphi_5 = p_5, \tag{57.5}$$

so that

$$f_{\substack{\min \\ x_3, x_4, x_5}\ \substack{\max \\ x_1, x_2}} = f_6 = -4. \tag{56.6}$$

The "minimaxing" points are given by formulas $x_5 = \varphi_5$, $x_4 = \varphi_4$, $x_3 = \varphi_3$, $x_2 = \bar{\varphi}_2$, $x_1 = \bar{\varphi}_1$; taking into account relations (57), we obtain

(58) $$x_5 = p_5, \quad x_4 = 0, \quad x_3 = 1, \quad x_2 = 1, \quad x_1 = 0.$$

In case that the variables are subject to certain constraints, we can proceed as follows. Let $\mathscr{F}_1, \ldots, \mathscr{F}_p$ be the families of solutions to the constraints (see Chapter IV). For each family $\mathscr{F}_k$, we introduce the fixed values of the variables into the objective function f, obtaining thus a new function $f^{(k)}$ with less variables. We solve the minimax problem for each function $f^{(k)}$ and select then

(59) $$\min_{k=1,\ldots,p} f^{(k)}_{\min\max}.$$

Example 7. Let us find

(60) $$\min_{x_4, x_5, x_6} \max_{x_1, x_2, x_3} (2x_1 + 3x_2 - 7x_3 - 5x_1 x_2 x_5 + 3x_2 x_4 + 8x_5 x_6 - 2\bar{x}_3 x_5)$$

under the constraints

(61.1) $$x_1 x_2 + 4\bar{x}_1 x_3 - 3x_2 x_3 x_5 + 6\bar{x}_2 x_4 x_6 \geqq -1,$$

(61.2) $$3x_2 x_4 - 5\bar{x}_1 \bar{x}_3 \bar{x}_5 + 4x_4 x_6 \geqq 1.$$

The families of solutions to the constraints were determined in Example 7 of Chapter VI. Applying the above indicated procedure, we obtain the following table:

Table 5

No.	Fam. of solutions x_1	x_2	x_3	x_4	x_5	x_6	$f^{(k)}$	The minimax of $f^{(k)}$ is reached for:	$\min_{x_4 x_5 x_6} \max_{x_1 x_2 x_3} f^{(k)}$
1	0	1	1	1	—	0	-1	*	-1
2	0	—	1	1	—	1	$-7 + 6x_2 + 8x_5$	$x_2 = 1,\ x_5 = 0$	-1
3	0	1	0	1	1	0	4	*	4
4	0	—	0	1	1	1	$6 + 6x_2$	$x_2 = 1$	12
5	0	1	0	1	0	1	6	*	6
6	1	0	—	1	—	1	$2 - 7x_3 + 8x_5 - 2\bar{x}_3 x_5$	$x_3 = 0,\ x_5 = 0$	2
7	1	1	0	1	—	—	$8 - 7x_5 + 8x_5 x_6$	$x_5 = 1,\ x_6 = 0$	1
8	1	1	1	1	0	—	1	*	1

Therefore

(62) $$\min_{x_4, x_5, x_6} \max_{x_1, x_2, x_3} f = f(0, 1, 1, 1, p, 0) = f(0, 1, 1, 1, 0, 1) = -1.$$

Application. The following typical minimax problem, studied by D. A. Pospelov [1], may be solved with the method indicated in this section.

A system of m digital computers $C_1, \ldots, C_m$ guided by a "director-computer", has to perform a rather complex program. At a certain stage, this program consists of n subprograms $S_1, \ldots, S_n$, which may be performed independently. Assuming that $m \geqq n$, one has to assign each subprogram S_j to a certain computer C_i, so that the corresponding stage of the original program should be performed as quickly as possible.

For each $i = 1, \ldots, m$ and each $j = 1, \ldots, n$, let t_{ij} be the time necessary for the computer C_i to perform the subprogram S_j, and let x_{ij} be a variable which takes the value 1 if the subprogram S_j is assigned to the computer C_i, and 0 in the opposite case.

If follows that the variables x_{ij} are subject to the constraints

$$\sum_{i=1}^{m} x_{ij} = 1 \qquad (j = 1, \ldots, n), \tag{63}$$

$$\sum_{j=1}^{n} \sum_{\substack{k=1 \\ k \neq j}}^{n} x_{ij} x_{ik} = 0 \qquad (i = 1, \ldots, m), \tag{64}$$

while the time necessary for the computer C_i to perform the assigned subprogram (if any) is

$$t_i = \sum_{j=1}^{n} t_{ij} x_{ij} \qquad (i = 1, \ldots, m). \tag{65}$$

Thus the problem is to find

$$\min_{x_{ij}} \max_{i} t_i \tag{66}$$

under the constraints (63) and (64).

Introducing m auxiliary bivalent variables $y_1, \ldots, y_m$, subject to the condition

$$\sum_{i=1}^{m} y_i = 1, \tag{67}$$

we can reformulate the problem as follows:

Find

$$\min_{x_{ij}} \max_{y_i} \sum_{i=1}^{m} y_i \left(\sum_{j=1}^{n} t_{ij} x_{ij} \right), \tag{68}$$

under the constraints (63), (64) and (67).

A combinational approach to the above problem was proposed in K. Maciščak and D. A. Pospelov [1].

§ 5. Fractional Pseudo-Boolean Programming

The problem of minimizing the cost of production of a certain item, i.e. the quotient of the (usually linear) function representing the total cost, by the (usually linear) function representing the produced amount of that item, is an outstanding example of what is termed "hyperbolic" or "fractional" programming. Methods for solving this problem were given by B. Martos [1], A. Charnes and W. W. Cooper [1] and W. Dinkelbach [1], etc.

We are now going to present a method for solving problems of this type in the case when the variables are bivalent. In other words, the problem is to minimize a function of the form

$$F = \frac{a_0 + a_1 x_1 + \cdots + a_n x_n}{b_0 + b_1 x_1 + \cdots + b_n x_n}, \tag{69}$$

where the variables x_h $(h = 1, \ldots, n)$ may take only the values 0 and 1.

We shall examine the case — frequently appearing in practical applications — when

$$b_h > 0 \qquad (h = 0, 1, \ldots, n). \tag{70}$$

As a matter of fact, it will be shown at the end of the section (Remark 4) that this restriction is not essential.

Let I and J be the sets of all indices $i > 0$ and $j > 0$ satisfying

$$\frac{a_i}{b_i} \leqq \frac{a_0}{b_0} \tag{71}$$

and

$$\frac{a_0}{b_0} < \frac{a_j}{b_j}, \tag{72}$$

respectively.

The following three lemmas are obvious.

Lemma 1. *Relations* (70) *and* $\bigcup_{j \in J} x_j = 1$ *imply*

$$\frac{a_0 + \sum_{i \in I} a_i x_i}{b_0 + \sum_{i \in I} b_i x_i} < \frac{a_0 + \sum_{i \in I} a_i x_i + \sum_{j \in J} a_j x_j}{b_0 + \sum_{i \in I} b_i x_i + \sum_{j \in J} b_j x_j}. \tag{73}$$

Lemma 2. *If relations* (70) *hold and*

$$\min_{i \in I} \frac{a_i}{b_i} = \frac{a_{i_1}}{b_{i_1}} < \frac{a_0}{b_0}, \tag{74}$$

then

$$\frac{a_0 + a_{i_1} + \sum_{i \in I - i_1} a_i x_i}{b_0 + b_{i_1} + \sum_{i \in I - i_1} b_i x_i} < \frac{a_0 + \sum_{i \in I - i_1} a_i x_i}{b_0 + \sum_{i \in I - i_1} b_i x_i}. \tag{75}$$

Lemma 3. *If, for a certain set K of indices, relations* (70) *and*

$$\frac{a_k}{b_k} = \frac{a_0}{b_0} \tag{76}$$

hold, then the identity

$$\frac{a_0 + \sum_{k \in K} a_k x_k}{b_0 + \sum_{k \in K} b_k x_k} = \frac{a_0}{b_0} \tag{77}$$

also holds.

Now consider the case when certain coefficients a_h, b_h may be zero. It is obvious that: if $a_h = 0$ and $b_h > 0$, then in the minimizing points $x_h^* = 1$; if $a_h > 0$ and $b_h = 0$, then $x_h^* = 0$.

We are lead to the following

Algorithm I. 1. Whenever $a_h = 0$, $b_h > 0$, put $x_h = 1$.
2. Whenever $a_h > 0$, $b_h = 0$, put $x_h = 0$.
3. Determine the sets I and J.
4. For each $j \in J$, put $x_j = 0$.
5. Determine the first index i_1 for which $\frac{a_{i_1}}{b_{i_1}} = \min_{i \in I} \frac{a_i}{b_i}$.
6. Repeat steps 3,4 and 5 until all variables are exhausted.

Case (α). If $\frac{a_{i_1}}{b_{i_1}} < \frac{a_0}{b_0}$, put $x_{i_1} = 1$, transform a_0 into $a_0 + a_{i_1}$, b_0 into $b_0 + b_{i_1}$. Case (β). If $\frac{a_{i_1}}{b_{i_1}} = \frac{a_0}{b_0}$, then for each $i \in I$, put $x_i = p_i$, where p_i is an arbitrary parameter.

The above discussion shows that the following theorem holds:

Theorem 7. *The above algorithm I determines all the minimizing points of the function* (69) *with non-negative coefficients* a_h, b_h.

Example 8. Minimize

(78)
$$F = \frac{3+2x_1+4x_2+x_3+2x_4+9x_5+6x_6+12x_7+8x_8+2x_9+3x_{10}+3x_{11}+x_{12}}{6+x_1+8x_2+3x_3+5x_4+15x_5+10x_6+25x_7+18x_8+6x_9+3x_{10}+7x_{11}}.$$

1) There is no h with $a_h = 0$, $b_h > 0$.

2) Since $a_{12} = 1$, $b_{12} = 0$, we take $x_{12} = 0$; there is no other h with $a_h > 0$, $b_h = 0$.

3) We have to determine the sets J and I for the function

$$F_1 = \frac{3+2x_1+4x_2+x_3+2x_4+9x_5+6x_6+12x_7+8x_8+2x_9+3x_{10}+3x_{11}}{6+x_1+8x_2+3x_3+5x_4+15x_5+10x_6+25x_7+18x_8+6x_9+3x_{10}+7x_{11}}.$$

We have $\frac{a_0}{b_0} = \frac{1}{2}$ and

$$J = \{1, 5, 6, 10\},$$
$$I = \{2, 3, 4, 7, 8, 9, 11\}.$$

4) We put $x_1 = x_5 = x_6 = x_{10} = 0$.

5) The first index i_1 for which $\frac{a_{i_1}}{b_{i_1}} = \min_{i \in I} \frac{a_i}{b_i}$ is $i_1 = 3$. Since $\frac{a_3}{b_3} = \frac{1}{3} < \frac{a_0}{b_0}$, we are in the case ($\alpha$) and so we take $x_3 = 1$. Now we are faced with the new function

$$F_2 = \frac{4+4x_2+2x_4+12x_7+8x_8+2x_9+3x_{11}}{9+8x_2+5x_4+25x_7+18x_8+6x_9+7x_{11}}.$$

3′) We have now $\frac{a_0'}{b_0'} = \frac{4}{9}$ and

$$J' = \{2, 7\},$$
$$I' = \{4, 8, 9, 11\}.$$

4′) We put $x_2 = x_7 = 0$.

5′) The single index i_1 for which $\frac{a_{i_1}}{b_{i_1}} = \min_{i \in I'} \frac{a_i}{b_i}$ is $i_1 = 9$. We have $\frac{a_9}{b_9} = \frac{1}{3} < \frac{a_0'}{b_0'}$, so that we are again in the case (α) and $x_9 = 1$. The new function is

$$F_3 = \frac{6+2x_4+8x_8+3x_{11}}{15+5x_4+18x_8+7x_{11}}.$$

3″) Now $\frac{a_0''}{b_0''} = \frac{2}{5}$ and

$$J'' = \{8, 11\},$$

$$I'' = \{4\}.$$

4″) We put $x_8 = x_{11} = 0$.

5″) Since $\frac{a_4}{b_4} = \min\limits_{i \in I''} \frac{a_i}{b_i} = \frac{2}{5} = \frac{a_0''}{b_0''}$, we are in the case ($\beta$) so that x_4 is an arbitrary parameter.

The sought minimum is

$$F_4 = f(0, 0, 1, x_4, 0, 0, 0, 0, 1, 0, 0, 0) = \frac{6 + 2x_4}{15 + 5x_4} = \frac{2}{5}. \tag{79}$$

We shall now prove that for the functions of the form (69), the locally minimizing points coincide with the globally minimizing points.

Let $P = (\xi_1, \ldots, \xi_k, \xi_{k+1}, \ldots, \xi_n)$ and $P' = (\bar{\xi}_1, \ldots, \bar{\xi}_k, \xi_{k+1}, \ldots, \xi_n)$ be two points in B_2^n so that $k \geqq 2$ and

$$F(P) > F(P'). \tag{80}$$

Let us consider the points

$$P_i = (\bar{\xi}_1, \ldots, \bar{\xi}_{i-1}, \xi_i, \bar{\xi}_{i+1}, \ldots, \bar{\xi}_k, \xi_{k+1}, \ldots, \xi_n) \quad (i = 1, \ldots, k). \tag{81}$$

We have

Lemma 4. *If relations* (70) *and* (80) *hold, then at least one of the relations*

$$F(P) > F(P_i) \qquad (i = 1, \ldots, k) \tag{82}$$

holds.

Proof. Let us put

$$a_0 + \sum_{j=1}^{n} a_j \xi_j = \alpha,$$

$$a_0 + \sum_{j=1}^{k} a_j \bar{\xi}_j + \sum_{h=k+1}^{n} a_h \xi_h = \beta,$$

$$b_0 + \sum_{j=1}^{n} b_j \xi_j = \gamma,$$

$$b_0 + \sum_{j=1}^{k} b_j \bar{\xi}_j + \sum_{h=k+1}^{n} b_h \xi_h = \delta.$$

Then relation (80) states that

$$\frac{\alpha}{\gamma} > \frac{\beta}{\delta}. \tag{83}$$

Now,

$$F(P_i) = \frac{\beta + a_i(\xi_i - \bar{\xi}_i)}{\delta + b_i(\xi_i - \bar{\xi}_i)}.$$

Let us suppose that the lemma is false; then

$$\frac{\alpha}{\gamma} \leqq \frac{\beta + a_i(\xi_i - \bar{\xi}_i)}{\delta + b_i(\xi_i - \bar{\xi}_i)}$$

for $i = 1, \ldots, k$. Hence

$$\alpha\,\delta + \alpha\, b_i(\xi_i - \bar{\xi}_i) - \gamma\,\beta - \gamma\, a_i(\xi_i - \bar{\xi}_i) \leqq 0$$

for $i = 1, \ldots, k$. Summing the above relations, we get

$$k\,\alpha\,\delta - k\,\gamma\,\beta + \alpha\Big(b_0 + \sum_{i=1}^{k} b_i\,\xi_i + \sum_{j=k+1}^{n} b_j\,\xi_j\Big) -$$

$$- \alpha\Big(b_0 + \sum_{i=1}^{k} b_i\,\bar{\xi}_i + \sum_{j=k+1}^{n} b_j\,\xi_j\Big) - \gamma\Big(a_0 + \sum_{i=1}^{k} a_i\,\xi_i + \sum_{j=k+1}^{n} a_j\,\xi_j\Big) +$$

$$+ \gamma\Big(a_0 + \sum_{i=1}^{k} a_i\,\bar{\xi}_i + \sum_{j=k+1}^{n} a_j\,\xi_j\Big) \leqq 0,$$

or,

$$k\,\alpha\,\delta - k\,\gamma\,\beta + \alpha\,\gamma - \alpha\,\delta - \gamma\,\alpha + \gamma\,\beta \leqq 0,$$

i.e.

$$(k-1)\,(\alpha\,\delta - \beta\,\gamma) \leqq 0$$

and, k being greater or equal to 2, we find

$$\alpha\,\delta - \beta\,\gamma \leqq 0,$$

in contradiction with (83).

Theorem 8. *If the function* (69) *satisfies the conditions* (70), *then any local minimum of the function F is a global minimum of it.*

Proof. Let $P = (\xi_1, \ldots, \xi_n)$ be a local minimum of (69) and let us suppose that it is not a global minimum of it. It results that a point $P' = (\bar{\xi}_1, \ldots, \bar{\xi}_k, \xi_{k+1}, \ldots, \xi_n)$ exists so that $F(P') < F(P)$. P being a locally minimizing point, k is greater or equal to 2. From Lemma 4, we see that there is a point, say $P^1 = (\bar{\xi}_1, \ldots, \bar{\xi}_{k-1}, \xi_k, \ldots, \xi_n)$, (differing in $k-1$ coordinates from P), so that $F(P^1) < F(P)$. If $k - 1 = 1$, then the definition of P, as a locally minimizing point is contradicted. If $k - 1 \geqq 2$, then applying again the lemma, we find a point, say $P^2 = (\bar{\xi}_1, \ldots, \bar{\xi}_{k-2}, \xi_{k-1}, \xi_k, \ldots, \xi_n)$, differing from P in at most $k-2$ coordinates, and so that $F(P^2) < F(P)$. Etc. Finally, we obtain a point, say $P^{k-1} = (\bar{\xi}_1, \xi_2, \ldots, \xi_n)$ with $F(P^{k-1}) < F(P)$, in contradiction with the definition of P.

This theorem shows, that for the determination of the minimizing points of (69), it suffices to determine its locally minimizing points. This can be done as follows.

Algorithm II. Start with any initial vector $(x_1^*, \ldots, x_n^*)$ (the initial vector

$$x_i^* = \begin{cases} 1, & \text{if } b_i > 0 \text{ and } \dfrac{a_i}{b_i} < \dfrac{a_0}{b_0} \\ 0, & \text{otherwise} \end{cases} \tag{84}$$

seems to give a good approximation of the minimizing point). Compute, for this vector, the left-hand sides of formulas (4. i), which become

$$T_i^* = (x_i^* - \overline{x_i^*})\,(a_i\,B - b_i A), \tag{85.0}$$

where

$$A = \sum_{j=1}^{n} a_j\, x_j^*, \tag{85.1}$$

$$B = \sum_{j=1}^{n} b_j\, x_j^*. \tag{85.2}$$

If all $T_i^* \leqq 0$, then we have obtained the sought minimizing point. If there are positive T_i^*'s, then choose the greatest one, say $T_{i_0}^*$, and change $x_{i_0}^*$ into $\overline{x_{i_0}^*}$, leaving $x_1^*, \ldots, x_{i_0-1}^*, x_{i_0+1}^*, \ldots, x_n^*$ unchanged. Compute again the new T_i^{**}'s, and check whether they are all non-positive or not, etc. This procedure is continued, until all T_i's become non-positive, showing that we have obtained a minimizing point $(\xi_1, \ldots, \xi_n)$.

Let K be the set of those indices i for which $T_i(\xi_1, \ldots, \xi_n) = 0$ and put

$$x_i = \begin{cases} \xi_i, & \text{if } i \notin K, \\ \text{arbitrary}, & \text{if } i \in K. \end{cases} \tag{86}$$

Theorem 9. *The points* $(x_1, \ldots, x_n)$ *given by formula* (86), *are all the* (*globally*!) *minimizing points of the function* (69) *which satisfy* (70).

Proof. (α) When changing $x_{i_0}^*$ into $\overline{x_{i_0}^*}$, the corresponding value of the objective function decreases with $T_{i_0}^*$; therefore after a finite number of steps, we do attain the (locally) minimizing point $(\xi_1, \ldots, \xi_n)$.

(β) Moreover, according to Theorem 8, $(\xi_1, \ldots, \xi_n)$ is even a globally minimizing point. The fact that $T_i = 0$ for all $i \in K$ shows that for all the points $(x_1, \ldots, x_n)$ satisfying (86), we have $F(x_1, \ldots, x_n) = F(\xi_1, \ldots, \xi_n) = \min F$.

(γ) But, Theorem 7 also states that the minimizing points are of the following form: x_i arbitrary for i belonging to a certain set K', while x_i have fixed values for i not in K'.

(δ) It results from (β) and (γ) that $K \subseteqq K'$. For $i \notin K$, we have $T_i < 0$, so that

$$F(\xi_1, \ldots, \xi_n) < F(\xi_1, \ldots, \xi_{i-1}, \bar{\xi}_i, \xi_{i+1}, \ldots, \xi_n),$$

implying that $i \notin K'$. Therefore $K = K'$, completing the proof.

Example 9. Applying the above algorithm to the function (78) studied in Example 8, we obtain the following table:

Table 6

No.	x_1	x_2	x_3	x_4	x_5	x_6	x_7	x_8	x_9	x_{10}	x_{11}	x_{12}	T_1	T_2	T_3	T_4	T_5	T_6	T_7	T_8	T_9	T_{10}	T_{11}	T_{12}	i_0
1	0	0	1	1	0	0	1	1	1	0	1	0	−	−	−	−	−	−	+	+	−	−	+	−	7
2	0	0	1	1	0	0	0	1	1	0	1	0	−	−	−	−	−	−	−	+	−	−	+	−	8
3	0	0	1	1	0	0	0	0	1	0	1	0	−	−	−	−	−	−	−	−	−	−	+	−	11
4	0	0	1	1	0	0	0	0	1	0	0	0	−	−	−	0	−	−	−	−	−	−	−	−	$\vee$

Hence, the minimizing points are

$$(87)\quad x_1 = x_2 = 0, \quad x_3 = 1, \quad x_4 \text{ arbitrary}, \quad x_5 = x_6 = x_7 = x_8 = 0,$$
$$x_9 = 1, \quad x_{10} = x_{11} = x_{12} = 0.$$

Theorem 4 of § 1 shows that in case of constraints, the minimization of a fractional pseudo-Boolean function (69) [satisfying (70)] can not be performed as in the restriction-free case, by finding the locally minimizing points.

In this case, we shall:

1) determine (as in Chapters III and IV) the families of solutions of the restraints;

2) for each family, minimize the fractional pseudo-Boolean function obtained from (69) by introducing the fixed values characterizing the family;

3) choose the smallest value(s) obtained at 2.

Example 10. Let us minimize the function

(78)
$$F = \frac{3 + 2x_1 + 4x_2 + x_3 + 2x_4 + 9x_5 + 6x_6 + 12x_7 + 8x_8 + 2x_9 + 3x_{10} + 3x_{11} + x_{12}}{6 + x_1 + 8x_2 + 3x_3 + 5x_4 + 15x_5 + 10x_6 + 25x_7 + 18x_8 + 6x_9 + 3x_{10} + 7x_{11}}$$

under the constraints

$$(88.1)\quad x_1 - 3x_2 + 12x_3 + x_5 - 7x_6 + x_7 - 3x_{10} + 5x_{11} + x_{12} - 6 \geqq 0,$$

$$(88.2)\quad -3x_1 + 7x_2 - x_4 - 6x_5 + 1 \geqq 0,$$

$$(88.3)\quad -11x_1 - x_3 + 7x_4 + x_6 - 2x_7 - x_8 + 5x_9 - 9x_{11} - 4 \geqq 0,$$

$$(88.4)\quad -5x_2 - 6x_3 + 12x_5 - 7x_6 - 3x_8 - x_9 + 8x_{10} - 5x_{12} + 8 \geqq 0,$$

$$(88.5)\quad 7x_1 + x_2 + 5x_3 - 3x_4 - x_5 + 8x_6 + 2x_8 - 7x_9 - x_{10} + 7x_{12} - 7 \geqq 0,$$

$$(88.6)\quad 2x_1 + 4x_4 + 3x_7 + 5x_8 + x_9 - x_{11} - x_{12} - 4 \geqq 0.$$

This system of inequalities was solved in Example 5 of Chapter III, and its families of solutions are:

Table 7

No.	x_1	x_2	x_3	x_4	x_5	x_6	x_7	x_8	x_9	x_{10}	x_{11}	x_{12}
1	0	1	1	1	1	0	0	1	0	—	0	1
2	0	1	1	1	1	0	1	0	0	—	0	1
3	0	—	1	1	0	0	1	0	0	1	0	1
4	0	0	1	1	0	0	0	1	0	1	0	1

Table 8

No.	Family of solutions x_1	x_2	x_3	x_4	x_5	x_6	x_7	x_8	x_9	x_{10}	x_{11}	x_{12}	Remaining function	Minimizing point x_1	x_2	x_3	x_4	x_5	x_6	x_7	x_8	x_9	x_{10}	x_{11}	x_{12}	Minimum
1	0	1	1	1	1	0	0	1	0	—	0	1	$\frac{28+3x_{10}}{55+3x_{10}}$	0	1	1	1	1	0	0	1	0	0	0	1	28/55
2	0	1	1	1	1	0	1	0	0	—	0	1	$\frac{32+3x_{10}}{62+3x_{10}}$	0	1	1	1	1	0	1	0	0	0	0	1	32/62
3	0	—	1	1	0	0	1	0	0	1	0	1	$\frac{22+4x_2}{42+8x_2}$	0	1	1	1	0	0	1	0	0	1	0	1	26/50
4	0	0	1	1	0	0	0	1	0	1	0	1	18/35	0	0	1	1	0	0	0	1	0	1	0	1	18/35

Proceeding as indicated above, we obtain the following Table 8, showing that

(89)
$$F_{\min} = F(0, 1, 1, 1, 1, 0, 0, 1, 0, 0, 0, 1) = \frac{28}{55}.$$

Remark 4. The above described procedures can be extended to the case when condition (70) is not satisfied. If the pseudo-Boolean function

(90)
$$h(x_1, \ldots, x_n) = b_0 + b_1 x_1 + \cdots + b_n x_n$$

takes only positive values, this can be simply done by performing the substitutions $x_i = 1 - \bar{x}_i$ for all i such that $b_i < 0$. If $h(x_1, \ldots, x_n)$ takes only negative values, we multiply both the numerator and the denominator of (69) by -1, and then apply the above indicated substitutions. If the function $h(x_1, \ldots, x_n)$ has not a constant sign, we find the families of solutions of the inequalities

$$h(x_1, \ldots, x_n) > 0 \text{ and } h(x_1, \ldots, x_n) < 0;$$

afterwards we introduce, in turn, each of these families of solutions into (69) and apply the above methods separately.

Remark 5. The accelerating methods described in § V.2 could perhaps be adapted to fractional progamming.

Part II

Chapter VIII

Integer Mathematical Programming

In this chapter we shall show that problems of:

1. all-integer (linear or nonlinear) programming,
2. mixed integer-continuous (linear or nonlinear) programming,

may be reduced to problems of pseudo-Boolean programming.

Also, a method of K. Maghout [2–6] using Boolean algebra for the solution of linear, parametric linear and quadratic continuous programming will be presented.

Non-Boolean techniques for integer programming are surveyed in the papers of M. L. Balinski [1], E. M. L. Beale [1], A. Ben-Israel and A. Charnes [1], ect. Whereas most of the research in this field is devoted to linear integer programming, there are also several papers (H. P. Künzi and W. Oettli [1], C. Witzgall [1], etc.) dealing with specific problems of nonlinear integer programming.

§ 1. All-Integer Mathematical Programming

Let us consider the following

Problem I. Minimize the function

$$P_0(x_1, \ldots, x_n) \tag{1}$$

under the conditions

$$P_j(x_1, \ldots, x_n)\, R_j\, 0 \qquad (j = 1, \ldots, m), \tag{2}$$

$$0 \leqq x_i \leqq M_i \qquad (i = 1, \ldots, n), \tag{3}$$

$$x_i \text{ integers} \qquad (i = 1, \ldots, n), \tag{4}$$

where $P_0, P_1, \ldots, P_m$ are polynomials with real coefficients, and each R_j is one of the relations $=, <, >, \leqq, \geqq, \neq$.

Remark 1. If, instead of the bounds (3), for certain indices i we are given the bounds

$$m_i \leqq x_i \leqq M_i, \tag{3'}$$

we replace these variables x_i by the new variables

$$x_i' = x_i - m_i, \tag{5}$$

satisfying the conditions $0 \leqq x_i' \leqq M_i - m_i$, obtaining thus a problem of the type I.

Remark 2. If, for certain indices i, the corresponding conditions (3′) are lacking, and if the polynomials $P_1, \ldots, P_m$ are linear, then we can try to determine the missing m_i's and M_i's by minimizing (maximizing) the variable x_i under the constraints (2), without taking into account the conditions (4) (so that these auxiliary extremum problems may be solved as common linear programs).

Now, Problem I may be reduced to a pseudo-Boolean program.

To do this, we introduce the binary developments of the variables x_i:

$$x_i = \sum_{j=0}^{r_i} 2^j x_{ij} \qquad (i = 1, \ldots, n), \tag{6}$$

where x_{ij} are binary $(0, 1)$ variables and r_i coincides with the greatest exponent in the binary development of M_i (i.e., r_i is the greatest integer for which $2^{r_i} \leqq M_i$). The pseudo-Boolean program equivalent to the original problem is obtained simply by performing the substitution (6) into formulas (1), (2) and (3).

Example 1 (R. Gomory [1]). Solve the following program: maximize

$$f(x_1, x_2, x_3) = 4x_1 + 5x_2 + x_3 \tag{7}$$

under the constraints

$$3x_1 + 2x_2 \leqq 10, \tag{8.1}$$

$$x_1 + 4x_2 \leqq 11, \tag{8.2}$$

$$3x_1 + 3x_2 + x_3 \leqq 13, \tag{8.3}$$

$$x_i \geqq 0 \qquad (i = 1, 2, 3), \tag{9}$$

$$x_i \text{ integers} \qquad (i = 1, 2, 3). \tag{10}$$

Maximizing the variable x_1 under the constraints (8), we find

$$x_1 \leqq \frac{10}{3} \tag{11.1}$$

and similarly

$$x_2 \leqq \frac{11}{4}, \tag{11.2}$$

$$x_3 \leqq 13. \tag{11.3}$$

From (10) and (11) we deduce

$$x_1 \leqq 3, \tag{12.1}$$

$$x_2 \leqq 2, \tag{12.2}$$

$$x_3 \leqq 13, \tag{12.3}$$

hence the binary developments of x_1, x_2 and x_3 are of the form

$$x_1 = y_1 + 2y_2, \tag{13.1}$$

$$x_2 = y_3 + 2y_4, \tag{13.2}$$

$$x_3 = y_5 + 2y_6 + 4y_7 + 8y_8. \tag{13.5}$$

It follows from (13) that the original problem is reduced to the following one: maximize

$$g(y_1, \ldots, y_8) = 4y_1 + 8y_2 + 5y_3 + 10y_4 + y_5 + 2y_6 + 4y_7 + 8y_8 \tag{14}$$

under the constraints

$$(15.1)\qquad 3y_1 + 6y_2 + 2y_3 + 4y_4 \leqq 10,$$

$$(15.2)\qquad y_1 + 2y_2 + 4y_3 + 8y_4 \leqq 11,$$

$$(15.3)\qquad 3y_1 + 6y_2 + 3y_3 + 6y_4 + y_5 + 2y_6 + 4y_7 + 8y_8 \leqq 13,$$

$$(16.1)\qquad y_3 + 2y_4 \leqq 2,$$

$$(16.2)\qquad y_5 + 2y_6 + 4y_7 + 8y_8 \leqq 13,$$

$$(17)\qquad y_j = 0 \text{ or } 1 \qquad (j = 1, \ldots, 8).$$

Using, for instance, the method described in § 1 of Chapter V, we find that the above pseudo-Boolean program has a single solution, namely

$$(18)\quad y_1 = 0,\ \ y_2 = 1,\ \ y_3 = 0,\ \ y_4 = 1,\ \ y_5 = 1,\ \ y_6 = 0,\ \ y_7 = 0,\ \ y_8 = 0.$$

From (18) and (13) we get the corresponding solution to the original problem:

$$(19)\qquad x_1 = 2,\quad x_2 = 2,\quad x_3 = 1,$$

hence the sought maximum of the function (7) is

$$(20)\qquad f(2, 2, 1) = 19.$$

Example 2. Let us minimize

$$(21)\qquad f(x_1, x_2, x_3, x_4) = 3x_1 x_2 - 4x_2 x_3 + 5x_3 x_4 - 8x_4^2$$

under the restrictions

$$(22.1)\qquad x_1^2 - 3x_1 x_4 \leqq 7,$$

$$(22.2)\qquad 2x_1 x_3 x_4^2 + 5x_2 x_3 \geqq 10,$$

$$(22.3)\qquad 2x_1 - 3x_2 + 4x_3 + 3x_4 \leqq 21,$$

$$(23.1)\qquad 0 \leqq x_1 \leqq 2,$$

$$(23.2)\qquad 0 \leqq x_2 \leqq 3,$$

$$(23.3)\qquad 0 \leqq x_3 \leqq 6,$$

$$(23.4)\qquad 0 \leqq x_4 \leqq 1,$$

$$(24)\qquad x_i \text{ integers} \qquad (i = 1, 2, 3, 4).$$

It results from relations (23) that the binary developments of x_1, x_2, x_3 and x_4 are:

$$(25.1)\qquad x_1 = y_1 + 2y_2,$$

$$(25.2)\qquad x_2 = y_3 + 2y_4,$$

$$(25.3)\qquad x_3 = y_5 + 2y_6 + 4y_7,$$

$$(25.4)\qquad x_4 = y_8,$$

where $y_1, \ldots, y_8$ are bivalent variables.

Hence the original program is equivalent to the following pseudo-Boolean program: minimize

$$(26)\quad g(y_1, \ldots, y_8) = 3y_1 y_3 + 6y_1 y_4 + 6y_2 y_3 + 12y_2 y_4 - 4y_3 y_5 - \\ - 8y_3 y_6 - 16y_3 y_7 - 8y_4 y_5 - 16y_4 y_6 - 32y_4 y_7 + \\ + 5y_5 y_8 + 10y_6 y_8 + 20y_7 y_8 - 8y_8$$

under the restrictions

(27.1) $$y_1 + 4y_1 y_2 + 4y_2 - 3y_1 y_8 - 6y_2 y_8 \leqq 7,$$

(27.2) $$2y_1 y_5 y_8 + 4y_1 y_6 y_8 + 8y_1 y_7 y_8 + 4y_2 y_5 y_8 + \\ + 8y_2 y_6 y_8 + 16y_2 y_7 y_8 + 5y_3 y_5 + 10y_3 y_6 + 20y_3 y_7 + \\ + 10y_4 y_5 + 20y_4 y_6 + 40y_4 y_7 \geqq 10,$$

(27.3) $$2y_1 + 4y_2 - 3y_3 - 6y_4 + 4y_5 + 8y_6 + 16y_7 + 3y_8 \leqq 21,$$

(28.1) $$y_1 + 2y_2 \leqq 2,$$

(28.2) $$y_5 + 2y_6 + 4y_7 \leqq 6,$$

(29) $$y_j = 0 \text{ or } 1 \quad (j = 1, \ldots, 8).$$

Using the methods given in Chapter VI, we find that the single solution to the above pseudo-Boolean program is

(30) $$y_1 = 0, \; y_2 = 0, \; y_3 = 1, \; y_4 = 1, \; y_5 = 0, \; y_6 = 1, \; y_7 = 1, \; y_8 = 0.$$

Formulas (25) and (30) determine the solution to the original problem:

(31) $$x_1 = 0, \quad x_2 = 3, \quad x_3 = 6, \quad x_4 = 0,$$

the minimum value being

(32) $$f(0, 3, 6, 0) = -72.$$

§ 2. Mixed Integer-Continuous Mathematical Programming

Several authors have proposed a simple idea for converting mixed integer-continuous programs into all-integer programs, the price being a certain increase in the number of binary components of the variables.

Namely, suppose that a continuous variable y_i is bounded by m_i and M_i:

(33) $$m_i \leqq y_i \leqq M_i$$

and that we want to determine an approximative value y_i' of y_i, so that

(34) $$|y' - y_i| \leqq 10^{-\alpha_i}$$

where α_i is a given positive integer. Then we can replace $10^{\alpha_i} y_i$ by a new variable z_i, taking only integer values and satisfying

(35) $$10^{\alpha_i} m_i \leqq z_i \leqq 10^{\alpha_i} M_i.$$

For starting the procedure, multiply the objective function as well as the constraints by suitable powers of 10 and introduce the new variables z_i instead of the continuous variables y_i.

Example 3 (M. SIMONNARD [1], 9.5.3). Maximize

$$f = x_1 - 3x_2 + 3x_3 \tag{36}$$

under the constraints

$$2x_1 + x_2 - x_3 \leqq 4, \tag{37.1}$$

$$4x_1 - 3x_2 \leqq 2, \tag{37.2}$$

$$-3x_1 + 2x_2 + x_3 \leqq 3, \tag{37.3}$$

$$x_i \geqq 0 \qquad (i = 1, 2, 3), \tag{38}$$

$$x_2 \text{ and } x_3 \text{ integers.} \tag{39}$$

Maximizing separately each x_i under the constraints (37) and (38), we find

$$x_1 \leqq 3, \tag{40.1}$$

$$x_2 \leqq \frac{10}{3}, \tag{40.2}$$

$$x_3 \leqq \frac{9}{2}. \tag{40.3}$$

From (38), (39) and (40.2), (40.3), we deduce

$$0 \leqq x_2 \leqq 3, \tag{41.2}$$

$$0 \leqq x_3 \leqq 4. \tag{41.3}$$

Let us suppose that we are interested in finding x_1 with an approximation of $\frac{1}{10}$; we shall replace $10x_1$ by an integer variable z_1 satisfying

$$0 \leqq z_1 \leqq 30. \tag{41.1}$$

Thus the original program is reduced to the following all-integer program: maximize

$$g = z_1 - 30x_2 + 30x_3 \tag{42}$$

under the constraints

$$2z_1 + 10x_2 - 10x_3 \leqq 40, \tag{43.1}$$

$$4z_1 - 30x_2 \leqq 20, \tag{43.2}$$

$$-3z_1 + 20x_2 + 10x_3 \leqq 30, \tag{43.3}$$

$$0 \leqq z_1 \leqq 30, \tag{41.1}$$

$$0 \leqq x_2 \leqq 3, \tag{41.2}$$

$$0 \leqq x_3 \leqq 4, \tag{41.3}$$

$$z_1, x_2 \text{ and } x_3 \text{ integers.} \tag{42}$$

By applying, for instance, the method described in the preceding section, we find that there exists a single maximizing point:

$$z_1 = 5, \quad x_2 = 0, \quad x_3 = 4, \tag{43}$$

hence

$$x_1 = \tfrac{1}{2}, \quad x_2 = 0, \quad x_3 = 4, \tag{44}$$

and the sought maximum is

$$f\left(\frac{1}{2}, 0, 4\right) = \frac{25}{2}. \tag{45}$$

Example 4. Let us solve the following nonlinear mixed integer-continuous program: maximize

$$f = x_1 x_2 x_3 + 2x_1 + 4x_2 + 3x_3 - 1 \tag{46}$$

under the constraints

$$4x_1 + x_1 x_2 \leqq 4, \tag{47.1}$$

$$x_1 x_3^2 \geqq 5, \tag{47.2}$$

$$0 \leqq x_1 \leqq 1, \tag{48.1}$$

$$0 \leqq x_2 \leqq 5, \tag{48.2}$$

$$0 \leqq x_3 \leqq 3, \tag{48.3}$$

$$x_2 \text{ and } x_3 \text{ integers}; \tag{49}$$

it suffices to determine x_1 with an approximation of $\frac{1}{10}$.

Then, as in the preceding example, the given program is reduced to the following one: maximize

$$g = z_1 x_2 x_3 + 2z_1 + 40x_2 + 30x_3 - 10 \tag{50}$$

under the constraints

$$4z_1 + z_1 x_2 \leqq 40, \tag{51.1}$$

$$z_1 x_3^2 \geqq 50, \tag{51.2}$$

$$0 \leqq z_1 \leqq 10, \tag{48.1'}$$

$$0 \leqq x_2 \leqq 5, \tag{48.2}$$

$$0 \leqq x_3 \leqq 3, \tag{48.3}$$

$$z_1, x_2, x_3 \text{ integers}, \tag{52}$$

(where $z_1 = 10x_1$).

This program, solved as in the preceding section, yields the solution

$$z_1 = 6, \quad x_2 = 2, \quad x_3 = 3, \tag{53}$$

hence

$$x_1 = \tfrac{3}{5}, \quad x_2 = 2, \quad x_3 = 3 \tag{54}$$

and the maximum is

$$f\left(\frac{3}{5}, 2, 3\right) = \frac{104}{5}. \tag{55}$$

§ 3. Linear and Quadratic Continuous Programming

In this section we shall briefly describe the results of K. MAGHOUT [2—4] on the application of Boolean algebra to linear continuous programming.

The authors have no informations about the power of Boolean methods for continuous programming on electronic computers. As to hand computation, the classical methods are more efficient. However, the underlying ideas of this section could be useful in solving special types of mathematical programs (as for instance, transportation or assignment problems).

Denote by $A = (a_{ij})$ a $m \times n$ matrix, by $b = (b_i)$ an $m \times 1$ vector and by $c = (c_j)$ an $1 \times n$ vector, the elements a_{ij}, b_i, c_j being real numbers. Consider the two problems of linear programming associated to these data:

Primal problem. Maximize

$$\sum_{j=1}^{n} c_j x_j \tag{56}$$

under the restrictions

$$\sum_{j=1}^{n} a_{ij} x_j \leqq b_i \qquad (i = 1, \ldots, m), \tag{57}$$

$$x_j \geqq 0 \qquad (j = 1, \ldots, n). \tag{58}$$

Dual problem. Minimize

$$\sum_{i=1}^{m} y_i b_i \tag{59}$$

under the restrictions

$$\sum_{i=1}^{m} y_i a_{ij} \geqq c_j \qquad (j = 1, \ldots, n), \tag{60}$$

$$y_i \geqq 0 \qquad (i = 1, \ldots, m). \tag{61}$$

It is well-known, that the primal problem has optimal solutions if and only if the dual problem has optimal solutions. Further, it is also known that the vectors $x = (x_j)$ and $y = (y_i)$ are optimal solutions to the primal and to the dual problem, respectively, if and only if relations (58), (61) and

$$\sum_{j=1}^{n} c_j x_j = \sum_{i=1}^{m} y_i b_i \tag{62}$$

hold.

But, if x and y are such solutions, then relations (57) and (61) imply

$$\sum_{i=1}^{m} y_i \left(b_i - \sum_{j=1}^{n} a_{ij} x_j\right) \geqq 0, \tag{63}$$

while from (60) and (58) one deduces

$$\sum_{j=1}^{n} \left(\sum_{i=1}^{m} y_i a_{ij} - c_j\right) x_j \geqq 0. \tag{64}$$

In view of (62), the sum of the left-hand sides of the inequalities (63) and (64) is equal to zero, so that, taking into account (57) and (60), we deduce

$$y_i \left(b_i - \sum_{j=1}^{n} a_{ij} x_j\right) = 0 \qquad (i = 1, \ldots, m), \tag{65}$$

$$\left(\sum_{i=1}^{m} y_i a_{ij} - c_j\right) x_j = 0 \qquad (j = 1, \ldots, n). \tag{66}$$

Conversely, from (65) and (66) we infer immediately (62).

Thus the above discussion has shown that the problem of solving both the primal and the dual linear programs is equivalent to that of solving the system of equations (65) and (66) under the conditions (58) and (61).

Setting

$$b_i - \sum_{j=1}^{n} a_{ij} x_j = v_i \qquad (i = 1, \ldots, m), \tag{67}$$

$$\sum_{i=1}^{m} y_i a_{ij} - c_j = u_j \qquad (j = 1, \ldots, n), \tag{68}$$

we see that the original problem is reduced to the following one: solve the system consisting of equations (67), (68),

$$\sum_{i=1}^{m} y_i v_i = 0, \tag{69}$$

and

$$\sum_{j=1}^{n} u_j x_j = 0, \tag{70}$$

under the conditions

$$y_i, v_i \geqq 0 \qquad (i = 1, \ldots, m), \tag{71}$$

$$u_j, x_j \geqq 0 \qquad (j = 1, \ldots, n). \tag{72}$$

But conditions (70) and (72) mean that for each i, at least one of the values u_j, x_j is equal to zero; in other words, we have

$$x_j = \xi_j X_j \qquad (j = 1, \ldots, n), \tag{73}$$

$$u_j = \bar{\xi}_j X_j \qquad (j = 1, \ldots, n), \tag{74}$$

where X_j are new variables satisfying

$$X_j \geqq 0 \qquad (j = 1, \ldots, n), \tag{75}$$

while ξ_j are Boolean (0, 1) variables. Similarly, conditions (69) and (71) are equivalent to

$$y_i = \bar{\eta}_i Y_i \qquad (i = 1, \ldots, m), \tag{76}$$

$$v_i = \eta_i Y_i \qquad (i = 1, \ldots, m), \tag{77}$$

where

$$Y_i \geqq 0 \qquad (i = 1, \ldots, m) \tag{78}$$

and η_i are Boolean variables.

We introduce the expressions (73), (74), (76) and (77) into the system (67) and (68), obtaining the system

$$\sum_{j=1}^{n} a_{ij} \xi_j X_j + \eta_i Y_i = b_i \qquad (i = 1, \ldots, m), \tag{79}$$

$$\sum_{i=1}^{m} a_{ij} \bar{\eta}_i Y_i - \bar{\xi}_j X_j = c_j \qquad (j = 1, \ldots, n). \tag{80}$$

Thus the original problem is equivalent to that of solving the system (79) and (80), where the unknowns X_j and Y_i must satisfy (75) and (78), respectively, while ξ_j and η_i are Boolean variables.

As it is well-known, the determination of all the optimal solutions of a problem of linear programming reduces to the determination of the so-called basic solutions*.

Now we shall assume, as it is generally done, that the given linear programs are not degenerate. In this case, it is also known that: (α) each basic solution $S = (x_1^*, \ldots, x_n^*, v_1^*, \ldots, v_m^*)$ of the primal problem has exactly m positive components; (β) if $H = \{j_1, \ldots, j_r, i_1, \ldots, i_s\}$ is a set of indices j and i, with $r + s = m$, then there exists at most one basic solution S such that $x_{j_1}^*, \ldots, x_{j_r}^*, v_{i_1}^*, \ldots, v_{i_s}^* > 0$, while for j and i not in H, $x_j^* = v_i^* = 0$. Similar results hold for the basic solutions $T = (y_1^*, \ldots, y_m^*, u_1^*, \ldots, u_n^*)$ of the dual problem.

But x_j^*, v_i^*, y_i^* and u_j^* are of the form $x_j^* = \xi_j^* X_j^*$, $v_i^* = \eta_i^* Y_i^*$, $y_i^* = \overline{\eta_i^*} Y_i^*$ and $u_j^* = \overline{\xi_j^*} X_j^*$, respectively. Hence the above results may be re-formulated as follows: (α') for each basic solution S, the corresponding vector $(\xi_1^*, \ldots, \xi_n^*, \eta_1^*, \ldots, \eta_m^*)$ has exactly m ones; (β') if $H = \{j_1, \ldots, j_r, i_1, \ldots, i_s\}$ is a set of indices j and i for which $r + s = m$, then there exists at most one basic solution S determined by the conditions $\xi_{j_1}^* = \cdots = \xi_{j_r}^* = \eta_{i_1}^* = \cdots = \eta_{i_s}^* = 1$, $X_{j_1}^*, \ldots, X_{j_r}^*$, $Y_{i_1}^*, \ldots, Y_{i_s}^* > 0$, while for j and i not in H, $\xi_j^* = \eta_i^* = 0$.

Therefore, we have to check the $\frac{(m+n)!}{m!\,n!}$ vectors $(\xi_1^*, \ldots, \xi_n^*, \eta_1^*, \ldots, \eta_m^*) \in B_2^{m+n}$ for which the number of ones is equal to m, in order to detect those for which the corresponding system (79) and (80) has a *unique* and *positive* solution in $X_1^*, \ldots, X_n^*, Y_1^*, \ldots, Y_m^*$.

But the determinant of the system (79) and (80) is

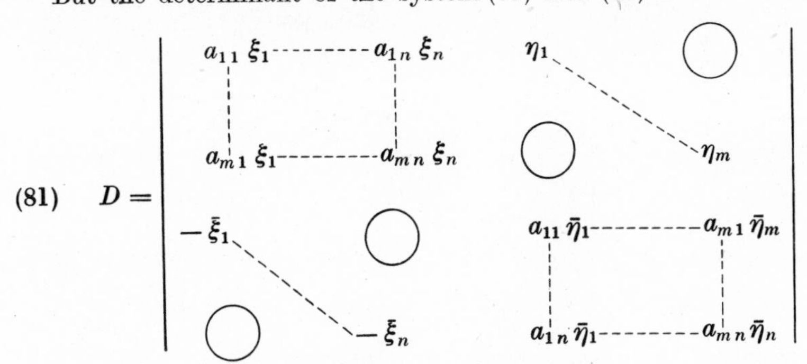

[and it is easy to see directly that if the vector $(\xi_1, \ldots, \xi_m, \eta_1, \ldots, \eta_n)$ has less or more than m ones, then $D = 0$].

* This classical concept is not to be confused with the basic solutions of a linear pseudo-Boolean inequality, as defined in Chapter III.

Let $J = \{j_1, \ldots, j_r\} = \{j \mid \xi_j^* = 1\}, \bar{J} = \{1, \ldots, n\} - J, I = \{i, \ldots, i_s\} = \{i \mid \eta_i^* = 1\}$ $(r + s = m)$, $\bar{I} = \{1, \ldots, m\} - I$. Further, a notation like $|a_{IJ}|$ will mean the determinant $|a_{ij}|$ where $i \in I$ and $j \in J$, while a notation like $|a_{IJ} - a_{Ij} + b_I|$ indicates the determinant obtained from $|a_{IJ}|$ by replacing column j by the column vector b_I. Finally, let $D_{j,0}$ and $D_{0,i}$ be the determinants corresponding to the variables X_j and Y_i by Cramer's rule, and let D^*, $D_{j,0}^*$, $D_{0,i}^*$ be the values of D, $D_{j,0}$ and $D_{0,i}$, respectively, corresponding to the vector $(\xi_1^*, \ldots, \xi_m^*, \eta_1^*, \ldots, \eta_n^*)$.

It is not difficult to establish the following formulas:

$$D^* = (-1)^{n(m+1)} |a_{\bar{I}J}|^2, \tag{82}$$

$$D_{j,0}^* = (-1)^{n(m+1)} |a_{\bar{I}J}| \cdot |a_{\bar{I}J} - a_{\bar{I}j} + b_{\bar{I}}| \quad \text{for} \quad j \in J, \tag{83}$$

$$D_{j,0}^* = -(-1)^{n(m+1)} |a_{\bar{I}J}| \cdot \begin{vmatrix} c_j & c_J \\ a_{\bar{I}j} & a_{\bar{I}J} \end{vmatrix} \quad \text{for} \quad j \notin J, \tag{84}$$

$$D_{0,i}^* = (-1)^{n(m+1)} |a_{\bar{I}J}| \cdot \begin{vmatrix} b_i & a_{iJ} \\ b_{\bar{I}} & a_{\bar{I}J} \end{vmatrix} \quad \text{for} \quad i \in I, \tag{85}$$

$$D_{0,i}^* = (-1)^{n(m+1)} |a_{\bar{I}J}| \cdot |a_{\bar{I}J} - a_{iJ} + c_J| \quad \text{for} \quad i \notin I \tag{86}$$

(implying, in particular, that for $\bar{I} = J = \varnothing$, we have $D^* = (-1)^{n(m+1)}$, $D_{j,0}^* = -(-1)^{n(m+1)} c_j$, $D_{0,i}^* = (-1)^{n(m+1)} b_i$).

From the above discussion, we deduce the following *algorithm* for solving the given linear program:

1) Form the set $B^* = \{Z^{1*}, \ldots, Z^{p*}\}$ of all the vectors $Z^{k*} = (\xi_1^{k*}, \ldots, \xi_n^{k*}, \eta_1^{k*}, \ldots, \eta_m^{k*})$ for which the number of ones is equal to m.

2) For each vector $Z^{k*} \in B^*$, compute the corresponding values $D_{j,0}^{k*}$ and $D_{0,i}^{k*}$ according to formulas (83)—(86). (α) If there exists either an index j for which $D_{j,0}^{k*}$ has not the sign $(-1)^{n(m+1)}$, or an index i for which $D_{0,i}^{k*}$ has not the sign $(-1)^{n(m+1)}$, then the vector Z^{k*} is to be dropped. (β) If all $D_{j,0}^{k*}$ and $D_{0,i}^{k*}$ have the sign $(-1)^{n(m+1)}$, then apply formulas

$$x_j^* = \xi_j^{k*} X_j^{k*} = \xi_j^{k*} \frac{D_{j,0}^{k*}}{D^{k*}} \qquad (j = 1, \ldots, n), \tag{87}$$

which determine a basic solution of the primal problem (while formulas

$$y_i^* = \bar{\eta}_i^* Y_i^* = \bar{\eta}_i^* \frac{D_{0,i}^{k*}}{D^*} \qquad (i = 1, \ldots, m) \tag{88}$$

determine a basic solution of the dual problem).

Similar treatments apply to the case when the c_j are of the form $c_j = \sum_{h=1}^{p} c_{jh} \lambda_h$, with λ_h arbitrary parameters, and to the case of a quadratic objective function with linear constraints.

Example 5 (K. MAGHOUT [3]). Maximize

(89) $$3x_1 + 2x_2 + x_3$$

under the constraints

(90.1) $$2x_1 + 3x_2 + 4x_3 \leqq 8,$$

(90.2) $$3x_1 + x_2 + 2x_3 \leqq 5,$$

(91) $$x_j \geqq 0 \qquad (j = 1, 2, 3).$$

Here $m = 2$, $n = 3$,

(92) $$(-1)^{n(m+1)} = -1,$$

(93) $$A = \begin{pmatrix} 2 & 3 & 4 \\ 3 & 1 & 2 \end{pmatrix},$$

(94) $$b = \begin{pmatrix} 8 \\ 5 \end{pmatrix},$$

(95) $$c = (3 \quad 2 \quad 1).$$

The set B consists of the following vectors Z^{k*}:

Table 1

	ξ_1^*	ξ_2^*	ξ_3^*	η_1^*	η_2^*
Z^{1*}	1	1	0	0	0
Z^{2*}	1	0	1	0	0
Z^{3*}	1	0	0	1	0
Z^{4*}	1	0	0	0	1
Z^{5*}	0	1	1	0	0
Z^{6*}	0	1	0	1	0
Z^{7*}	0	1	0	0	1
Z^{8*}	0	0	1	1	0
Z^{9*}	0	0	1	0	1
Z^{10*}	0	0	0	1	1

We begin step 2 with the vector Z^{1*} for which $J_1 = \{j \mid \xi_j^{1*} = 1\} = \{1, 2\}$, $I_1 = \{i \mid \eta_i^{1*} = 1\} = \varnothing$, so that $\bar{J}_1 = \{3\}$, $\bar{I}_6 = \{1, 2\}$. It follows, via (83)—(86), that

$$D_{1,0}^{1*} = -|a_{\{1,2\}\{1,2\}}| \cdot |a_{\{1,2\}\{1,2\}} - a_{\{1,2\}1} + b_{\{1,2\}}| = -\begin{vmatrix} 2 & 3 \\ 3 & 1 \end{vmatrix} \cdot \begin{vmatrix} 8 & 3 \\ 5 & 1 \end{vmatrix} = -49,$$

$$D_{2,0}^{1*} = -|a_{\{1,2\}\{1,2\}}| \cdot |a_{\{1,2\}\{1,2\}} - a_{\{1,2\}2} + b_{\{1,2\}}| = -\begin{vmatrix} 2 & 3 \\ 3 & 1 \end{vmatrix} \cdot \begin{vmatrix} 2 & 8 \\ 3 & 5 \end{vmatrix} = -98,$$

$$D_{3,0}^{1*} = +|a_{\{1,2\}\{1,2\}}| \cdot \begin{vmatrix} c_3 & c_{\{1,2\}} \\ a_{\{1,2\}3} & a_{\{1,2\}\{1,2\}} \end{vmatrix} = \begin{vmatrix} 2 & 3 \\ 3 & 1 \end{vmatrix} \cdot \begin{vmatrix} 1 & 3 & 2 \\ 4 & 2 & 3 \\ 2 & 3 & 1 \end{vmatrix} = -100,$$

$$D_{0,1}^{1*} = -|a_{\{1,2\}\{1,2\}}| \cdot |a_{\{1,2\}\{1,2\}} - a_{1\{1,2\}} + c_{\{1,2\}}| = -\begin{vmatrix} 2 & 3 \\ 3 & 1 \end{vmatrix} \cdot \begin{vmatrix} 3 & 2 \\ 3 & 1 \end{vmatrix} = -21,$$

$$D_{0,2}^{2*} = -|a_{\{1,2\}\{1,2\}}| \cdot |a_{\{1,2\}\{1,2\}} - a_{2\{1,2\}} + c_{\{1,2\}}| = -\begin{vmatrix} 2 & 3 \\ 3 & 1 \end{vmatrix} \cdot \begin{vmatrix} 2 & 3 \\ 3 & 2 \end{vmatrix} = -35,$$

so that we are in the case $2(\beta)$.

Hence we have to compute

$$D^{1*} = -\left| a_{\{1,2\}\{1,2\}} \right|^2 = -\begin{vmatrix} 2 & 3 \\ 3 & 1 \end{vmatrix}^2 = -49$$

[by (82)], so that formulas (87) give the basic solution

$$(96) \quad x_1^{1*} = \xi_1^{1*} \frac{D_{1,0}^{1*}}{D^{1*}} = 1, \quad x_2^{1*} = \xi_2^{1*} \frac{D_{2,0}^{1*}}{D^{1*}} = 2, \quad x_3^{1*} = \xi_3^{1*} \frac{D_{3,0}^{1*}}{D^{1*}} = 0,$$

while from (88) we obtain a basic solution of the dual problem:

$$(96') \qquad y_1^{1*} = \overline{\eta_1^{1*}} \frac{D_{0,1}^{1*}}{D^{1*}} = \frac{3}{7}, \quad y_2^{1*} = \overline{\eta_2^{1*}} \frac{D_{0,2}^{1*}}{D^{1*}} = \frac{5}{7}.$$

We have now to check $Z^{2*}, \ldots, Z^{10*}$ in the same way.

The computations may be summarized in Tables 1 and 2.

Table 2

	$D_{1,0}^*$	$D_{2,0}^*$	$D_{0,3}^*$	$D_{0,1}^*$	$D_{0,2}^*$	D^*	x_1^*	x_2^*	x_3^*	y_1^*	y_2^*
Z^{1*}	−49	−98	−100	−21	−35	−49	1	2	0	3/7	5/7
Z^{2*}	−32	+120					No basic solution				
Z^{3*}	−15	+9					Idem				
Z^{4*}	−16	−10	−20	−6	+28		Idem				
Z^{5*}	+30						Idem				
Z^{6*}	−3	−5	−3	+7			Idem				
Z^{7*}	+15						Idem				
Z^{8*}	+6						Idem				
Z^{9*}	+40						Idem				
Z^{10*}	+3						Idem				

Thus the single basic solution of the primal(dual)problem is given by formulas (96.1) [(96′), respectively].

§ 4. Discrete-Variable Problems

In certain practical applications we are faced with problems, the variables x_j $(j = 1, \ldots, n)$ of which have to take values in a specified finite set $S = \{s_1, s_2, \ldots, s_p\}$ of real numbers.

For instance, when planning to build warehouses in different points, we can choose in each of them, one of the possible sizes $s_1, s_2, \ldots, s_p$.

In such cases, the problem may be converted into one with bivalent variables (as indicated by G. B. Dantzig [3]) in the following way:

Put

$$(97) \qquad x_j = y_{j1} s_1 + y_{j2} s_2 + \cdots + y_{jp} s_p \qquad (j = 1, \ldots, n),$$

where the bivalent variables y_{jk} are subject to

$$(98) \qquad \sum_{k=1}^{p} y_{jk} = 1 \qquad (j = 1, \ldots, n).$$

Thus the problem is reduced to one with bivalent variables.

Using this device, a nonlinear function $f_{ij} = f_{ij}(x_j)$ occuring in a system

$$\sum_{j=1}^{n} f_{ij}(x_j) = a_i \qquad (i = 1, \ldots, m) \tag{99}$$

of constraints, may be replaced by the values of the function corresponding to a sample of values $x_j = x_j^r$ $(r = 1, \ldots, q)$; in this way, we obtain

$$\begin{aligned} &\sum_{j=1}^{n} \sum_{r=1}^{q} f_{ij}(x_j^r)\, y_{ijr} = a_i \qquad (i = 1, \ldots, m), \\ &\sum_{i=1}^{m} \sum_{r=1}^{q} y_{ijr} = 1, \qquad (j = 1, \ldots, n), \end{aligned} \tag{100}$$

where y_{ijr} are bivalent variables, while $f_{ij}(x_j^r)$ are specified values.

Chapter IX

Connectedness and Path Problems in Graphs

This chapter is concerned with the Boolean matrix approach to problems concerning the existence and determination of paths in directed graphs.

The presentation is not exhaustive. After the completion of the manuscript we became aware of the very interesting papers of D. ROSENBLATT [1—3], which are closely related to the topics discussed in this chapter. It may turn out that unfortunately we also had ignored other relevant papers.

§ 1. Basic Concepts in Graph Theory: Paths, Circuits, Connectedness*

Definition 1. By a *graph* we shall mean a couple $G = (N, \Gamma)$, where $N = \{1, 2, \ldots, n\}$ is a finite set consisting of n elements called *nodes*, or *vertices*, or *points*, and Γ is a mapping which associates to each vertex $i \in N$ a subset $\Gamma i \subseteqq N$. A couple of points (i, j) so that $j \in \Gamma$ is said to be an *arc*.

A graph may be depicted as a diagram in which the vertices are represented as points in the plane, while the arcs (i, j) are indicated by arrows with starting point i and end point j.

The diagram of a graph is not to be confused with the graph itself. Let us consider the graph $G = (N, \Gamma)$, with $N = \{1, 2, 3, 4\}$ and the mapping Γ defined as follows: $\Gamma 1 = \{2\}$, $\Gamma 2 = \{3, 4\}$, $\Gamma 3 = \{4\}$, $\Gamma 4 = \{1, 4\}$. As the relative position of the points (representing the vertices) is immaterial, as well as the form of the arrows, we see that the same graph G may be represented in several ways; see for instance, Fig. 1a, b, c. Notice also that in Fig. 1c the intersection of the arcs $(1, 4)$ and $(2, 3)$ does not stand for a vertex. However, the diagram representation of a graph has the advantage of being intuitive and proves to be useful in many problems.

---

* For detailed presentations of graph theory, see, for instance, C. BERGE [2], C. BERGE and A. GHOUILA-HOURI [1], F. HARARY, R. NORMAN and D. CARTWRIGHT [1], A. KAUFMANN [1], D. KÖNIG [1], O. ORE [1], B. ROY [1].

Many phenomena of real interest may be visualised as graphs. Communication networks (roads, telephone nets, systems of TV relay stations, etc.), electrical networks, pipe line systems, may be quoted

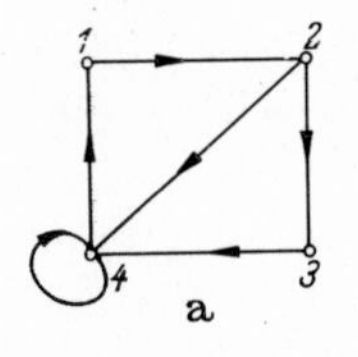

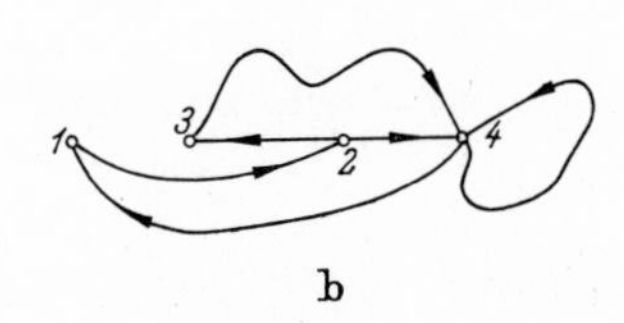

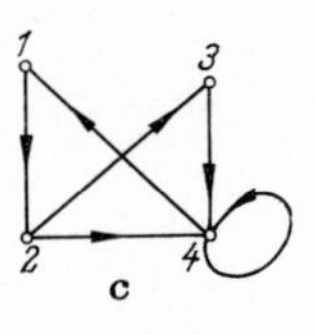

Fig. 1

as typical examples of graphs. Numerous graph-theoretical models are encountered in operations research, psychology, biology, etc. (see e.g. F. Harary, R. Norman and D. Cartwright [1]).

Definition 2. A *path* from the vertex i_0 to the vertex i_p is a sequence

$$(i_0, i_1, \ldots, i_p) \tag{1}$$

of vertices $i_k \in N$ such that $i_k \in \Gamma i_{k-1}$ for $k = 1, 2, \ldots, p$. The *length* of the path (1) is p. The sequence (1) is said to be an *elementary path*, if the vertices $i_1, \ldots, i_p$ are pairwise distinct.

For instance, in the graph of Fig. 1, $(1, 2, 4, 1, 2, 3)$, $(1, 2, 3, 4)$ and $(1, 2, 3, 4, 1)$ are paths, the last two being elementary.

Remark 1. If there is a path from i_0 to i_p, there is also an elementary one joining these two vertices.

In the above example, the vertices 1 and 3 are joined by the path $(1, 2, 4, 1, 2, 3)$, but the "subpath" of it formed by the vertices $(1, 2, 3)$ is an elementary path which also links 1 to 3.

Definition 3. A path (1) is called an (*elementary*) *circuit*, if $i_0 = i_p$ (and if the path $(i_0, \ldots, i_{p-1})$ is elementary). We identify the circuit $(i_0 = i_p, i_1, i_2, \ldots, i_p)$ with each of the circuits $(i_{k-1}, i_k, i_{k+1}, \ldots, i_p, i_1, i_2, \ldots, i_{k-1})$ $(k = 1, 2, \ldots, p)$. In case the circuit is made up of a single arc, then it is called a *loop*.

For instance, in Fig. 1, $(1, 2, 4, 1, 2, 3, 4, 1)$ is a circuit, $(1, 2, 3, 4, 1)$ is an elementary one, while $(4, 4)$ is a loop. Notice also that $(1, 2, 3, 4, 1)$, $(2, 3, 4, 1, 2)$, $(3, 4, 1, 2, 3)$ and $(4, 1, 2, 3, 4)$ represent in fact the same circuit. This circuit is Hamiltonian, in the following sense:

Definition 4. An elementary path (circuit) running through all the vertices of the graph is said to be a *Hamiltonian path* (*circuit*).

Definition 5. A graph $G = (N, \Gamma)$ is called *strongly connected*, (or, *total*) if for every two vertices $i, j \in N$, there is a path from i to j.

Remark 2. According to the above definition, there is also a path from j to i, so that Definition 5 may be re-formulated as follows: a graph is *strongly connected,* if for every two vertices, $i, j \in N$, with $i \neq j$, there exists a circuit containing them.

Definition 6. A graph $G = (N, \Gamma)$ is termed *connected,* if for every two vertices $i, j \in N$, with $i \neq j$, there is either a path from i to j, or a path from j to i (or both).

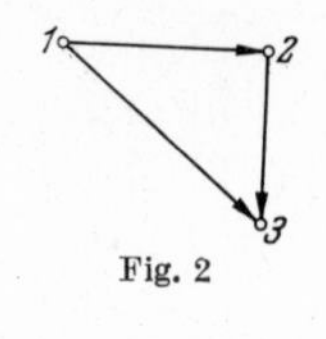

Fig. 2

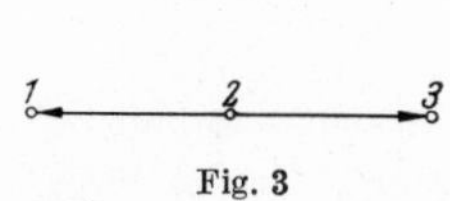

Fig. 3

Obviously, every strongly connected graph is connected. It is also at hand to see that a graph containing a Hamiltonian circuit is strongly connected.

The graph of Fig. 1 is hence strongly connected. The graph of Fig. 2 is connected but not strongly connected, for there is no path from 3 to 1, while the graph of Fig. 3 is not connected, for there is no path between 1 and 3.

Besides Γ, two other mappings, $\check{\Gamma}$ and $\hat{\Gamma}$, from N to the set of all subsets of N, are defined, describing the connectedness of the graph.

Definition 7. The *algebraic transitive closure**, or, simply termed, *algebraic closure,* of a graph $G = (N, \Gamma)$ is the graph $\check{G} = (N, \check{\Gamma})$, where $\check{\Gamma}$ is the mapping which associates to each vertex $i \in N$ the set of all vertices $j \in N$ so that there is a path from i to j.

We see that $i \in \check{\Gamma} i$ if and only if there is a circuit (possibly a loop) passing through i.

Definition 8. A graph $G = (N, \Gamma)$ is called *strongly complete* (respectively *complete*), if for every two vertices $i, j \in N$, with $i \neq j$, relation $j \in \Gamma i$ holds (respectively, if at least one of the two relations $j \in \Gamma i$ and $i \in \Gamma j$ holds).

Remark 3. We see that a strongly complete graph is strongly connected. Further, G is strongly connected if and only if $\check{G}$ is strongly complete.

Definition 9. The *transitive closure* of a graph $G = (N, \Gamma)$ is the graph $\hat{G} = (N, \hat{\Gamma})$, where $\hat{\Gamma}$ is the mapping which associates to each vertex $i \in N$ the set consisting of the vertex i itself and of all vertices j so that there is a path from i to j.

We see that

$$\check{\Gamma} i = \Gamma i \cup \Gamma^2 i \cup \cdots \cup \Gamma^n i \tag{2}$$

and that

$$\hat{\Gamma} i = \{i\} \cup \check{\Gamma} i, \tag{3}$$

where n is the number of elements of N and $\Gamma^p i$ is defined by induction as follows: 1) $\Gamma^1 i = \Gamma i$; 2) $\Gamma^p i$ is the set of all $k \in N$ with the property that there exists a $j \in \Gamma^{p-1} i$ so that $k \in \Gamma j$.

* Called "μ-closure" by B. Roy [1].

Definition 10. Given two vertices, i and j, of a graph, the *distance* $d(i, j)$ from i to j, is defined as follows:

1) $d(i, i) = 0$;

2) if $i \neq j$ and there is no path from i to j, then $d(i, j) = \infty$;

3) in the other cases, $d(i, j)$ is the smallest length of a path linking i to j.

Notice that a path from i to j so that $d(i, j)$ equals its length, must be an elementary one. Notice also that $d(i, j) \geqq 0$, $d(i, j) = 0$ if and only if $i = j$, $d(i, k) \leqq d(i, j) + d(j, k)$, but it may happen that $d(i, j) \neq d(j, i)$. We also see that G is strongly connected if and only if all $d(i, j)$ are finite, and it is connected if and only if for each $i, j \in N$, at least one of $d(i, j)$ and $d(j, i)$ is finite.

Definition 11. The *elongation* $e(i)$ of a vertex $i \in N$ is defined by

$$e(i) = \max_{j \in N} d(i, j); \tag{4}$$

the *radius* ϱ of the graph is

$$\varrho = \min_{i \in N} e(i), \tag{5}$$

while the *diameter* δ of the graph is

$$\delta = \max_{i, j \in N} d(i, j); \tag{6}$$

a point i_c such that

$$e(i_c) = \varrho \tag{7}$$

is called a *center* of the graph, provided that ϱ is finite, while a vertex i_p satisfying

$$e(i_p) = \delta \tag{8}$$

is said to be a *peripheral point.*

The graph is strongly connected if and only if δ is finite.

§ 2. Unitary Incidence Matrices

A graph $G = (N, \Gamma)$ is completely characterized by its *incidence matrix* $A = (a_{ij})$, having n rows and n columns, and defined by

$$a_{ij} = \begin{cases} 1, & \text{if } j \in \Gamma i, \\ 0, & \text{if } j \notin \Gamma i. \end{cases} \tag{9}$$

Incidence matrices are usually termed, in the American literature, „adjacency matrices". See, for instance, F. Harary, R. Z. Norman and D. Cartwright [1], where in Chap. 5 a thorough study of these matrices is given.

For instance, the incidence matrices of the graphs of Fig. 1, 2, 3, are

(10)
$$A_1 = \begin{pmatrix} 0 & 1 & 0 & 0 \\ 0 & 0 & 1 & 1 \\ 0 & 0 & 0 & 1 \\ 1 & 0 & 0 & 1 \end{pmatrix},$$

(11)
$$A_2 = \begin{pmatrix} 0 & 1 & 1 \\ 0 & 0 & 1 \\ 0 & 0 & 0 \end{pmatrix},$$

(12)
$$A_3 = \begin{pmatrix} 0 & 0 & 0 \\ 1 & 0 & 1 \\ 0 & 0 & 0 \end{pmatrix},$$

respectively.

The *unitary incidence matrix* $A' = (a'_{ij})$ of the graph G is obtained from its incidence matrix (9) by putting all diagonal elements equal to 1, that is

(13)
$$a'_{ij} = \begin{cases} 1, & \text{if } i = j, \\ a_{ij}, & \text{if } i \neq j, \end{cases}$$

or, equivalently,

(14)
$$a'_{ij} = a_{ij} \cup \delta_{ij},$$

where δ_{ij} is the Kronecker symbol (i.e. $\delta_{ii} = 1$, and $\delta_{ij} = 0$ for $i \neq j$); we may also write

(15)
$$A' = A \cup E,$$

where E is the unit matrix (I.57).

The existence of loops is shadowed in A' by the obligatory 1's on the diagonal (while $a_{ii} = 1$ in A if and only if there is a loop in i). Nevertheless, the unitary incidence matrices are useful in handling different path problems, as we shall point out in the sequel, following K. MAGHOUT [6].

In the sequel, the powers p of a matrix have the meaning

(I.58)
$$A^0 = E, \quad A^p = A^{p-1} \times A \quad (p = 1, 2, \ldots),$$

where $\times$ is the product

(I.44)
$$(a_{ij}) \times (b_{jk}) = \left(\bigcup_{j=1}^{n} a_{ij}\, b_{jk}\right).$$

Theorem 1 (Main). *Let a'^p_{ij} denote the elements of the p-th power $(A')^p$ of the unitary incidence matrix A'. Assume $i \neq j$; then $a'^p_{ij} = 1$ if and only if there is a path of length at most p from i to j.*

Proof. We shall use the development

(16)
$$a'^p_{ij} = \bigcup_{i_1, \ldots, i_{p-1}} a'_{ii_1}\, a'_{i_1 i_2} \ldots a'_{i_{p-1} j}$$

[see (I.59)], which shows that $a'^{p}_{ij} = 1$ if and only if there exist $p-1$ indices $i_1, i_2, \ldots, i_{p-1}$ such that $a'_{i i_1} = a'_{i_1 i_2} = \cdots = a'_{i_{p-1} j} = 1$. If every two successive indices i_k, i_{k+1} are distinct, then we have the path $(i, i_1, i_2, \ldots, i_{p-1}, j)$ of length p; in the opposite case, the length is $< p$.

Corollary 1. Assume $i \neq j$; relation $a'^{p}_{ij} = 1$ holds if and only if there is an elementary path of length at most p from i to j (see Remark 1).

Corollary 2. Relations $a'^{p}_{ij} = a'^{p}_{ji} = 1$ $(i \neq j)$ hold if and only if there is a circuit of length $2p$ passing through i and j (or, equivalently, if and only if there is an elementary circuit of length $\leqq 2p$ passing through i and j).

Corollary 3. Assume $i \neq j$; $d(i, j)$ is the first integer p for which $a'^{p}_{ij} = 1$, if any; $d(i, j) = \infty$ if $a'^{p}_{ij} = 0$ for all p.

Remark 4. Let $\hat{G}$ be the transitive closure of a graph G. The incidence matrix $\hat{A}$ of $\hat{G}$ coincides with the unitary incidence matrix $(\hat{A})'$ of $\hat{G}$.

Theorem 2. *The (unitary) incidence matrix $\hat{A}$ of the transitive closure $\hat{G} = (N, \hat{\Gamma})$ of the graph $G = (N, \Gamma)$ is given by*

$$\hat{A} = (A')^e \tag{17}$$

where e is the characteristic exponent of the matrix A' associated to G.

We recall (see I., § 3) that e is the first integer for which

$$(A')^e = (A')^{e+1} = \cdots = (A')^n = \cdots \tag{18}$$

Proof. It follows from Definition 9 that for every i, j with $i \neq j$, $\hat{a}_{ij} = 1$ if and only if there is a path from i to j. Hence

$$\hat{a}_{ij} = a'_{ij} \cup a'^{2}_{ij} \cup a'^{3}_{ij} \cup \cdots \tag{19}$$

by Theorem 1; now Theorem I.3 shows that the right-hand side of (19) equals $a'_{ij} \cup a'^{2}_{ij} \cup \cdots \cup a'^{e}_{ij}$.

Computational remarks. 1) The above theorem provides us with a quickly converging procedure for finding the transitive closure $\hat{G}$ of a given graph G. Indeed, instead of computing all the powers A', $(A')^2$, $(A')^3, \ldots$, it suffices to compute

$$(A')^2, (A')^4, \ldots, (A')^{2^m} \tag{20}$$

where m is the first integer so that $2^m \geqq n-1$ (for, $e \leqq n-1$ cf. Theorem I.3).

2) If we are interested in finding the distance $d(i, j)$, $(i \neq j)$ via Corollary 3, then two situations may occur. In case $\hat{a}'_{ij} = 0$, then $d(i, j) = \infty$. In case $\hat{a}_{ij} = 1$, we seek in (20) the first r for which $a'^{2^r}_{ij} = 0$ and $a'^{2^{r+1}}_{ij} = 1$, showing that $2^r < d(i, j) \leqq 2^{r+1}$; now, it obviously suffices to check the interval by successively halving it, until we reach the integer p with $a'^{p-1}_{ij} = 0$, $a'^{p}_{ij} = 1$ (of course, $p \leqq e$).

3) The same halving procedure enables us to find the characteristic exponent e.

Example 1. The graph of Fig. 4 has the unitary incidence matrix

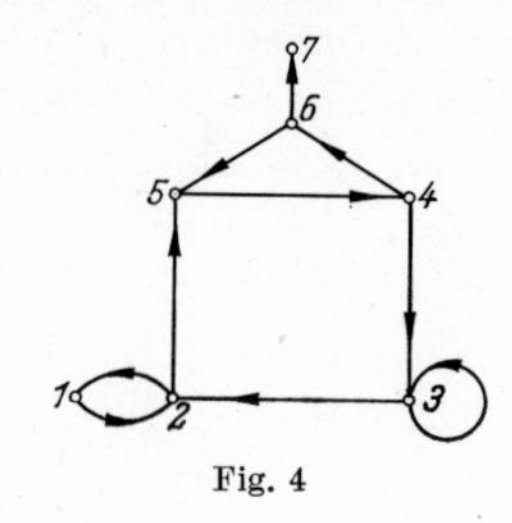

Fig. 4

$$A' = \begin{bmatrix} 1 & 1 & 0 & 0 & 0 & 0 & 0 \\ 1 & 1 & 0 & 0 & 1 & 0 & 0 \\ 0 & 1 & 1 & 0 & 0 & 0 & 0 \\ 0 & 0 & 1 & 1 & 0 & 1 & 0 \\ 0 & 0 & 0 & 1 & 1 & 0 & 0 \\ 0 & 0 & 0 & 0 & 1 & 1 & 1 \\ 0 & 0 & 0 & 0 & 0 & 0 & 1 \end{bmatrix} \tag{21}$$

hence

$$(A')^2 = \begin{bmatrix} 1 & 1 & 0 & 0 & 1 & 0 & 0 \\ 1 & 1 & 0 & 1 & 1 & 0 & 0 \\ 1 & 1 & 1 & 0 & 1 & 0 & 0 \\ 0 & 1 & 1 & 1 & 1 & 1 & 1 \\ 0 & 0 & 1 & 1 & 1 & 1 & 0 \\ 0 & 0 & 0 & 1 & 1 & 1 & 1 \\ 0 & 0 & 0 & 0 & 0 & 0 & 1 \end{bmatrix} \tag{22}$$

$$(A')^4 = \begin{bmatrix} 1 & 1 & 1 & 1 & 1 & 1 & 0 \\ 1 & 1 & 1 & 1 & 1 & 1 & 1 \\ 1 & 1 & 1 & 1 & 1 & 1 & 0 \\ 1 & 1 & 1 & 1 & 1 & 1 & 1 \\ 1 & 1 & 1 & 1 & 1 & 1 & 1 \\ 0 & 1 & 1 & 1 & 1 & 1 & 1 \\ 0 & 0 & 0 & 0 & 0 & 0 & 1 \end{bmatrix} \tag{23}$$

$$\hat{A} = (A')^8 = \begin{bmatrix} 1 & 1 & 1 & 1 & 1 & 1 & 1 \\ 1 & 1 & 1 & 1 & 1 & 1 & 1 \\ 1 & 1 & 1 & 1 & 1 & 1 & 1 \\ 1 & 1 & 1 & 1 & 1 & 1 & 1 \\ 1 & 1 & 1 & 1 & 1 & 1 & 1 \\ 1 & 1 & 1 & 1 & 1 & 1 & 1 \\ 0 & 0 & 0 & 0 & 0 & 0 & 1 \end{bmatrix} \tag{24}$$

Now we must compute $(A')^6$; but $(A')^6 = (A')^{n-1} = \hat{A}$, which is readily computed in (24). Further, we compute $(A')^5 = (A')^4 \times A'$ and find $(A')^5 = \hat{A}$. Since $(A')^4 \neq \hat{A}$, it follows that $e = 5$.

If we want to find $d(5,7)$, we notice that $a_{57}'^2 = 0$, while $a_{57}'^4 = 1$. Computing $a_{57}'^3 = \bigcup_{i=1}^{7} a_{5i}'^2 a_{i7}'$, we find $a_{57}'^3 = 1$ so that $d(5,7) = 3$.

Computational remark. Since $a_{ij}'^p = 1$ implies $a_{ij}'^{p+q} = 1$ for all q, the determination of $(A')^{2p}$ reduces to the computation of those $a_{ij}'^{2p}$ for which $a_{ij}'^p = 0$.

From Theorem 2 and Corollary 3 we deduce the following

Corollary 4. The elongation $e(i)$ of a vertex i is equal to the first integer p for which *all* $a_{ij}'^p = 1$ $(j = 1, \ldots, n)$, if any, and is ∞ if there is no p with this property. In other words, if $\prod_j \hat{a}_{ij} = 0$, then $e(i) = \infty$; in the opposite case, $e(i)$ is the first p for which $\prod_j a_{ij}'^p = 1$.

It follows from Definition 11 that the elongation $e(i)$ of a center is finite, hence, taking into account Corollary 4, we deduce

Corollary 5. The graph G has a center if and only if

$$\bigcup_{i=1}^{n} \prod_{j=1}^{n} \hat{a}_{ij} = 1; \tag{25}$$

the first r for which

$$\bigcup_{i=1}^{n} \prod_{j=1}^{n} a_{ij}'^r = 1 \tag{26}$$

equals the radius ϱ of the graph; the centers are all those vertices i for which

$$\prod_{j=1}^{n} a_{ij}'^\varrho = 1. \tag{27}$$

We recall that a graph is strongly connected if and only if the diameter δ (Definition 11) is finite, i.e. if and only if all $e(i)$ are finite. Therefore:

Corollary 6. The graph G is strongly connected if and only if $\hat{A} = I$ [see (I.48)]. In this case $\delta = e$ [see (17)] and the peripheral points are those i for which $\prod_{j=1}^{n} a_{ij}'^{e-1} = 0$. If $\hat{A} \neq I$, then G is not strongly connected, $\delta = \infty$, and the peripheral points are those i for which $\prod_{j=1}^{n} \hat{a}_{ij} = 0$.

Example 2. Let us consider the graph of Fig. 4.

1) The elongation $e(1) = 5$ because $a_{17}'^4 = 0$, while $\prod_{j=1}^{7} a_{1j}'^5 = 1$. The elongation $e(7) = \infty$ because $\prod_{j=1}^{7} \hat{a}_{7j} = 0$.

2) The graph has a center because relation (25) is fulfilled.

3) The radius ϱ of the graph is 3, because relation (26) is not fulfilled for $r = 2$, while $\prod_{j=1}^{7} a_{4j}'^{3} = 1$, showing also that the vertex 4 is a center of the graph. Since $\prod_{j=1}^{7} a_{ij}'^{3} = 0$ for any $i \neq 4$, it follows that 4 is the single center of G.

4) Since $\hat{A} \neq I$, the graph is not strongly connected, $\delta = \infty$, an the single peripheral point is 7, as it may be seen from (24).

The study of circuits in graphs is facilitated by operations with the transpose A^T of a given matrix A.

Theorem 3. Let

$$B'^{(p)} = \bigcup_{r=1}^{p-1} (A')^{p-r} \cdot [(A')^r]^T = (b_{ij}'^{p}) \tag{28}$$

[see (I.41) and (I.42)]; then $b_{ij}'^{p} = 1$ $(i \neq j)$ *if and only if there is a circuit of length at most* p *passing through the vertices* i *and* j.

Proof. Follows from relation

$$b_{ij}'^{p} = \bigcup_{r=1}^{p-1} a_{ij}'^{p-r}\, a_{ji}'^{r}. \tag{29}$$

This theorem is a generalization of Corollary 2. Similarly, one proves

Theorem 4. *There is a circuit passing through the nodes* i *and* j $(i \neq j)$ *if and only if* $\hat{a}_{ij}\,\hat{a}_{ji} = 1$; *the graph* G *is circuit-free if and only if*

$$\hat{B} = \hat{A}(\hat{A})^T = E. \tag{30}$$

Definition 12. The *strongly connected component* of a vertex i of the graph $G = (N, \Gamma)$ is the graph $G_i = (N_i, \Gamma_i)$, where N_i is the set consisting of i and of those vertices $j \in N$ for which there exists a circuit in G passing through i and j, while $\Gamma_i j$ is defined for $j \in N_i$ as being $\Gamma_i j = \Gamma j \cap N_i$, i.e. the set of those $k \in \Gamma i$ which belong also to N_i.

We notice that a strongly connected component of a vertex i of G is a strongly connected graph. We also notice that G decomposes into pairwise disjoint strongly connected components

Corollary 7. The nodes of the strongly connected component of the vertex i are those points j for which the element $\hat{b}_{ij}$ of the matrix $\hat{B}$ equals 1.

Example 3. For the graph of Fig. 4 we see from (30) and (24) that

$$\hat{B} = \begin{bmatrix} 1 & 1 & 1 & 1 & 1 & 1 & 0 \\ 1 & 1 & 1 & 1 & 1 & 1 & 0 \\ 1 & 1 & 1 & 1 & 1 & 1 & 0 \\ 1 & 1 & 1 & 1 & 1 & 1 & 0 \\ 1 & 1 & 1 & 1 & 1 & 1 & 0 \\ 1 & 1 & 1 & 1 & 1 & 1 & 0 \\ 0 & 0 & 0 & 0 & 0 & 0 & 1 \end{bmatrix} \tag{31}$$

so that the graph contains circuits passing through any couple of vertices (i, j), with $i = 1, \ldots, 6$; $j = 1, \ldots, 6$; $i \neq j$; and there is no circuit passing through the vertex 7. The strongly connected component of any vertex i with $1 \leqq i \leqq 6$ is determined by the set $\{1, 2, \ldots, 6\}$, while the strongly connected component of 7 is the degenerate graph reduced to the single vertex 7.

§ 3. Incidence Matrices

In this section we shall describe several applications of the incidence matrices $A = (a_{ij})$, which, unlike the unitary incidence matrices, have a diagonal element $a_{ii} = 1$ if and only if there is a loop in vertex i.

This feature of the incidence matrices allows the argument of Theorem 1 to be strengthened as follows: if $i_1, i_2, \ldots, i_{p-1}$ are $p - 1$ indices such that $a_{i i_1} = a_{i_1 i_2} = \cdots = a_{i_{p-1} j} = 1$, then there exists a path having *exactly* the length p and joining the vertex i to the vertex j. Hence:

Theorem 5. *The element*

$$a_{ij}^{(p)} = \sum_{i_1, \ldots, i_{p-1}} a_{i i_1} a_{i_1 i_2} \ldots a_{i_{p-1} j} \tag{32}$$

of the p-th ordinary power of the matrix A is equal to the number of paths of length p joining i to j.

Now, we shall establish analogues of the results presented in § 2.

Taking into account the above remark and reasoning as in Theorem 1, we obtain:

Theorem 6 (Main). *Let a_{ij}^p denote the elements of the p-th power A^p of the incidence matrix A. Then $a_{ij}^p = 1$ if and only if there is a path of length (exactly!) p from i to j.*

Corollary 8. Relation $a_{ij}^p = 1$ is equivalent to the existence of an elementary path of length at most p from i to j. In particular, relation $a_{ii}^p = 1$ is equivalent to the existence of an (elementary) circuit of length (at most) p passing through i.

Corollary 9. Relations $a_{ij}^p = a_{ji}^p = 1$ hold if and only if there is an (elementary) circuit of length (at most) $2p$ passing through i and j.

Corollary 10. Assume $i \neq j$; $d(i, j)$ is equal to the first integer p for which $a_{ij}^p = 1$, if any; $d(i, j) = \infty$ if $a_{ij}^p = 0$ for all p.

Corollary 11.

$$\hat{A} = E \cup A \cup A^2 \cup \cdots \cup A^n. \tag{33}$$

Corollary 12. The elongation $e(i)$ of a vertex i is equal to the first integer p for which

$$\prod_{\substack{j=1 \\ j \neq i}}^{n} \bigcup_{r=1}^{p} a_{ij}^r = 1, \tag{34}$$

if any, and is ∞ if there is no p with this property.

Corollary 13. The radius ϱ of a graph equals the first integer r for which

$$\bigcup_{i=1}^{n} \prod_{\substack{j=1\\ j\neq i}}^{n} \bigcup_{s=1}^{r} a_{ij}^{s} = 1, \tag{35}$$

while the centers of the graph are all those vertices i for which

$$\prod_{\substack{j=1\\ j\neq i}}^{n} \bigcup_{s=1}^{\varrho} a_{ij}^{s} = 1. \tag{36}$$

Corollary 14. Let

$$B^{(p)} = \bigcup_{r=1}^{p-1} A^{p-r} \cdot (A^{r})^{T} = (b_{ij}^{p}); \tag{37}$$

then $b_{ij}^{p} = 1$ if and only if there is a circuit of length (exactly!) p passing through the vertices i and j.

§ 4. The Algebraic Closure of a Graph

We have defined in § 1 the algebraic closure $\check{G} = (N, \check{\Gamma})$ of a graph $G = (N, \Gamma)$ by

$$\check{\Gamma}\, i = \{j \in N \mid \text{there is a path from } i \text{ to } j\} \tag{38}$$

and we have noticed that

$$\check{\Gamma}\, i = \Gamma i \cup \Gamma^{2} i \cup \cdots \cup \Gamma^{n} i \tag{2}$$

and that

$$\hat{\Gamma}\, i = \{i\} \cup \check{\Gamma}\, i. \tag{3}$$

If A, $\check{A}$ and $\hat{A}$ are the incidence matrices of the graphs $G = (N, \Gamma)$, $\check{G} = (N, \check{\Gamma})$ and $\hat{G} = (N, \hat{\Gamma})$ respectively, then the above relations may be translated by

$$\check{A} = A \cup A^{2} \cup \cdots \cup A^{n} \tag{39}$$

and by

$$\hat{A} = E \cup \check{A}, \tag{40}$$

respectively.

The reader may easily verify that the mapping $A \to \check{A}$ is a *closure operator* in the sense of K. Kuratowski, i.e.

$$\check{0} = 0, \tag{41}$$

$$A \leqq \check{A}, \tag{42}$$

$$\check{\check{A}} = \check{A}, \tag{43}$$

$$\text{if } A \leqq B, \text{ then } \check{A} \leqq \check{B} \tag{44}$$

(these properties are also obvious from the graph-theoretical interpretation of A).

It is clear from the discussions of the preceding sections, that the graph $\breve{G}$ offers us a description of the path properties of G. A thorough investigation of the properties of $\breve{A}$ was done by B. Roy [1] and some of his results will be presented below.

The basic idea here is that of computing $\breve{A}$ with the aid of n simple transformations T_i $(i = 1, \ldots, n)$, defined as follows. If

$$A = (a_{hk}) \tag{45}$$

is a Boolean $n \times n$ matrix, then

$$T_i A = (\alpha_{hk}), \tag{46}$$

where

$$\alpha_{hk} = a_{hk} \cup a_{hi} a_{ik}. \tag{47}$$

In other words, in order to compute the row h of $T_i A$, we shall proceed as follows. If $a_{hi} = 0$, then the row h of $T_i A$ coincides with row h of A: $\alpha_{hk} = a_{hk}$ $(k = 1, \ldots, n)$. If $a_{hi} = 1$, then the row h of $T_i A$ is the disjunction (logical sum) of the rows i and h of A: $\alpha_{hk} = a_{hk} \cup a_{ik}$ $(i = 1, \ldots, n)$.

Lemma 1. *The transformations T_i are idempotent and commutative:*

$$T_i T_i A = T_i A, \tag{48}$$

$$T_i T_j A = T_j T_i A. \tag{49}$$

Proof. In view of formula (47), $\alpha_{hi} = a_{hi}$, $\alpha_{ik} = a_{ik}$, so that $\alpha_{hi} \alpha_{ik} = a_{hi} a_{ik} \leqq \alpha_{hk}$. Hence the elements β_{hk} of $T_i T_i A$ are

$$\beta_{hk} = \alpha_{hk} \cup \alpha_{hi} \alpha_{ik} = \alpha_{hk}.$$

Now, let γ_{hk} denote the elements of $T_j T_i A$. Then

$$\gamma_{hk} = \alpha_{hk} \cup \alpha_{hj} \alpha_{jk} = a_{hk} \cup a_{hi} a_{ik} \cup (a_{hj} \cup a_{hi} a_{ij})(a_{jk} \cup a_{ji} a_{ik})$$
$$= a_{hk} \cup a_{hi} a_{ik} \cup a_{hj} a_{jk} \cup a_{hj} a_{ji} a_{ik} \cup a_{hi} a_{ij} a_{jk}$$

(since the term $a_{hi} a_{ij} a_{ji} a_{ik}$ was absorbed by $a_{hi} a_{ik}$). The last expression of γ_{hk} is symmetric in i and j, proving thus (49).

Lemma 2. *For every A and i,*

$$\breve{A} = \widebreve{T_i A}. \tag{50}$$

Proof. Formula (47) shows that $A \leqq T_i A \leqq A \cup A^2$, whence, in view of (39), we have

$$A \leqq T_i A \leqq \breve{A}. \tag{51}$$

Applying (44) and (43), we get (50).

Theorem 7. *Relation*

$$A = \breve{A} \tag{52}$$

is equivalent to

$$A = T_i A \quad \text{for all} \quad i = 1, \ldots, n. \tag{53}$$

Proof. If $A = \check{A}$, we see from (51) that $A = T_i A$. Conversely, if $A \neq \check{A}$, in the graph G having A as its incidence matrix there exists a path, say $\{1, 2, \ldots, p\}$, with $p \notin \Gamma 1$. Let r be the first vertex of this path for which $r \notin \Gamma 1$. Then $r - 1 \in \Gamma 1$, $r \in \Gamma(r-1)$ and the element α_{1r} of the matrix $T_{r-1} A$ is $\alpha_{1r} = a_{1r} \cup a_{1,r-1}\, a_{r-1,r} = 1 \neq 0 = a_{1r}$, so that $T_{r-1} A \neq A$.

Remark 5. An important class of graphs is that of the so-called *transitive graphs*. These have the property that $j \in \Gamma i$ and $k \in \Gamma j$ imply $k \in \Gamma i$. In other words, whenever there exists a path joining two vertices i and j, there is also an arc from i to j. There is a one-to-one correspondence between the class of all (finite) partially ordered sets and the class of all (finite) transitive, loop-free graphs.

Theorem 7 offers a quick procedure for checking the transitivity of a graph (because this property means nothing but $A = \check{A}$).

Theorem 8. *For every $n \times n$ Boolean matrix A,*

$$\check{A} = T_n T_{n-1} \ldots T_1 A. \tag{54}$$

Proof. Let A_n denote the right-hand side of (54). According to Lemma 1, $T_i A_n = A_n$ for $i = 1, \ldots, n$; therefore

$$\check{A}_n = A_n, \tag{55}$$

by Theorem 7. The repeated application of Lemma 2 shows that

$$\check{A} = \check{A}_n; \tag{56}$$

from (55) and (56) we get (54).

As we shall see below, Theorem 8 provides a simple procedure for computing $\check{A}$, whose importance was already shown. This procedure furnishes us the matrix $\hat{A}$ as well, via relation (40).

Computational remarks. 1) In order to avoid useless recopying of matrices when computing $\check{A}$, it seems convenient to leave empty those cells in which A and its forthcoming transforms $T_1 A, T_2 T_1 A, \ldots$ contain zeros, and to perform the successive transformations on the *same* tableau.

2) It may be useful to improve the procedure by replacing each 1 of the matrix A through $n + 1$ and each 1 introduced in stage i through i. On the one hand, this trick avoids certain errors or allows their simple detection; on the other hand, it permits to determine, for each couple of vertices (i, j) corresponding to a non-empty cell of A, a path joining i to j in G. Namely, if the cell (i, j) of the tableau contains the number k, then there is a path from i to j passing through k. Inspecting now the cells (i, k) and (k, j), we determine further points of the path.

Example 4. If A is the matrix of the graph in Fig. 4, then we have $\check{A}$ (computed as at point 2) equal to:

$$\check{A} = \begin{bmatrix} 2 & 8 & 5 & 5 & 2 & 5 & 6 \\ 8 & 1 & 5 & 5 & 8 & 5 & 6 \\ 2 & 8 & 8 & 5 & 2 & 5 & 6 \\ 3 & 3 & 8 & 5 & 3 & 8 & 6 \\ 4 & 4 & 4 & 8 & 4 & 4 & 6 \\ 5 & 5 & 5 & 5 & 8 & 5 & 8 \\ 0 & 0 & 0 & 0 & 0 & 0 & 0 \end{bmatrix}$$

If we are interested in an elementary path joining 3 to 6, for instance, then we see that the cell $(3, 6)$ contains the number 5. Hence there is an elementary path from 3 to 5 and one from 5 to 6. The cell $(3, 5)$ contains the number 2 hence there are elementary paths from 3 to 2 and from 2 to 5. But both $(3, 2)$ and $(2, 5)$ contain the number 8, hence 3 and 2 as well as 2 and 5 are joined by arcs. The cell $(5, 6)$ contains 4, the cells $(5, 4)$ and $(4, 6)$ contain 8; hence $5, 4$ and $4, 6$ are joined by arcs. We have hence determined the path $(3, 2, 5, 4, 6)$.

§ 5. Decompositions of Graphs

We recall that the strongly connected components of a graph $G = (N, \Gamma)$ (Definition 12), determine a partition of the set N of vertices, i.e. a decomposition of N into pairwise disjoint subsets $N_1, N_2, \ldots, N_m$. With these sets, taken as vertices, we construct a new graph, according to the following:

Definition 13. *The reduced graph* of a graph $G = (N, \Gamma)$ is a graph $G^R = (N^R, \Gamma^R)$, where $N^R = \{N_1, N_2, \ldots, N_m\}$ is the set of all strongly connected components of G, and $N_j \in \Gamma^R N_i$ if and only if there exist two vertices, $h \in N_i$ and $k \in N_j$ such that $k \in \Gamma h$.

While in a strongly connected graph, for every two vertices there is a circuit passing through them, in a reduced graph there are no circuits (and, a fortiori, no loops). In fact, each graph may be regarded as a reduced graph whose vertices are strongly connected graphs.

B. Roy [1] has shown the importance of the reduced graph G^R in solving many problems concerning the graph G. The determination of the strongly connected components of a graph and of its reduced graph, is nothing but the determination of the two following matrices:

1) the *decomposition matrix* $D = (d_{ij})$ having n rows and m columns, where

$$d_{ij} = \begin{cases} 1, & \text{if} \quad i \in N_j, \\ 0, & \text{if} \quad i \notin N_j; \end{cases} \tag{57}$$

2) the incidence matrix A^R of G^R.

To do this job, let us introduce the following two transformations, $A \to A^V$ and $A \to A^W$, defined for square (say, $n \times n$) Boolean matrices A.

1) The matrix A^V is obtained in $m (m \leqq n)$ steps, each one of them consisting in the deletion of certain columns. More precisely, in step p $(p = 1, 2, \ldots, m)$ we consider the first column h which remained undeleted after the previous steps, and delete all those columns j for which $j > h$ and $a_{jh} = 1$.

2) A^W is the matrix which has only zeros on the diagonal, and coincides with A in the other cells.

Theorem 9 (Roy). *With the above notations, we have*

$$D = [\hat{A}(\hat{A}^T)]^V \tag{58}$$

and

$$A^R = (D^T \times A \times D)^W. \tag{59}$$

Proof. We have already noticed in Corollary 7 that the elements $\hat{b}_{ij}$ of the matrix $\hat{B} = \hat{A}(\hat{A}^T)$ have the property that $\hat{b}_{ij} = 1$ if and only if i and j are in the same strongly connected component.

Therefore, the meaning of the transformation $\hat{B} \to (\hat{B})^V$ is, in fact, to preserve one vertex from each strongly connected component and to delete the columns corresponding to the other vertices. Hence the vertices corresponding to the columns h which remained in $(\hat{B})^V$ form a complete system of representatives of the strongly connected components. Moreover, if $\hat{b}_{jh}$ is an element of $(\hat{B})^V$, then $\hat{b}_{jh} = 1$ means that j belongs to the class represented by h; hence $(\hat{B})^V = D$.

An element e_{ij} of the right-hand side of (59) is given by

$$e_{ij} = \begin{cases} \bigcup_{h=1}^{m} \bigcup_{k=1}^{n} d_{hi}\, a_{hk}\, d_{kj}, & \text{if } i \neq j, \\ 0, & \text{if } i = j, \end{cases} \tag{60}$$

where d_{ij} are the elements of D. Therefore, $e_{ij} = 1$ if and only if $i \neq j$ and there exist indices h and k so that $d_{hi} = a_{hk} = d_{kj} = 1$, expressing the fact that:

1) h and i belong to the same strongly connected component N_i;
2) j and k belong to the same strongly connected component N_j;
3) $k \in \Gamma h$.

According to Definition 13, this means that $N_j \in \Gamma^R N_i$, thus completing the proof.

Example 5. Let us consider the graph in Fig. 5

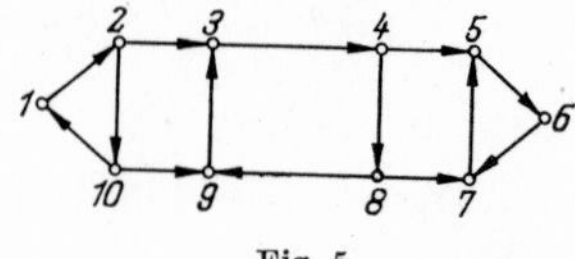

Fig. 5

It is easy to see that for this graph,

$$\hat{A} = \begin{bmatrix} 1 & 1 & 1 & 1 & 1 & 1 & 1 & 1 & 1 & 1 \\ 1 & 1 & 1 & 1 & 1 & 1 & 1 & 1 & 1 & 1 \\ 0 & 0 & 1 & 1 & 1 & 1 & 1 & 1 & 1 & 0 \\ 0 & 0 & 1 & 1 & 1 & 1 & 1 & 1 & 1 & 0 \\ 0 & 0 & 0 & 0 & 1 & 1 & 1 & 0 & 0 & 0 \\ 0 & 0 & 0 & 0 & 1 & 1 & 1 & 0 & 0 & 0 \\ 0 & 0 & 0 & 0 & 1 & 1 & 1 & 0 & 0 & 0 \\ 0 & 0 & 1 & 1 & 1 & 1 & 1 & 1 & 1 & 0 \\ 0 & 0 & 1 & 1 & 1 & 1 & 1 & 1 & 1 & 0 \\ 1 & 1 & 1 & 1 & 1 & 1 & 1 & 1 & 1 & 1 \end{bmatrix}$$

Hence

$$\hat{A}(\hat{A}^T) = \begin{bmatrix} 1 & 1 & 0 & 0 & 0 & 0 & 0 & 0 & 0 & 1 \\ 1 & 1 & 0 & 0 & 0 & 0 & 0 & 0 & 0 & 1 \\ 0 & 0 & 1 & 1 & 0 & 0 & 0 & 1 & 1 & 0 \\ 0 & 0 & 1 & 1 & 0 & 0 & 0 & 1 & 1 & 0 \\ 0 & 0 & 0 & 0 & 1 & 1 & 1 & 0 & 0 & 0 \\ 0 & 0 & 0 & 0 & 1 & 1 & 1 & 0 & 0 & 0 \\ 0 & 0 & 0 & 0 & 1 & 1 & 1 & 0 & 0 & 0 \\ 0 & 0 & 1 & 1 & 0 & 0 & 0 & 1 & 1 & 0 \\ 0 & 0 & 1 & 1 & 0 & 0 & 0 & 1 & 1 & 0 \\ 1 & 1 & 0 & 0 & 0 & 0 & 0 & 0 & 0 & 1 \end{bmatrix}$$

We deduce

$$D = \begin{bmatrix} 1 & 0 & 0 \\ 1 & 0 & 0 \\ 0 & 1 & 0 \\ 0 & 1 & 0 \\ 0 & 0 & 1 \\ 0 & 0 & 1 \\ 0 & 0 & 1 \\ 0 & 1 & 0 \\ 0 & 1 & 0 \\ 1 & 0 & 0 \end{bmatrix}$$

Hence, the three strongly connected components of this graph are respectively containing the vertices $\{1, 2, 10\}$, $\{3, 4, 8, 9\}$, $\{5, 6, 7\}$.

Finally, we deduce

$$A^R = \begin{pmatrix} 0 & 1 & 1 \\ 0 & 0 & 1 \\ 0 & 0 & 0 \end{pmatrix}$$

§ 6. Latin Multiplication

The methods studied in §§ 2, 3 enable us to decide, for every two vertices i and j of a graph, whether there is a path from i to j, or not; it was noticed that the procedure of § 4 offers also a means for actually determining one of those paths.

A. KAUFMANN and Y. MALGRANGE [1] (see also A. KAUFMANN [1]) have developed an efficient procedure (the "Latin multiplication") for an irredundant and successive listing of all elementary paths of lengths $1, 2, \ldots, n-1$; it is the aim of this section to present their method.

Definition 14. The *reduced* of an elementary path $\pi = (i_1, i_2, \ldots, i_p)$ is the elementary path $\pi' = (i_2, \ldots, i_p)$. The *Latin product*, $\pi_1 * \pi_2$, of two elementary (possibly empty: $\varnothing$) paths $\pi_1 = (i_1, i_2, \ldots, i_p)$ and $\pi_2 = (j_1, j_2, \ldots, j_q)$ is defined as follows:

1) if $\pi_1 \neq \varnothing$, $\pi_2 \neq \varnothing$, $i_p = j_1$ and $\pi_1 \pi_2' = (i_1, i_2, \ldots, i_p, j_2, \ldots, j_q)$ is an elementary path, then

$$\pi_1 * \pi_2 = \pi_1 \pi_2'; \tag{61.1}$$

2) if $\pi_1 \neq \varnothing$, $\pi_2 \neq \varnothing$, but $i_p \neq j_1$ or $\pi_1 \pi_2'$ is not an elementary path (or both), then

$$\pi_1 * \pi_2 = \varnothing; \tag{61.2}$$

3) if $\pi_1 = \varnothing$, or $\pi_2 = \varnothing$ (or both), then

$$\pi_1 * \pi_2 = \varnothing. \tag{61.3}$$

For example, consider again the graph of Fig. 4. Let us consider here the elementary paths $\pi_1 = (2, 5, 4)$, $\pi_2 = (4, 6, 7)$, $\pi_3 = (4, 3, 2, 1)$, $\pi_4 = (1, 2, 5)$. According to the above definition:

$$\pi_1 * \pi_2 = \pi_1 \pi_2' = (2, 5, 4, 6, 7),$$

$$\pi_1 * \pi_3 = \pi_1 * \pi_4 = \pi_2 * \pi_3 = \pi_2 * \pi_4 = \pi_3 * \pi_4 = \varnothing.$$

Definition 15. Let $P_{ij}^q = \{\pi_1, \pi_2, \ldots, \pi_u\}$ be the set of all elementary paths of length q from i to j, and $P_{jk}^r = \{\omega_1, \omega_2, \ldots, \omega_v\}$, the set of all elementary paths of length r from j to k. The *Latin product* of the sets P_{ij}^q, P_{jk}^r is the set

$$P_{ij}^q * P_{jk}^r = \{\pi_1 * \omega_1, \pi_1 * \omega_2, \ldots, \pi_u * \omega_v\} \tag{62}$$

of all possible Latin products of the paths in P_{ij}^q with the paths in P_{jk}^r.

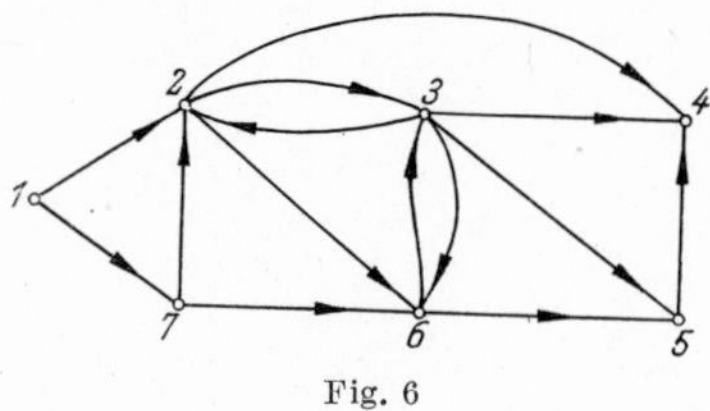

Fig. 6

For example, in the case of the graph of Fig. 6, the set of elementary paths of length 3 joining 1 to 3, is

$$P_{13}^3 = \{(1, 2, 6, 3), (1, 7, 6, 3), (1, 7, 2, 3)\},$$

the set of elementary paths of length 2 joining 3 to 4, is

$$P_{34}^2 = \{(3,2,4), (3,5,4)\}$$

and hence, the set

$$P_{13}^3 * P_{34}^2 =$$

$$= \{(1,2,6,3,5,4), (1,7,6,3,2,4), (1,7,6,3,5,4), (1,7,2,3,5,4)\},$$

because $(1,2,6,3) * (3,2,4) = (1,7,2,3) * (3,2,4) = \varnothing$.

Remark 6. The set P_{ik}^s of all elementary paths of length s from i to k may be expressed as a set-theoretical join:

$$P_{ik}^s = \bigcup_{j=1}^{m} P_{ij}^q * P_{jk}^r, \tag{63}$$

where q and r are any two natural numbers such that $q + r = s$.

Definition 16. The *Latin matrix* L^s of order s of a graph G with n vertices is the $n \times n$ matrix whose elements are the sets P_{ij}^s. The *Latin product* of two Latin matrices L^s and L^t is the Latin matrix $L^s * L^t$, the elements of which are

$$P_{ik}^{s+t} = \bigcup_{j=1}^{n} P_{ij}^s * P_{jk}^t. \tag{64}$$

Remark 7. We have

$$L^s * L^t = L^{s+t}. \tag{65}$$

Remark 8. Since there is no elementary path of length $> n - 1$,

$$L^n = L^{n+1} = \cdots = (\varnothing). \tag{66}$$

Remark 9. The halving procedure described in § 2 (Theorem 2, computational remarks 2 and 3) is also suitable for computing the Latin matrices of different orders.

Example 6.
For the graph in Fig. 6, we have

$$L^1 = \begin{bmatrix} 0 & 12 & 0 & 0 & 0 & 0 & 17 \\ 0 & 0 & 23 & 24 & 0 & 26 & 0 \\ 0 & 32 & 0 & 34 & 35 & 36 & 0 \\ 0 & 0 & 0 & 0 & 0 & 0 & 0 \\ 0 & 0 & 0 & 54 & 0 & 0 & 0 \\ 0 & 0 & 63 & 0 & 65 & 0 & 0 \\ 0 & 72 & 0 & 0 & 0 & 76 & 0 \end{bmatrix}$$

$L^2 = L^1 * L^1 =$

0	172	123	124	0	126 176	0
0	0	263	234	235 265	236	0
0	0	0	324 354	365	326	0
0	0	0	0	0	0	0
0	0	0	0	0	0	0
0	632	0	634 654	635	0	0
0	0	723 763	724	765	726	0

$L^3 = L^1 * L^2 =$

0	0	1263 1723 1763	1234 1724	1235 1265 1765	1236 1726	0
0	0	0	2354 2634 2654	2365 2635	0	0
0	0	0	3654	3265	0	0
0	0	0	0	0	0	0
0	0	0	0	0	0	0
0	0	0	6324 6354	0	0	0
0	7632	7263	7234 7634 7654	7235 7265 7635	7236	0

$L^4 = L^2 * L^2 =$

0	17632	17263	12354 12634 12654 17234 17634 17654	12365 12635 17235 17265 17635	17236	0
0	0	0	23654 26354	0	0	0
0	0	0	32654	0	0	0
0	0	0	0	0	0	0
0	0	0	0	0	0	0
0	0	0	0	0	0	0
0	0	0	72354 72634 72654 76324 76354	72365 72635	0	0

$L^5 = L^2 * L^3 =$

0	0	0	126354 123654 172354 172634 172654 176324 176354	172365 172635	0	0
0	0	0	0	0	0	0
0	0	0	0	0	0	0
0	0	0	0	0	0	0
0	0	0	0	0	0	0
0	0	0	0	0	0	0
0	0	0	723654 726354	0	0	0

$L^6 = L^3 * L^3 =$

0	0	0	1723654 1726354	0	0	0
0	0	0	0	0	0	0
0	0	0	0	0	0	0
0	0	0	0	0	0	0
0	0	0	0	0	0	0
0	0	0	0	0	0	0
0	0	0	0	0	0	0

Obviously, $L^7 = L^8 = \cdots = (0)$, because no elementary path in the graph of Fig. 6 may have a length greater than 6.

Computational remark. The determination of $L^s = L^r * L^{s-r}$ becomes easier if the matrix L^r has many zeros. Thus, in the above example, it is convenient to compute L^6 as $L^5 * L^1$.

Elementary circuits. A slight modification of the above discussion also permits the detection of all circuits.

Namely, if in Definition 14 we split rule 2) into rules 2′) and 2″) (see below), then the diagonal elements of the matrix L^s will be of the form $(h_1, h_2, \ldots, h_s)$, with the property that $(h_s, h_1, h_2, \ldots, h_s)$ is an elementary circuit; each elementary circuit of length s is obtained in this way.

2′) If $\pi_1 \neq \varnothing$, $\pi_2 \neq \varnothing$, $i_1 = j_q$, $i_p = j_1$, and $\pi_1' \pi_2'$ is an elementary path, then

(61.2′) $$\pi_1 * \pi_2 = \pi_1' \pi_2' = (i_2, i_3, \ldots, i_p, j_2, j_3, \ldots, j_q);$$

2″) if $\pi_1 \neq \varnothing$, $\pi_2 \neq \varnothing$ and we are in one of the following situations:

— $i_p \neq j_1$

or

$$- \; i_1 \neq j_q \text{ and } \pi_1 \pi_2' \text{ is not elementary,}$$

or

$$- \; \pi_1' \pi_2' \text{ is not elementary,}$$

then:

(61.2″)
$$\pi_1 * \pi_2 = \varnothing .$$

Example 7. If we apply this new rule of computing the elementary paths and circuits, we shall obtain the matrices L'^1, L'^2, L'^3, L'^4, L'^5, L'^6 which can differ from $L_1, L_2, \ldots, L_6$, respectively only on the diagonal elements. More precisely, the diagonal elements of these matrices, written here, for the sake of simplicity, horizontally, are:

L'^1:	0	0	0	0	0	0	0
L'^2:	0	32	23 63	0	0	36	0
L'^3:	0	632	263	0	0	326	0
L'^4:	0	0	0	0	0	0	0
L'^5:	0	0	0	0	0	0	0
L'^6:	0	0	0	0	0	0	0

showing that the elementary circuits of this graph are: $(2, 3, 2)$, $(3, 2, 3)$, $(3, 6, 3)$, $(6, 3, 6)$, $(2, 6, 3, 2)$, $(3, 2, 6, 3)$, $(6, 3, 2, 6)$, which, in fact, reduce to the following three circuits:

(67)
$$(2, 3, 2), \quad (3, 6, 3), \quad (2, 6, 3, 2).$$

§ 7. Hamiltonian Paths and Circuits

We recall that a *Hamiltonian path* was defined as a path which runs through all the vertices of a given graph, so that no vertex is touched twice (the latter condition means that the path is *elementary*, cf. Definition 3). If this path is a circuit, then it is called a *Hamiltonian circuit* (Definition 4).

The problem of finding the Hamiltonian circuits of a graph occurs in many economic contexts (see e.g. A. Kaufmann [1]). A frequently quoted example is the *Traveling-Salesman Problem.*

A traveling salesman has to visit n places so that the total distance covered by him should be minimal. If we represent the n places as the n vertices of a graph, the arcs of which are the direct routes connecting the places, then the problem reduces to that of finding a Hamiltonian circuit of minimal length*.

There are various methods for finding all the Hamiltonian paths and circuits of a graph. However, as in the other sections, our attention will be confined to the Boolean procedures for solving this problem.

Since any Hamiltonian path is characterized by the property of being an elementary path of length $n - 1$, the elements P_{ij}^{n-1} of the Latin matrix L^{n-1} of order $n - 1$ (see Definition 16) give us a complete listing of the Hamiltonian paths. More precisely,

$$(68) \qquad \bigcup_{i=1}^{n} \bigcup_{j=1}^{n} P_{ij}^{n-1}$$

is the set of all Hamiltonian paths.

The finding of the Hamiltonian circuits reduces to the finding of the diagonal elements P_{ii}^{n-1} of L^{n-1}, computed by using formulas (61.2′) and (61.2″). As $(i_1, i_2, \ldots, i_{n-1}, i_n, i_1) = (i_2, \ldots, i_{n-1}, i_n, i_1, i_2) = \cdots$, it follows that all the sets P_{ii}^{n-1} are equal; therefore, it suffices to calculate one of them.

Example 8.

As we have seen previously (Example 7) for the graph in Fig. 6, we have

$$L'^{(n-1)} = L'^{6} = \begin{bmatrix} 0 & 0 & 0 & \begin{matrix}1723654\\1726354\end{matrix} & 0 & 0 & 0 \\ 0 & 0 & 0 & 0 & 0 & 0 & 0 \\ 0 & 0 & 0 & 0 & 0 & 0 & 0 \\ 0 & 0 & 0 & 0 & 0 & 0 & 0 \\ 0 & 0 & 0 & 0 & 0 & 0 & 0 \\ 0 & 0 & 0 & 0 & 0 & 0 & 0 \\ 0 & 0 & 0 & 0 & 0 & 0 & 0 \end{bmatrix}$$

showing us that this graph has no Hamiltonian circuits (all the diagonal elements being equal to zero) and that it has two Hamiltonian paths:

$$(1, 7, 2, 3, 6, 5, 4)$$

and

$$(1, 7, 2, 6, 3, 5, 4).$$

Another approach to the problem of determining the Hamiltonian circuits and the Hamiltonian paths of a graph was suggested by

* For details the reader is referred to G. B. Dantzig, D. R. Fulkerson and S. M. Johnson [1], M. M. Flood [1], A. W. Tucker [1] and to any standard textbook on mathematical programming. R. M. Karp [1] has shown that this problem also arises in digital computer programming.

R. Fortet [1] and developed by P. Camion [1], and S. Rudeanu [6]; it consists in reducing the original problem to that of solving certain *Boolean* or *diophantine equations*.

A path (and, more generally, a collection of arcs) of a given graph may be characterized by a system of bivalent variables in the following way: to each arc (i, j) of the graph a variable x_{ij} is associated, defined by

$$(69) \qquad x_{ij} = \begin{cases} 1, & \text{if the arc } (i, j) \text{ belongs to the path,} \\ 0, & \text{if the arc } (i, j) \text{ does not belong to the path;} \end{cases}$$

if $j \notin \Gamma i$, x_{ij} is not defined.

Now consider an arbitrary but fixed Hamiltonian circuit H; for each vertex k of the graph there exists exactly one arc of H of the form (i, k) and exactly one arc of H of the form (k, j). Therefore, the variables x_{ij} associated to H must satisfy

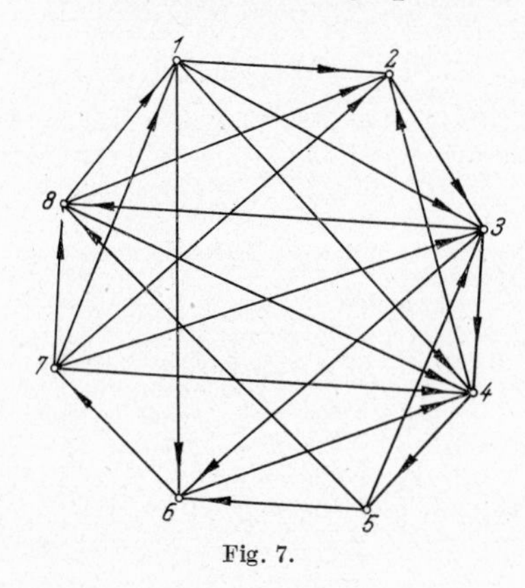

Fig. 7.

$$(70) \qquad \sum_j x_{ij} = 1 \qquad (i = 1, \ldots, n)$$

and

$$(71) \qquad \sum_i x_{ij} = 1 \qquad (j = 1, \ldots, n),$$

where $\sum_j$ means arithmetic sum over those variables $x_{ij_1}, x_{ij_2}, x_{ij_3}, \ldots,$ which are defined.

Equations (70) and (71) are necessary, but not sufficient conditions in order that the collection of arcs determined by the x_{ij}'s be a Hamiltonian circuit. Thus, in the graph of Fig. 7, equations (70) and (71) become

$$(72.1) \qquad x_{12} + x_{13} + x_{14} + x_{16} = 1,$$

$$(72.2) \qquad x_{23} = 1,$$

$$(72.3) \qquad x_{34} + x_{36} + x_{38} = 1,$$

$$(72.4) \qquad x_{42} + x_{45} = 1,$$

$$(72.5) \qquad x_{53} + x_{56} + x_{58} = 1,$$

$$(72.6) \qquad x_{64} + x_{67} = 1,$$

$$(72.7) \qquad x_{71} + x_{72} + x_{73} + x_{74} + x_{78} = 1,$$

$$(72.8) \qquad x_{81} + x_{82} + x_{84} = 1,$$

and

(73.1) $$x_{71} + x_{81} = 1,$$

(73.2) $$x_{12} + x_{42} + x_{72} + x_{82} = 1,$$

(73.3) $$x_{13} + x_{23} + x_{53} + x_{73} = 1,$$

(73.4) $$x_{14} + x_{34} + x_{64} + x_{74} + x_{84} = 1,$$

(73.5) $$x_{45} = 1,$$

(73.6) $$x_{16} + x_{36} + x_{56} = 1,$$

(73.7) $$x_{67} = 1,$$

(73.8) $$x_{38} + x_{58} + x_{78} = 1,$$

respectively. If we take $x_{12} = x_{23} = x_{38} = x_{81} = 1$, $x_{45} = x_{56} = x_{67} = x_{74} = 1$ and the other $x_{ij} = 0$, then equations (72) and (73) are satisfied, but the corresponding collection of arcs is not a Hamiltonian circuit: it is the set-theoretical join $\{(1,2),(2,3),(3,8),(8,1)\} \cup \cup \{(4,5),(5,6),(6,7),(7,4)\}$ of two disjoint elementary circuits, covering all the vertices of the graph.

Therefore, after solving the equations (70) and (71), it is necessary to check all the solutions, in order to eliminate those corresponding to *fragmentary Hamiltonian circuits* (as in the above example).

It is convenient to find various necessary conditions for the Hamiltonian circuits, in order to add new equations to the system (70) and (71), so that many (if possible, all) fragmentary Hamiltonian circuits be eliminated from the solutions of the augmented system.

To do this job, we can try to find elementary but non-Hamiltonian circuits and express the condition that the searched circuit should not include them. It seems convenient to look for such elementary circuits starting from the "obligatory" arcs corresponding to equations of the form $x_{ij} = 1$ in the system (70), (71).

To illustrate these ideas let us resume the above example. As we must take $x_{23} = 1$ (72.2), let us seek some elementary circuits containing the arc $(2,3)$; we easily find $(2,3,8,2)$, $(2,3,8,1,2)$ and $(2,3,6,7,2)$. Similarly, we easily find the circuit $(4,5,8,4)$ containing the obligatory arc $(4,5)$ [see (73.5)] and the circuit $(6,7,2,3,6)$, containing the arc $(6,7)$ [see (73.7)]. Now we express the conditions that the searched Hamiltonian circuit include none of the above circuits; we obtain

(74.1) $$x_{38}\, x_{82} = 0$$

[since, by (72.2), we know that $x_{23} = 1$] and similarly

$$x_{38}\, x_{81}\, x_{12} = 0, \tag{74.2}$$

$$x_{36}\, x_{72} = 0, \tag{74.3}$$

$$x_{58}\, x_{84} = 0, \tag{74.4}$$

$$x_{72}\, x_{36} = 0, \tag{74.5}$$

respectively. The augmented system (72), (73) and (74) coincides with the system (IV.91), solved in Example IV.15, whose solutions were

$x_{12} = x_{23} = x_{38} = x_{84} = x_{45} = x_{56} = x_{67} = x_{71} = 1$, the other $x_{ij} = 0$;

$x_{16} = x_{67} = x_{72} = x_{23} = x_{34} = x_{45} = x_{58} = x_{81} = 1$, the other $x_{ij} = 0$;

$x_{14} = x_{45} = x_{56} = x_{67} = x_{72} = x_{23} = x_{38} = x_{81} = 1$, the other $x_{ij} = 0$;

$x_{14} = x_{45} = x_{58} = x_{82} = x_{23} = x_{36} = x_{67} = x_{71} = 1$, the other $x_{ij} = 0$;

$x_{12} = x_{23} = x_{34} = x_{45} = x_{56} = x_{67} = x_{78} = x_{81} = 1$, the other $x_{ij} = 0$;

$x_{12} = x_{23} = x_{36} = x_{67} = x_{74} = x_{45} = x_{58} = x_{81} = 1$, the other $x_{ij} = 0$;

corresponding to the Hamiltonian circuits $(1, 2, 3, 8, 4, 5, 6, 7, 1)$, $(1, 6, 7, 2, 3, 4, 5, 8, 1)$, $(1, 4, 5, 6, 7, 2, 3, 8, 1)$, $(1, 4, 5, 8, 2, 3, 6, 7, 1)$, $(1, 2, 3, 4, 5, 6, 7, 8, 1)$ and $(1, 2, 3, 6, 7, 4, 5, 8, 1)$. As a matter of fact, the conditions (72), (73) and (74) were necessary and sufficient, since we have obtained no fragmentary Hamiltonian circuit.

P. Camion [1], following the same idea of solving Boolean and diophantine equations which represent necessary conditions, has solved some problems of the following type: find all Hamiltonian paths satisfying certain supplementary conditions.

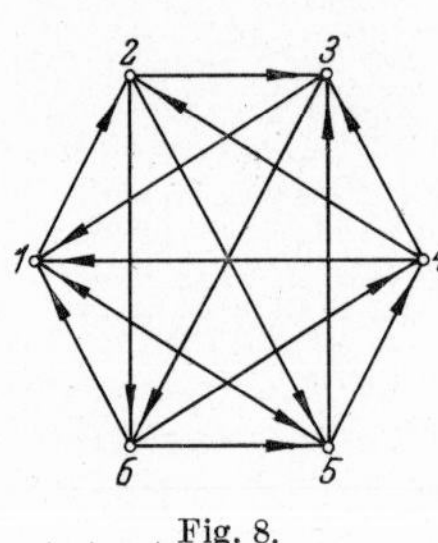

Fig. 8.

For instance, in the graph of Fig. 8, let us determine all Hamiltonian paths either from 1 to 2, or from 1 to 3. The following conditions are necessary (but not sufficient):

1) each vertex $\neq 1$ is the terminal point of a single arc;
2) vertex 1 is not the terminal point of any arc;
3) each vertex $\neq 2, 3$ is the starting point of at least an arc.

These conditions are translated respectively by the following equations:

$$\sum_i x_{ij} = 1 \qquad (j = 2, 3, 4, 5, 6), \tag{75}$$

$$\bigcup_i x_{i1} = 0 \qquad (i = 2, 3, 4, 5, 6), \tag{76}$$

$$\prod_{k \in K} \bigcup_j x_{kj} = 1, \quad \text{where} \quad K = \{1, 4, 5, 6\}, \tag{77}$$

or, explicitly,

(75.2) $x_{12} + x_{42} = 1,$

(75.3) $x_{23} + x_{43} + x_{53} = 1,$

(75.4) $x_{54} + x_{64} = 1,$

(75.5) $x_{15} + x_{25} + x_{65} = 1,$

(75.6) $x_{26} + x_{36} = 1,$

(76′) $x_{31} = x_{41} = x_{51} = x_{61} = 0,$

(77′) $(x_{12} \cup x_{15})(x_{41} \cup x_{42} \cup x_{43})(x_{51} \cup x_{53} \cup x_{54})(x_{61} \cup x_{64} \cup x_{65}) = 1.$

Equations (75) are easily solved:

(75′.2) $x_{12} = z_1, \quad x_{42} = \bar{z}_1;$

(75′.3) $x_{23} = z_2 z_3, \quad x_{43} = z_2 \bar{z}_3, \quad x_{53} = \bar{z}_2;$

(75′.4) $x_{54} = z_4, \quad x_{64} = \bar{z}_4;$

(75′.5) $x_{15} = z_5 z_3, \quad x_{25} = z_5 \bar{z}_6, \quad x_{65} = \bar{z}_5;$

(75′.6) $x_{26} = z_7, \quad x_{36} = \bar{z}_7;$

where $z_1, \ldots, z_6$ are new variables. We introduce the values (75′) and (76′) into (77′) and obtain

(77″) $(z_1 \cup z_5 z_6)(\bar{z}_1 \cup z_2 \bar{z}_3)(\bar{z}_2 \cup z_4)(\bar{z}_4 \cup \bar{z}_5) = 1.$

Taking $z_1 = 1$, equation (77″) becomes $z_2 \bar{z}_3 (\bar{z}_2 \cup z_4)(\bar{z}_4 \cup \bar{z}_5) = 1$, hence $z_2 = 1$, $z_3 = 0$ and (77″) becomes $z_4(\bar{z}_4 \cup \bar{z}_5) = 1$, hence $z_4 = 1$ and $z_5 = 0$. Taking $z_1 = 0$, equation (77″) becomes $z_5 z_6 (\bar{z}_2 \cup z_4) \cdot$ $\cdot (\bar{z}_4 \cup \bar{z}_5) = 1$, hence $z_5 = z_6 = 1$ and (77″) becomes $(\bar{z}_2 \cup z_4)\bar{z}_4 = 1$, hence $z_4 = 0$ and $z_2 = 0$.

We have thus obtained the following 8 solutions:

(78.1) $z_1 = z_2 = z_4 = 1, \quad z_3 = z_5 = 0, \quad z_6$ and z_7 arbitrary;

(78.2) $z_5 = z_6 = 1, \quad z_1 = z_2 = z_4 = 0, \quad z_3$ and z_7 arbitrary;

in terms of the x_{ij} we have 4 solutions:

(78′.1) $x_{12} = x_{43} = x_{54} = x_{65} = 1, \; x_{26} = u, \; x_{36} = \bar{u}$, the other $x_{ij} = 0$;

(78′.2) $x_{42} = x_{53} = x_{64} = x_{15} = 1, \; x_{26} = u, \; x_{36} = \bar{u}$, the other $x_{ij} = 0$;

where u is an arbitrary parameter.

Taking $u = 1$ in (78′.1), we obtain the Hamiltonian path $(1, 2, 6, 5, 4, 3)$, while for $u = 0$ we obtain the set of arcs $\{(1, 2), (4, 3), (3, 6), (6, 5), (5, 4)\}$, to be rejected. Taking $u = 1$ in (78′.2) we obtain the set of arcs $\{(4, 2), (2, 6), (6, 4), (1, 5), (5, 3)\}$ to be rejected, while for $u = 0$ we obtain the Hamiltonian path $(1, 5, 3, 6, 4, 2)$.

Other examples, making use of certain graph-theoretical properties which facilitate the computations, can be found in P. Camion [1].

§ 8. Using Free Boolean Algebras

In § I.4 we have studied the general concept of Boolean algebra (not the two-element one) and we have noticed (Example I.2.3) that the set F_n of all Boolean functions of n variables $x_1, \ldots, x_n$ becomes a Boolean algebra, if we define, for each $f \in F_n$, $g \in F_n$, the elements $f \cup g \in F_n$, $f g \in F_n$ and $\bar{f}$ by the following rules:

$$(I.79) \qquad (f \cup g)(x_1, \ldots, x_n) = f(x_1, \ldots, x_n) \cup g(x_1, \ldots, x_n),$$

$$(I.80) \qquad (f g)(x_1, \ldots, x_n) = f(x_1, \ldots, x_n)\, g(x_1, \ldots, x_n),$$

$$(I.81) \qquad \bar{f}(x_1, \ldots, x_n) = \overline{f(x_1, \ldots, x_n)},$$

the elements $0 \in F_n$ and $1 \in F_n$ being the functions identically equal to 0 and to 1, respectively.

The Boolean algebra F_n has the following special properties:

1) it contains n elements, $x_1, \ldots, x_n$, which satisfy no identity other than the postulates of Definition I.19;

2) it contains each "Boolean expression" formed with $x_1, \bar{x}_1, \ldots, x_n, \bar{x}_n$ (see Definition I.3);

3) it contains no other elements.

It can be proved that, for each n, F_n is essentially the single Boolean algebra satisfying the above three properties (more precisely, each Boolean-algebra satisfying 1), 2) and 3) is isomorphic with F_n).

F_n is called the *free Boolean algebra* with n generators $x_1, \ldots, x_n$. Of course, besides the above mentioned peculiar properties, F_n satisfies the properties common to all Boolean algebras. Thus, (I.4)—(I.11) and (I.13)—(I.24) are also valid in F_n; moreover, the matrices with elements in F_n satisfy all the properties studied in I. § 3 (see Theorem I.5). In particular, the following theorems will play a fundamental role in this section:

Theorem I.3. If $A = (a_{ij})$ is a square Boolean matrix of order n (here, with $a_{ij} \in F_n$) such that $a_{ii} = 1$ $(i = 1, \ldots, n)$, then

$$(I.60) \qquad E \leqq A \leqq A^2 \leqq \cdots \leqq A^{n-1} = A^n = A^{n+1} = \cdots$$

where $E = (\delta_{ij})$, $A = (a_{ij}) \leqq B = (b_{ij})$ means $a_{ij} \leqq b_{ij}$ for all i, j, and $A^0 = E$, $A^p = A^{p-1} \times A$ $(p = 1, 2, \ldots)$, the multiplication $\times$ being defined by

$$(I.44) \qquad (a_{ij}) \times (b_{ij}) = \left(\bigcup_{k=1}^{n} a_{ik}\, b_{kj} \right).$$

Theorem I.4. With the same hypotheses and notations as before, we have

$$(I.70) \qquad \operatorname{adj} A = A^e,$$

where e is the characteristic exponent of A (i.e., the first integer ($\leqq n-1$) for which $A^e = A^{e+1}$), and $\operatorname{adj} A = (\alpha_{ij})$ is defined by $\alpha_{ij} = |A_{ji}|$ (Definitions I.17 and I.18).

A. G. LUNC [1, 2] has proved the above theorems and used them in order to solve the *analysis* and the *synthesis* problems for multi-terminal switching circuits. These problems can be stated as follows:

1) describe the working of a given switching circuit,

and

2) design a switching circuit with a prescribed working program. It turns out that part of LUNC's research can be re-formulated as a method for finding the elementary paths of a graph, and this will be done in the sequel*. In particular, we shall point out that the main results in §§ 2, 3 and 4 are, in fact, consequences of the more general results of this section.

We hope that this section will contribute to the obliteration of the prejudice that only the two-element Boolean algebra is used in applied mathematics.

Definition 17. Let: $G = (N, \Gamma)$ be a graph, $1, 2, \ldots, n$ its vertices, and let F_{n^2} be the free Boolean algebra with n^2 generators z_{ij} $(i, j = 1, \ldots, n)$. The *free incidence matrix* of G is the Boolean $n \times n$ matrix $A = (\alpha_{ij})$ defined by

$$\alpha_{ij} = \begin{cases} z_{ij}, & \text{if } j \in \Gamma i, \\ 0, & \text{if } j \notin \Gamma i, \end{cases} \tag{79}$$

while the *free unitary incidence matrix* of G is the Boolean $n \times n$ matrix $A' = (\alpha'_{ij})$ defined by

$$\alpha'_{ij} = \begin{cases} 1, & \text{if } i = j, \\ z_{ij}, & \text{if } i \neq j \text{ and } j \in \Gamma i, \\ 0, & \text{if } i \neq j \text{ and } j \notin \Gamma i. \end{cases} \tag{80}$$

The following result is the analogue of Theorem 6 in § 3:

Theorem 10. *Let*

$$\alpha_{ij}^{p} = \bigcup_{i_1, \ldots, i_{p-1}} \alpha_{i i_1} \alpha_{i_1 i_2} \cdots \alpha_{i_{p-1} j} \tag{81}$$

be the elements of the p-th power A^p of the free incidence matrix A of a graph G. Then: 1) There is a one-to-one correspondence between the paths $(i, i_1, i_2, \ldots, i_{p-1}, j)$ of length (exactly!) p from vertex i to vertex j in G and those terms $\alpha_{i i_1} \alpha_{i_1 i_2} \cdots \alpha_{i_{p-1} j}$ of the right-hand side of (81) that are not equal to 0, i.e. that are of the form

$$z_{i i_1} z_{i_1 i_2} \cdots z_{i_{p-1} j}; \tag{82}$$

* For further research concerning this problem, the reader is refered to GR. C. MOISIL [6] and to P. CONSTANTINESCU [1—5], S. RUDEANU [1], C. DRĂGUȘIN [1], S. MITITELU [1], I. TOMESCU [1—4].

2) *The terms* (*82*) *which remain in the right-hand side of* (*81*) *after the performance of all possible absorptions, correspond to set-theoretically minimal paths.* (*A* path *P* from *i* to *j* will be called set-theoretically minimal, if there is no other path *P'* from *i* to *j* such that the set of arcs of *P'* be included in the set of arcs of *P*).

Proof. 1) is obvious; 2) since the elements z_{hk} are generators of a free Boolean algebra, the absorption

$$z_{i i_1} z_{i_1 i_2} \ldots z_{i_{p-1} j} \cup z_{i h_1} z_{h_1 h_2} \ldots z_{h_{p-1} j} = z_{i h_1} z_{h_1 h_2} \ldots z_{h_{p-1} j} \tag{83}$$

holds if and only if the set $\{z_{i h_1}, z_{h_1 h_2}, \ldots, z_{h_{p-1} j}\}$ is included in the set $\{z_{i i_1}, z_{i_1 i_2}, \ldots, z_{i_{p-1} j}\}$ (the indices occurring in (82) are not necessarily pairwise disjoint).

Remark. The method of Latin multiplication is, in fact based on Theorem 10.1).

Thus, in contrast with Theorem 6 which just points out the *existence* of paths, the above Theorem 10 *actually determines* them. As a matter of fact, we can immediately deduce Theorem 6 from Theorem 10, using the homomorphism f of F_{n^2} onto B_2 which is uniquely defined by

$$f(z_{hk}) = \begin{cases} 1, & \text{if } z_{hk} \text{ appears in } A, \\ 0, & \text{in the opposite case.} \end{cases} \tag{84}$$

We establish now the analogue of (Corollary 1) of Theorem 1 in § 2.

Theorem 11. *Let*

$$\alpha'^{p}_{ij} = \bigcup_{i_1, \ldots, i_{p-1}} \alpha'_{i i_1} \alpha'_{i_1 i_2} \ldots \alpha'_{i_{p-1} j} \tag{85}$$

be the elements of the p-th power A'^p *of the free unitary incidence matrix* A' *of a graph* G. *For* $i \neq j$ *there is a one-to-one correspondence between the elements*

$$z_{i i_1} z_{i_1 i_2} \ldots z_{i_{p-r-1} j} \tag{86}$$

($r \geqq 0$) *which do appear in the right-hand side of* (85) *after the performance of all possible absorptions, and the elementary paths of length at most* p *joining vertex* i *to vertex* j *in* G.

Proof. Assuming the theorem true for $p-1$, we shall prove it for p (since for $p=1$ the theorem is obvious).

Let us write relation (85) in the form

$$\alpha'^{p}_{ij} = \alpha'^{p-1}_{ij} \cup \bigcup_{\substack{k=1 \\ k \neq j}}^{n} \alpha'^{p-1}_{ik} \alpha'_{kj}, \tag{87}$$

where n is the number of vertices, and assume that in α'^{p-1}_{ik} $(k \neq j)$ all possible absorptions have been performed. Then the terms of α'^{p-1}_{ij}

correspond to all elementary paths of length at most $p-1$ from i to j, while each term of

(88) $$\bigcup_{\substack{k=1\\k\neq j}}^{n} \alpha_{ik}'^{p-1}\,\alpha_{kj}'$$

corresponds either to an elementary path of length at most p from i to j, or to a path of the form

(89) $$(i, i_1, \ldots, i_{s-1}, i_s = j, i_{s+1}, \ldots, i_t, k, j),$$

where $(i, i_1, \ldots, i_s, \ldots, i_t, k)$ is an elementary path of length at most $p-1$. It follows that each term

(90) $$z_{i\,i_1}\, z_{i_1 i_2} \ldots z_{i_t k}\, z_{kj},$$

which appears in (88) and corresponds to a non-elementary path (89), is absorbed in formula (87) by the term $z_{i\,i_1}\, z_{i_1 i_2} \ldots z_{i_{s-1} j}$ which appears in the development of $\alpha_{ij}'^{p-1}$. Thus, after the performance of all possible absorptions, the terms appearing in the right-hand side of (87) correspond to elementary paths of length at most p from i to j; moreover, the terms which remain in (88) (if any) correspond to elementary paths of length exactly p.

Conversely, let $(i, i_1, \ldots, i_{p-r-1}, j)$ be an elementary path of lenght $p-r$. If $r>0$, then the corresponding term (86) does appear in $\alpha_{ij}'^{p-1}$, by the inductive hypothesis. If $r=0$, it is easy to see that (86) does appear in (88).

The proof is complete.

Now, using the same homomorphism defined by (84) as before, we see that Theorem 1 in § 2 and its Corollary 1 are consequences of the above Theorem 11.

Using Theorems 10 and 11, which are extensions of Theorems 6 and 1, we may easily generalize the other theorems of §§ 2, 3 and 4.

Thus, Theorem 2 is also valid for free unitary incidence matrices. The analogue of Theorem 3 states that, after performing all possible absorptions, the development of the element $b_{ij}'^{p}$ of the matrix (28) (where A' is the free unitary incidence matrix) gives us all the elementary circuits of length at most p passing through the vertices i and j. The analogue of Theorem 4 is similarly formulated. Theorems 7 and 8 are also valid for free incidence matrices.

As concerns the computational aspects, Theorem I.3 and the halving procedure described in § 2, applied to the free unitary incidence matrix A', provides us with a suitable method for finding all elementary paths of length at most p from i to j, for each given integer p and for each i and j in G.

Alternatively, we can determine all elementary paths from i to j, using Theorem I.4, applied also to A'.

The above two methods determine all the paths of the graph. A third procedure, which may be called the *"vertex-elimination method"* and will be described below, seems to be very suitable in the case we want to determine all elementary paths joining a single given couple of vertices. This method is based on the following

Theorem 12. *Let* $A' = (\alpha'_{hk})$ *be a Boolean* $n \times n$ *matrix with* $\alpha'_{hh} = 1$ $(h = 1, \ldots, n)$ *and characteristic exponent* e, *and* $B' = (\beta_{ij})$ $(i, j = 1, \ldots, n-1)$ *the Boolean matrix defined by*

$$\beta'_{ij} = \alpha'_{ij} \cup \alpha'_{in}\,\alpha'_{nj} \qquad (i, j = 1, \ldots, n-1); \tag{91}$$

then

$$\alpha'^{e}_{ij} = \beta'^{e}_{ij} \qquad \text{for all} \quad i, j = 1, \ldots, n-1. \tag{92}$$

Proof. Relation (91) implies

$$\alpha'_{ij} \leqq \beta'_{ij} \leqq \alpha'^{2}_{ij};$$

further, it follows by induction that

$$\alpha'^{p}_{ij} \leqq \beta'^{p}_{ij} \leqq \alpha'^{2p}_{ij} \quad \text{for all } p, \tag{93}$$

because

$$\alpha'^{p}_{ij} = \bigcup_{k=1}^{n-1} \alpha'^{p-1}_{ik}\,\alpha'_{kj} \cup \alpha'^{p-1}_{in}\,\alpha'_{nj} \leqq$$

$$\leqq \bigcup_{k=1}^{n-1} \beta'^{p-1}_{ik}\,\beta'_{kj} \cup \bigcup_{i_1, \ldots, i_{p-2}} \alpha'_{ii_1}\,\alpha'_{i_1 i_2} \cdots \alpha'_{i_{p-2} n}\,\alpha'_{nj} \leqq$$

$$\leqq \beta'^{p}_{ij} \cup \bigcup_{i_1, \ldots, i_{p-2}} \beta'_{ii_1}\,\beta'_{i_1 i_2} \cdots \beta'_{i_{p-3} i_{p-2}}\,\beta'_{i_{p-2} j} = \beta'^{p}_{ij} \cup \beta'^{p-1}_{ij}$$

$$= \beta'^{p}_{ij} = \bigcup_{k=1}^{n-1} \beta'^{p-1}_{ik}\,\beta'_{kj} \leqq \bigcup_{k=1}^{n-1} \alpha'^{2p-2}_{ik}\,\alpha'^{2}_{kj} \leqq \alpha'^{2p}_{ij}.$$

Since $\alpha'^{2e}_{ij} = \alpha'^{e}_{ij}$, the inequalities (93) imply (92), completing the proof.

We illustrate the three procedures suggested in this section for determining the paths, by the following

Example 9. Let us determine the elementary paths

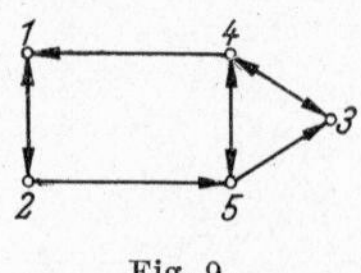

Fig. 9.

in the graph of Fig. 9. We have

$$A' = \begin{pmatrix} 1 & z_{12} & 0 & 0 & 0 \\ z_{21} & 1 & 0 & 0 & z_{25} \\ 0 & 0 & 1 & z_{34} & 0 \\ z_{41} & 0 & z_{43} & 1 & z_{45} \\ 0 & 0 & z_{53} & z_{54} & 1 \end{pmatrix}$$

$$A'^2 = \begin{pmatrix} 1 & z_{12} & 0 & 0 & z_{12}\, z_{25} \\ z_{21} & 1 & z_{25}\, z_{53} & z_{25}\, z_{54} & z_{25} \\ z_{34}\, z_{41} & 0 & 1 & z_{34} & z_{34}\, z_{45} \\ z_{41} & z_{41}\, z_{12} & z_{43} \cup z_{45}\, z_{53} & 1 & z_{45} \\ z_{54}\, z_{41} & 0 & z_{53} \cup z_{54}\, z_{43} & z_{53}\, z_{34} \cup z_{54} & 1 \end{pmatrix}$$

while the elements of A'^4 are:

$$\alpha'^4_{11} = 1,$$
$$\alpha'^4_{12} = z_{12},$$
$$\alpha'^4_{13} = z_{12}\, z_{25}\, z_{53} \cup z_{12}\, z_{25}\, z_{54}\, z_{43},$$
$$\alpha'^4_{14} = z_{12}\, z_{25}\, z_{54} \cup z_{12}\, z_{25}\, z_{53}\, z_{34},$$
$$\alpha'^4_{15} = z_{12}\, z_{25},$$
$$\alpha'^4_{21} = z_{21} \cup z_{25}\, z_{53}\, z_{34}\, z_{41} \cup z_{25}\, z_{54}\, z_{41},$$
$$\alpha'^4_{22} = 1,$$
$$\alpha'^4_{23} = z_{25}\, z_{53} \cup z_{25}\, z_{54}\, z_{43}$$

(here the term $z_{25}\, z_{54}\, z_{45}\, z_{53}$, corresponding to a non-elementary path, was absorbed by $z_{25}\, z_{53}$; similar remarks hold for α'^4_{25}, α'^4_{31}, α'^4_{34}, α'^4_{41}, α'^4_{43}, α'^4_{45}, α'^4_{51}, α'^4_{53}, α'^4_{54});

$$\alpha'^4_{24} = z_{25}\, z_{54} \cup z_{25}\, z_{53}\, z_{34},$$
$$\alpha'^4_{25} = z_{25},$$
$$\alpha'^4_{31} = z_{34}\, z_{41},$$
$$\alpha'^4_{32} = z_{34}\, z_{41}\, z_{12},$$
$$\alpha'^4_{33} = 1,$$
$$\alpha'^4_{34} = z_{34},$$
$$\alpha'^4_{35} = z_{34}\, z_{41}\, z_{12}\, z_{25} \cup z_{34}\, z_{45},$$
$$\alpha'^4_{41} = z_{41},$$
$$\alpha'^4_{42} = z_{41}\, z_{12},$$
$$\alpha'^4_{43} = z_{41}\, z_{12}\, z_{25}\, z_{53} \cup z_{43} \cup z_{45}\, z_{53},$$
$$\alpha'^4_{44} = 1,$$
$$\alpha'^4_{45} = z_{41}\, z_{12}\, z_{25} \cup z_{45},$$
$$\alpha'^4_{51} = z_{54}\, z_{41} \cup z_{53}\, z_{34}\, z_{41},$$
$$\alpha'^4_{52} = z_{54}\, z_{41}\, z_{12} \cup z_{53}\, z_{34}\, z_{41}\, z_{12},$$
$$\alpha'^4_{53} = z_{53} \cup z_{54}\, z_{43},$$
$$\alpha'^4_{54} = z_{53}\, z_{34} \cup z_{54},$$
$$\alpha'^4_{55} = 1;$$

we have thus obtained the list of all the elementary paths of the graph.

Now, let us determine the elementary paths from 1 to 3, that is α'^4_{13}, by the second procedure, which uses Theorem I.4. We have to compute the determinant of the complement A'_{31} of α'_{31} in A':

$$\alpha'^4_{13} = |A'_{31}| = \begin{vmatrix} z_{12} & 0 & 0 & 0 \\ 1 & 0 & 0 & z_5 \\ 0 & z_{43} & 1 & z_{45} \\ 0 & z_{53} & z_{54} & 1 \end{vmatrix} = z_{12} \begin{vmatrix} 0 & 0 & z_{25} \\ z_{43} & 1 & z_{45} \\ z_{53} & z_{54} & 1 \end{vmatrix} = z_{12} z_{25} \begin{vmatrix} z_{43} & 1 \\ z_{53} & z_{54} \end{vmatrix}$$

$$= z_{12} z_{25} (z_{43} z_{54} \cup z_{53}) = z_{12} z_{25} z_{53} \cup z_{12} z_{25} z_{54} z_{43}.$$

Alternatively, we can compute α'^4_{13} by eliminating successively the vertices $5, 4, 2$. We obtain thus the matrices

$$A'_{(5)} = \begin{pmatrix} 1 & z_{12} & 0 & 0 \\ z_{21} & 1 & z_{25} z_{53} & z_{25} z_{54} \\ 0 & 0 & 1 & z_{34} \\ z_{41} & 0 & z_{43} \cup z_{45} z_{53} & 1 \end{pmatrix},$$

$$A'_{(5,4)} = \begin{pmatrix} 1 & z_{12} & 0 \\ z_{21} \cup z_{25} z_{54} z_{41} & 1 & z_{25} z_{53} \cup z_{25} z_{54} z_{43} \\ z_{34} z_{41} & 0 & 1 \end{pmatrix}$$

$$A'_{(5,4,2)} = \begin{pmatrix} 1 & z_{12}(z_{25} z_{53} \cup z_{25} z_{54} z_{43}) \\ z_{34} z_{41} & 1 \end{pmatrix}$$

hence $\alpha'^4_{13} = z_{12} z_{25} z_{53} \cup z_{12} z_{25} z_{54} z_{43}$ and $\alpha'^4_{31} = z_{34} z_{41}$.

§ 9. Boolean Proofs of Graph-Theoretical Properties

All the applications studied in the preceding sections were of the following type: find one (all) mathematical object(s) having a certain property (if any).

Another type of applications was suggested by K. MAGHOUT [6], who showed that the demonstrations of several graph-theoretical properties reduce to the proofs of certain Boolean identities.

Let $A = (a_{ij})$ and $B = (b_{ij})$ be two Boolean (0,1) matrices, with $i, j = 1, \ldots, n$, and consider an arbitrary, but fixed integer i_0, with $1 \leqq i_0 \leqq n$. We shall compute first the expression

$$e_0 = \bigcup_{i=1}^{n} \bar{b}_{i_0 i} \prod_{j=1}^{n} \bigcup_{k=1}^{n} a_{ik} b_{kj}. \tag{94}$$

For each i with $1 \leqq i \leqq n$, let j_i be an integer such that

$$(95) \qquad \bigcup_{k=1}^{n} a_{ik} b_{kj_i} \leqq \bigcup_{k=1}^{n} a_{ik} b_{kj} \quad \text{for all} \quad j = 1, \ldots, n;$$

then

$$(96) \qquad e_0 = \bigcup_{i=1}^{n} \bar{b}_{i_0 i} \bigcup_{k=1}^{n} a_{ik} b_{kj_i} = \bigcup_{k=1}^{n} \bigcup_{i=1}^{n} \bar{b}_{i_0 i} a_{ik} b_{kj_i}.$$

Consider an arbitrary index i_1 $(1 \leqq i_1 \leqq n)$ and put $j_{i_1} = i_2$, $j_{i_2} = i_3, \ldots, j_{i_n} = i_{n+1}$; then (96) implies

$$(97) \qquad e_0 \geqq \bigcup_{k=1}^{n} \bigcup_{h=1}^{n} \bar{b}_{i_0 i_h} a_{i_h k} b_{k i_{h+1}}.$$

Relation (97) will be used in order to demonstrate:

Theorem 13. *Let* $A = (a_{ij})$ *and* $B = (b_{ij})$ *be two arbitrary Boolean (0,1)* $n \times n$ *matrices and* i_0 *an integer with* $1 \leqq i_0 \leqq n$. *Then*

$$(98) \quad \bigcup_{i=1}^{n} \bigcup_{j=1}^{n} (\bar{a}_{ij} \bar{b}_{ji} \cup \bar{a}_{ii} \cup \bar{b}_{jj}) \cup \bigcup_{i=1}^{n} \prod_{j=1}^{n} b_{ij} \cup \bigcup_{i=1}^{n} \bar{b}_{i_0 i} \prod_{j=1}^{n} \bigcup_{k=1}^{n} a_{ik} b_{kj} = 1.$$

Proof. Let i_1 be an integer such that

$$(99) \qquad b_{i_0 i_1} \leqq b_{i_0 j} \quad \text{for all} \quad j = 1, \ldots, n;$$

implying thus

$$(100) \qquad \bigcup_{i=1}^{n} \prod_{j=1}^{n} b_{ij} \geqq \prod_{j=1}^{n} b_{i_0 j} = b_{i_0 i_1}.$$

Let e denote the left-hand side of (98). Using (100) and (94), we deduce

$$(101) \qquad e \geqq \bigcup_{i=1}^{n} \bigcup_{j=1}^{n} \bar{a}_{ij} \bar{b}_{ji} \cup \bigcup_{i=1}^{n} (\bar{a}_{ii} \cup \bar{b}_{ii}) \cup b_{i_0 i_1} \cup e_0.$$

Now we replace e_0 by its expression (97), [written for an index i_1 satisfying (99)]; then (101) implies

$$(102) \quad e \geqq \bigcup_{h=1}^{n} \bigcup_{k=1}^{n} \bar{a}_{i_h k} \bar{b}_{k i_h} \cup \bigcup_{i=1}^{n} (\bar{a}_{ii} \cup \bar{b}_{ii}) \cup b_{i_0 i_1} \cup \\ \cup \bigcup_{h=1}^{n} \bigcup_{k=1}^{n} \bar{b}_{i_0 i_h} a_{i_h k} b_{k i_{h+1}} = e_1.$$

Notice that the right-hand side (e_1) of relation (102) satisfies

$$(103) \quad e_1 \geqq \bigcup_{h=1}^{n} (\bar{a}_{i_h i_0} \bar{b}_{i_0 i_h} \cup \bar{b}_{i_0 i_h} a_{i_h i_0} b_{i_0 i_{h+1}}) \\ = \bigcup_{h=1}^{n} \bigcup_{k=1}^{n} (\bar{a}_{i_h i_0} \bar{b}_{i_0 i_h} \cup \bar{b}_{i_0 i_h} a_{i_h i_0} b_{i_0 i_{h+1}}),$$

so that the right-hand side of (103) may be added to e_1 without altering it; hence (102) becomes

$$e \geqq \bigcup_{i=1}^{n} (\bar{a}_{ii} \cup \bar{b}_{ii}) \cup b_{i_0 i_1} \cup \bigcup_{h=1}^{n} \bigcup_{k=1}^{n} e_{h,k}, \tag{104}$$

where

$$e_{h,k} = \bar{a}_{i_h k} \bar{b}_{k i_h} \cup \bar{b}_{i_0 i_h} a_{i_h k} b_{k i_{h+1}} \cup \bar{a}_{i_h i_0} \bar{b}_{i_0 i_h} \cup \bar{b}_{i_0 i_h} a_{i_h i_0} b_{i_0 i_{h+1}}. \tag{105}$$

But it is well-known that $\bar{a}\, b \cup a\, c \geqq \bar{a}\, b\, c \cup a\, b\, c = b\, c$; using this inequality, with $a = a_{i_h k}$, $b = \bar{b}_{k i_h}$, and $c = \bar{b}_{i_0 i_h} b_{k i_{h+1}}$, we deduce from (105) that

$$\begin{aligned} e_{hk} &\geqq \bar{b}_{k i_h} \bar{b}_{i_0 i_h} b_{k i_{h+1}} \cup \bar{b}_{i_0 i_h} a_{i_h k} b_{k i_{h+1}} \cup \\ &\cup \bar{a}_{i_h i_0} \bar{b}_{i_0 i_h} \cup \bar{b}_{i_0 i_h} a_{i_h i_0} b_{i_0 i_{h+1}} \\ &= \bar{b}_{i_0 i_h} (\bar{b}_{k i_h} b_{k i_{h+1}} \cup a_{i_h k} b_{k i_{h+1}} \cup \bar{a}_{i_h i_0} \cup b_{i_0 i_{h+1}}) \geqq \\ &\geqq \bar{b}_{i_0 i_h} (\bar{b}_{k i_h} b_{k i_{h+1}} \cup a_{i_h k} b_{k i_{h+1}} \cup b_{i_0 i_{h+1}}). \end{aligned} \tag{106}$$

It follows from (104) and (106) that

$$\begin{aligned} e \geqq & \bigcup_{i=1}^{n} (\bar{a}_{ii} \cup \bar{b}_{ii}) \cup b_{i_0 i_1} \cup \\ & \cup \bigcup_{k=1}^{n} \bigcup_{h=1}^{n} \bar{b}_{i_0 i_h} (\bar{b}_{k i_h} b_{k i_{h+1}} \cup a_{i_h k} b_{k i_{h+1}} \cup b_{i_0 i_{h+1}}). \end{aligned} \tag{107}$$

But the indices $i_0, i_1, \ldots, i_n, i_{n+1}$ cannot be all distinct; let s denote the first integer for which there exists an $r < s$ with $i_r = i_{s+1}$. Relation (107) implies

$$\begin{aligned} e &\geqq \bigcup_{i=1}^{n} (\bar{a}_{ii} \cup \bar{b}_{ii}) \cup b_{i_0 i_1} \cup \bigcup_{k=1}^{n} \bigcup_{h=1}^{s} \bar{b}_{i_0 i_h} (b_{i_0 i_{h+1}} \cup a_{i_h k} b_{k i_{h+1}} \cup \bar{b}_{k i_h} b_{k i_{h+1}}) \\ &= \bigcup_{i=1}^{n} (\bar{a}_{ii} \cup \bar{b}_{ii}) \cup \bigcup_{h=1}^{s+1} b_{i_0 i_h} \cup \bigcup_{k=1}^{n} \bigcup_{h=1}^{s} (a_{i_h k} b_{k i_{h+1}} \cup \bar{b}_{k i_h} b_{k i_{h+1}}) \geqq \\ &\geqq \bar{a}_{i_r i_r} \cup \bar{b}_{i_r i_r} \cup \bigcup_{h=r}^{s} (a_{i_h i_r} b_{i_r i_{h+1}} \cup \bar{b}_{i_r i_h} b_{i_r i_{h+1}}) = \\ &= \bar{a}_{i_r i_r} \cup \bar{b}_{i_r i_r} \cup b_{i_r i_{r+1}} \cup \bigcup_{h=r+1}^{s} \bar{b}_{i_r i_h} b_{i_r i_{h+1}} \geqq \bar{b}_{i_r i_r} \cup \bigcup_{h=r}^{s} b_{i_r i_{h+1}} \geqq \\ &\geqq \bar{b}_{i_r i_r} \cup b_{i_r i_{s+1}} = \bar{b}_{i_r i_r} \cup b_{i_r i_r} = 1, \end{aligned}$$

thus completing the proof.

Now, several graph-theoretical results can be deduced from the above Theorem 13. We introduce first the following:

Definition 18. A graph $G = (N, \Gamma)$ is called (p, q)-*total*, if for every two vertices $i, j \in N$, either $d(i, j) \leqq p$, or $d(j, i) \leqq q$, or both (see Definition 10).

It follows that a (p, q)-total graph is also (q, p)-total. A $(1, 1)$-total graph is complete, in the sense of Definition 8. A (p, q)-total graph is characterized by relation

(108) $$A'^p \cup A'^q = I$$

(see § 2); compare to the obvious characterization

(109) $$\hat{A} \cup \hat{A}^T = I$$

of connected graphs.

Theorem 14. *In a* (p, q)*-total graph, one of the following two alternatives holds: either the radius* $\varrho \leqq q$, *or* $q < \varrho \leqq p + q$ *and for each vertex* i_0 *there exists a vertex* i_1 *such that* $d(i_0, i_1) > q$ *and* $e(i_1) \leqq p + q$.

Proof. We apply Theorem 13, with the powers A'^p and A'^q of the unitary incidence matrix A (see § 2) of the graph, in the roles of A and B, respectively. Taking into account that $a'^p_{ii} = b'^q_{jj} = 1$ for all i, j, we get

$$\bigcup_{i=1}^{n} \bigcup_{j=1}^{n} \overline{a'^p_{ij}}\, \overline{a'^q_{ji}} \cup \bigcup_{i=1}^{n} \prod_{j=1}^{n} a'^q_{ij} \cup \bigcup_{i=1}^{n} \overline{a'^q_{i_0 i}} \prod_{j=1}^{n} \bigcup_{k=1}^{n} a'^p_{ik}\, a'^q_{kj} = 1 .$$

This means that relation

(110) $$\bigcup_{i=1}^{n} \bigcup_{j=1}^{n} \overline{a'^p_{ij}}\, \overline{a'^q_{ji}} = 0$$

implies either

(111) $$\bigcup_{i=1}^{n} \prod_{j=1}^{n} a'^q_{ij} = 1$$

or

(112) $$\bigcup_{i=1}^{n} \overline{a'^q_{i_0 i}} \prod_{j=1}^{n} \bigcup_{k=1}^{n} a'^p_{ik}\, a'^q_{kj} = 1 .$$

But relation (110) does hold, because it is equivalent to the equality

$$\prod_{i=1}^{n} \prod_{j=1}^{n} (a'^p_{ij} \cup a'^q_{ji}) = 1$$

which expresses the condition (108) that the graph is (p, q)-total. Therefore, at least one of the two relations (111) and (112) holds also.

1) According to Corollary 4, relation (111) is equivalent to the existence of a vertex i with $e(i) \leqq q$; in other words, it is equivalent to relation $\varrho \leqq q$.

2) On the other hand, relation (112) is equivalent to the existence of an index i_1 such that: $i_0 \neq i_1$,

$$\overline{a'^{q}_{i_0 i_1}} = 1 \tag{113}$$

and

$$\prod_{j=1}^{n} \bigcup_{k=1}^{n} a'^{p}_{i_1 k}\, a'^{q}_{kj} = 1. \tag{114}$$

But relation (113) may be written $a'^{q}_{i_0 i_1} = 0$ and is equivalent to $d(i_0, i_1) > q$, while (114) means that to each vertex $j \neq i_1$ corresponds (at least) a vertex k_j such that $a'^{p}_{i_1 k_j} = a'^{q}_{k_j j} = 1$; in other words, relation (114) expresses the property that for each vertex $j \neq i_1$, there is a path of length at most $p + q$ from i_1 to j.

We have thus proved that relation (112) is equivalent to the existence of a vertex i_1 such that: α) $d(i_0, i_1) > q$ and β) $d(i_1, j) \leqq p + q$ for all $j \neq i_1$. Since β) is equivalent to $e(i_1) \leqq p + q$, this completes the proof.

Corollary 15. The radius of a (p, q)-total graph is $\leqq p + q$. In particular, the radius of a complete graph is $\leqq 2$.

It can also be inferred from Theorem 14 (by purely graph-theoretical means) that a (p, q)-total graph of radius $\varrho = p + q$, where $0 < p \leqq q$, has at least $[q/p] + 2$ centers ($[x]$ being the integer part of x). In particular, a complete graph ($p = q = 1$) has at least 3 centers.

Chapter X

Stable Sets, Kernels and Chromatic Decompositions of Graphs

In this chapter we shall be concerned with the problem of determining the (maximal) internally and (minimal) externally stable subsets, the kernels and the chromatic decomposition of a graph. Possible practical interpretations of these concepts are sketched in the first section, along with the necessary definitions.

§ 1. Definitions and Interpretations

Let us consider a graph $G = (N, \Gamma)$, where $N = \{1, 2, \ldots, n\}$ is the set of vertices and Γ is the mapping which associates with each vertex $i \in N$ the set of all those vertices $j \in N$ so that (i, j) is an arc.

Definition 1. A subset A of N will be called *internally stable* (or, *independent*), if there is no arc joining two vertices of A, or, equivalently, if

(1) $$A \cap \Gamma A = \emptyset .$$

Obviously, every subset of an internally stable set has also this property; in particular, the empty set $\emptyset$ is internally stable.

Definition 2. An internally stable set A is called *maximal,* if each set $A' \subseteqq N$ such that $A \subset A'$, ceases to be internally stable.

Thus a set is internally stable if and only if it is included in a maximal internally stable set.

Definition 3. An internally stable set A will be said to be *absolutely maximal,* if there is no internally stable set A' having more vertices than A. The number $\alpha(G)$ of vertices of each absolutely maximal internally stable set is called the *number of internal stability* (or, the *independence number*) of the graph.

Obviously, each absolutely maximal internally stable set is also maximal.

Example 1. (C. R. SHANNON [1], quoted from C. BERGE [2] and O. ORE [1]). A transmission code is given, having the symbols $s_1, \ldots, s_n$ as basic signals; certain couples of these signals may be confused by the receiver. In order to obtain an errorless code, it is desired to restrict the basic signals to a subset

$A = \{s_{i_1}, s_{i_2}, \ldots, s_{i_t}\}$ of the set $S = \{s_1, \ldots, s_n\}$, having the property that there is no possibility of confusing two signals s_{i_k} and s_{i_v} of A, and so that the number t of elements of A should be as large as possible. If we introduce the graph $G = (S, \Gamma)$, where for any $s', s'' \in S$, relation $s'' \in \Gamma s'$ holds if and only if the signals s' and s'' may be confused, then the above problem is nothing but that of determining an absolutely maximal internally stable set A.

Definition 4. A subset B of N will be called *externally stable* (or, *converse dominating*), if, for each vertex i not in B, there is a vertex j in B, so that (i, j) is an arc of the graph; or, equivalently, if

$$(2) \qquad i \notin B \quad \text{implies} \quad \Gamma i \cap B \neq \emptyset.$$

Obviously, if B is an externally stable set and $B \subseteqq C$, then C is also externally stable; in particular, the set N has this property.

Definition 5. An externally stable set B is called *minimal*, if each set $B' \subseteqq N$ such that $B' \subset B$, ceases to be externally stable.

Thus a set is externally stable if and only if it includes a minimal externally stable set.

Definition 6. An externally stable set B will be said to be *absolutely minimal*, if there is no externally stable set B' having fewer vertices than B. The number $\beta(G)$ of vertices of each absolutely minimal externally stable set is called the *number of external stability* (or, the *converse domination number*) of the graph.

Obviously, each absolutely minimal externally stable set is also minimal.

Example 2. A certain product is produced in several points $1, 2, \ldots, n$. From the point i, the product can be transported to certain other points, to which i is related. If we want to build warehouses or processing centers, then, from an economic point of view, we are faced with the following problem: Determine m points $i_1, i_2, \ldots, i_m$, such that from each point $1, 2, \ldots, n$, the product may be transported to at least one of the points $i_1, i_2, \ldots, i_m$, and that the number m be minimal. If we introduce the graph $G = (N, \Gamma)$, where $N = \{1, 2, \ldots, n\}$ and for each $i \in N, \Gamma i$ is the set of those points $j \in N$ to which i is related, then the above problem is translated as the problem of determining an absolutely minimal externally stable subset of N.

Definition 7. A subset K of N which is both internally and externally stable is called a *kernel* of the graph.

A graph may have one or more kernels, but it may also happen that it has no kernel.

In Example 3 of the Introduction it was shown that a certain problem in game theory leads to the finding of a certain subset K of the set $S = \{s_1, \ldots, s_l\}$ of strategies; conditions (4) and (5) of the example show clearly that K is nothing but a kernel of the graph $G = (S, \Gamma)$, where $b \in \Gamma a$ if and only if $a \prec b$.

Definition 8. By a *chromatic decomposition* of the graph $G = (N, \Gamma)$, we mean a family of disjoint internally stable sets

$C_1, C_2, \ldots, C_m$ of vertices covering the set N; that is

$$C_i \cap C_j = \emptyset \quad \text{if} \quad i \neq j, \tag{3}$$

$$\bigcup_{i=1}^{m} C_i = N. \tag{4}$$

Definition 9. A chromatic decomposition with the smallest number $\gamma(G)$ of sets is termed a *minimal chromatic decomposition,* while $\gamma(G)$ is said to be the *chromatic number* of G.

Example 3. 1) The map colouring problem, quoted in every treatise on graph theory, requires the colouring of a map in such way that any two neighbouring countries get different colours, and the number of colours employed be minimal. If we represent the countries as vertices of a graph, two vertices being linked if and only if the corresponding countries are neighbouring,then the above problem is in fact that of determining a minimal chromatic decomposition of the graph.

Consider an example given by J. WEISSMAN [1]: find the

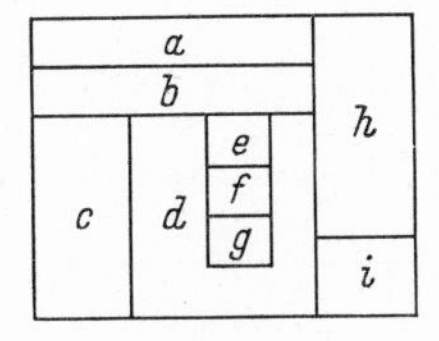

Fig. 1.

minimal colouring of the map in Fig. 1. The associated (undirected*) graph is given in Fig. 2. The maximal internally stable sets

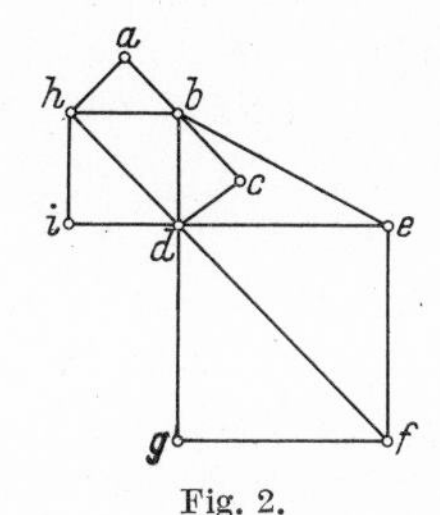

Fig. 2.

of this graph will be determined in Examples 4, 5.1 and 6, while the minimal chromatic decomposition will be given in Example 16.1.

2) The same interpretation may be given to the following problem. A telephone switching center is made up of many blocks, certain couples of blocks being joined by wires of various colours. In order to facilitate error detection, it is desired that any two wires having a common endpoint be coloured differently, and the number of colours be minimal.

* An undirected graph is a pair (N, ϱ), where N is a set of vertices, while ϱ is a symmetric binary relation on the set N. An undirected graph is depicted using lines instead of arrows (as in Fig. 2).

Various generalizations of the above concepts will be studied in the subsequent sections.

The following definition will permit us to translate problems concerning subsets of the vertex set of a graph into the language of Boolean algebra.

Definition 10. To each subset S of the vertex set $N = \{1, 2, \ldots, n\}$ of a graph $G = (N, \Gamma)$, we associate the *characteristic vector* $(x_1, \ldots, x_n)$, where

$$x_i = \begin{cases} 1, & \text{if } i \in S, \\ 0, & \text{if } i \notin S, \end{cases} \tag{5}$$

for each $i = 1, 2, \ldots, n$.

§ 2. Internally Stable Sets

We recall that in § 2 of the preceding chapter, the *incidence matrix* $A = (a_{ij})$ of the graph $G = (N, \Gamma)$ was defined by

$$a_{ij} = \begin{cases} 1, & \text{if } j \in \Gamma i, \\ 0, & \text{if } j \notin \Gamma i. \end{cases} \tag{IX.9}$$

Therefore, it results from Definitions 1 and 10 that a set S is internally stable if and only if its characteristic vector $(x_1, \ldots, x_n)$ satisfies the following condition: for any $i, j \in N$, if $a_{ij} = 1$, then either x_i, or x_j (or both) must be zero. This condition is obviously equivalent to the relation $a_{ij}\, x_i\, x_j = 0$. Hence:

Theorem 1. (K. Maghout [1, 6], J. Weissman [1], P. L. Hammer and I. Rosenberg [1]). *A set $S \subseteqq N$ is internally stable if and only if its characteristic vector $(x_1, \ldots, x_n)$ satisfies the Boolean equation*

$$\bigcup_{i=1}^{n} \bigcup_{j=1}^{n} a_{ij}\, x_i\, x_j = 0, \tag{6}$$

or, equivalently, the pseudo-Boolean equation

$$\sum_{i=1}^{n} \sum_{j=1}^{n} a_{ij}\, x_i\, x_j = 0. \tag{7}$$

This theorem provides us with a method for finding all internally stable sets of a graph, including all (absolutely) maximal internally stable sets.

K. Maghout and J. Weissman indicate the following way of solving equation (6), which is suggested by the classical Quine-McCluskey technique for simplifying truth functions. Let us consider the dual form of equation (6):

$$\prod_{i=1}^{n} \prod_{j=1}^{n} (\bar{a}_{ij} \cup \bar{x}_i \cup \bar{x}_j) = 1. \tag{8}$$

In fact, the left-hand side of (8) reduces to the product of those factors $(\bar{x}_i \cup \bar{x}_j)$ for which $a_{ij} = 1$; putting $\prod'$ for this product, equation (8) becomes

$$\prod{}' (\bar{x}_i \cup \bar{x}_j) = 1. \tag{8'}$$

Now, performing the necessary multiplications (using the distributive law) and all possible absorptions, we get

$$\bigcup \bar{x}_{h_1} \bar{x}_{h_2} \ldots \bar{x}_{h_{k(h)}} = 1. \tag{9}$$

Theorem 2. *If $\bar{x}_{h_1} \bar{x}_{h_2} \ldots \bar{x}_{h_{k(h)}}$ is one of the conjunctions appearing in the left-hand side of equation* (9), *then the vector $(x_1^*, \ldots, x_n^*)$ defined by*

$$x_i^* = \begin{cases} 0, & \text{if } i = h_1, h_2, \ldots, h_{k(h)}, \\ 1, & \text{otherwise}, \end{cases} \tag{10}$$

is the characteristic vector of a maximal internally stable set, and all maximal internally stable sets may be obtained in this way.

Proof. By expanding the left-hand side of (9) into the disjunctive canonical form, we see that: (α) each vector (10) is a solution of (9); (β) each vector $(x_1^{**}, \ldots, x_n^{**})$ with the property that there exists a conjunction $\bar{x}_{h_1} \bar{x}_{h_2} \ldots \bar{x}_{h_{k(h)}}$ in the left-hand side of (9), so that $x_{h_1}^{**} = x_{h_2}^{**} = \cdots = x_{h_{k(h)}}^{**} = 0$ is a solution of (9); (γ) each solution of (9) may be obtained as in (α) or (β). In other words, the solutions of equation (9) may be described as the characteristic vectors of all the subsets of the sets having the characteristic vectors (10).

Since, on the other hand, Theorem 1 states that the solutions of equation (9) may be described as the characteristic vectors of the internally stable sets, it follows that the maximal internally stable sets are precisely those having (10) as characteristic vectors, thus completing the proof.

Computational remarks. 1) Whenever $a_{ij} = a_{ji} = 1$, we shall maintain in equation (9) only one of the terms $(\bar{x}_i \cup \bar{x}_j)$ and $(\bar{x}_j \cup \bar{x}_i)$.

2) The multiplications in equation (8′) are facilitated by applying property (I.8′).

Example 4. Find the maximal internally stable sets of the graph in Fig. 2. Equation (6) becomes

$$(\bar{x}_a \cup \bar{x}_b)(\bar{x}_a \cup \bar{x}_h)(\bar{x}_b \cup \bar{x}_c)(\bar{x}_b \cup \bar{x}_d)(\bar{x}_b \cup \bar{x}_e)(\bar{x}_b \cup \bar{x}_h)(\bar{x}_c \cup \bar{x}_d)(\bar{x}_d \cup \bar{x}_e)(\bar{x}_d \cup \bar{x}_f) \cdot$$
$$\cdot (\bar{x}_d \cup \bar{x}_g)(\bar{x}_d \cup \bar{x}_h)(\bar{x}_d \cup \bar{x}_i)(\bar{x}_e \cup \bar{x}_i)(\bar{x}_f \cup \bar{x}_g)(\bar{x}_h \cup \bar{x}_i) = 1.$$

After performing the multiplications and absorptions, we get

$$\bar{x}_b \bar{x}_d \bar{x}_f \bar{x}_h \cup \bar{x}_a \bar{x}_b \bar{x}_d \bar{x}_f \bar{x}_i \cup \bar{x}_b \bar{x}_d \bar{x}_e \bar{x}_g \bar{x}_h \cup \bar{x}_a \bar{x}_b \bar{x}_d \bar{x}_e \bar{x}_g \bar{x}_i \cup$$
$$\cup \bar{x}_b \bar{x}_c \bar{x}_e \bar{x}_f \bar{x}_g \bar{x}_h \bar{x}_i \cup \bar{x}_a \bar{x}_c \bar{x}_d \bar{x}_e \bar{x}_g \bar{x}_h \cup \bar{x}_a \bar{x}_c \bar{x}_d \bar{x}_e \bar{x}_f \bar{x}_h = 1,$$

hence the maximal internally stable sets are

$$M_1 = \{a, c, e, g, i\}, \quad M_2 = \{c, e, g, h\}, \quad M_3 = \{a, c, f, i\},$$
$$M_4 = \{c, f, h\}, \quad M_5 = \{a, d\}, \quad M_6 = \{b, f, i\}, \quad M_7 = \{b, g, i\}.$$

Remark 1. The determination of the maximal internally stable sets (including, in particular, the absolutely maximal ones) is, in fact, an indirect way of determining all internally stable sets, because any internally stable set is characterized by the property of being a subset of a maximal internally stable set.

Another procedure for finding the maximal internally stable sets is given by the following

Theorem 3. (S. RUDEANU [9]; see also K. MAGHOUT [1]). *A set $S \subseteqq N$ is maximal internally stable if and only if its characteristic vector $(x_1, \ldots, x_n)$ satisfies the system of Boolean equations*

$$x_i = \overline{a^*_{ii}} \prod_{\substack{j=1 \\ j \neq i}}^{n} (\overline{a^*_{ij}} \cup \bar{x}_j) \qquad (i = 1, \ldots, n), \tag{11}$$

where

$$a^*_{ij} = a_{ij} \cup a_{ji} \qquad (i, j = 1, \ldots, n). \tag{12}$$

Proof*. We shall first give another form to the characteristic equation (6) of the internally stable sets. Since $a_{ij}\, x_i\, x_j \cup a_{ji}\, x_j\, x_i = (a_{ij} \cup a_{ji})\, x_i\, x_j = a^*_{ij}\, x_i\, x_j = a^*_{ji}\, x_j\, x_i = a^*_{ij}\, x_i\, x_j \cup a^*_{ji}\, x_j\, x_i$, equation (6) may be written in the form

$$\bigcup_{i=1}^{n} \bigcup_{j=1}^{n} a^*_{ij}\, x_i\, x_j = 0$$

and thus, is equivalent to the system

$$x_i \left(a^*_{ii} \cup \bigcup_{\substack{j=1 \\ j \neq i}}^{n} a^*_{ij}\, x_j \right) = 0 \qquad (i = 1, \ldots, n),$$

which, by (I.23″), may be written in the form

$$a^*_{ii} \cup \bigcup_{\substack{j=1 \\ j \neq i}}^{n} a^*_{ij}\, x_j \leqq \bar{x}_i \qquad (i = 1, \ldots, n). \tag{13}$$

Now, the condition of being maximal is the following: for any i, if $x_i = 0$, then either $a_{ii} = 1$, or there exists a $j \neq i$ so that $x_j = a^*_{ij} = 1$, or both. In other words, this means that $\bar{x}_i = 1$ implies $a^*_{ii} = a_{ii} = 1$, or $\bigcup_{\substack{j=1 \\ j \neq i}}^{n} a^*_{ij}\, x_j = 1$, or both. This condition can also be translated through

* For another proof see Corollary VII.4.

the system of inequalities

$$(14) \qquad \bar{x}_i \leqq a_{ii}^* \cup \bigcup_{\substack{j=1 \\ j \neq i}}^{n} a_{ij}^* x_j \qquad (j = 1, \ldots, n).$$

We have thus proved that the maximal internally stable sets are characterized by the inequalities (13) and (14), which are equivalent to the system of Boolean equations

$$(15) \qquad \bar{x}_i = a_{ii}^* \cup \bigcup_{\substack{j=1 \\ j \neq i}}^{n} a_{ij}^* x_j^* \qquad (i = 1, \ldots, n),$$

which is the dual form of (11).

Example 5.

1) Apply Theorem 3 in order to obtain the maximal internally stable sets of the graph in Fig. 2.

The system of equations (11) becomes

$$x_a = \bar{x}_b \bar{x}_h,$$

$$x_b = \bar{x}_a \bar{x}_c \bar{x}_d \bar{x}_e \bar{x}_h,$$

$$x_c = \bar{x}_b \bar{x}_d,$$

$$x_d = \bar{x}_b \bar{x}_c \bar{x}_e \bar{x}_f \bar{x}_g \bar{x}_h \bar{x}_i,$$

$$x_e = \bar{x}_b \bar{x}_d \bar{x}_f,$$

$$x_f = \bar{x}_d \bar{x}_e \bar{x}_g,$$

$$x_g = \bar{x}_d \bar{x}_f,$$

$$x_h = \bar{x}_a \bar{x}_b \bar{x}_d \bar{x}_i,$$

$$x_i = \bar{x}_d \bar{x}_h,$$

which was solved in Example II.8, and whose solutions coincide with those found in the above Example 4.

2) S. Marcus and Em. Vasiliu [1] introduced a graph, whose vertices are the 20 consonants of the Romanian language ($W, L, R, Y, S, N, P, K, F, M, \check{S}, T, Z, V, B, G, D, H, \check{Z}, Ţ$) and whose arcs are defined as follows: for any consonant c, the set Γc consists of all consonants c', so that there exists a Romanian word whose first consonantal group includes the sequence $c c'$. The 79 arcs are:

SF, ZB, ŠF, DW, ŽW, ZW, SB, ŠP, ŽV, BL, FL, ML, PL, VL, BY, PY, VY, FY, MY, ZD, SL, ŠL, ZL, ŽN, DY, LY, NY, SY, TY, LW, NW, RW, SM, ŠM, ZM, ŢW, SW, ŠW, TW, KL, GL, HL, KN, GN, KR, GR, HR, KT, MN, PN, BR, FR, MR, PR, VR, PS, FT, PT, BW, FW, MW, PW, VW, SK, ZG, SN, ŠN, DR, ŠR, TR, ST, ŠT, ŢY, ZY, KF, KW, GW, HW, KV.

For some linguistic reasons, it is interesting to determine a chromatic decomposition of this graph. As it will be shown in § 5, the first step in solving this problem is the determination of all maximal internally stable sets.

To do this, we write down the system (11), which becomes

$$W = \bar{L}\,\bar{R}\,\bar{S}\,\bar{N}\,\bar{P}\,\bar{K}\,\bar{F}\,\bar{M}\,\bar{\check{S}}\,\bar{T}\,\bar{Z}\,\bar{V}\,\bar{B}\,\bar{G}\,\bar{D}\,\bar{H}\,\bar{\check{Z}}\,\bar{Ţ},$$

$$L = \bar{W}\,\bar{Y}\,\bar{S}\,\bar{P}\,\bar{K}\,\bar{F}\,\bar{M}\,\bar{\check{S}}\,\bar{Z}\,\bar{V}\,\bar{B}\,\bar{G}\,\bar{H},$$

$$R = \bar{W}\,\bar{P}\,\bar{K}\,\bar{F}\,\bar{M}\,\bar{\check{S}}\,\bar{T}\,\bar{V}\,\bar{B}\,\bar{G}\,\bar{D}\,\bar{H},$$

$$Y = \bar{L}\,\bar{S}\,\bar{N}\,\bar{P}\,\bar{F}\,\bar{M}\,\bar{T}\,\bar{Z}\,\bar{V}\,\bar{B}\,\bar{D}\,\bar{Ţ},$$

$$S = \bar{W}\,\bar{L}\,\bar{Y}\,\bar{N}\,\bar{P}\,\bar{K}\,\bar{F}\,\bar{M}\,\bar{T},$$

$$N = \bar{W}\,\bar{Y}\,\bar{S}\,\bar{P}\,\bar{K}\,\bar{M}\,\bar{\check{S}}\,\bar{G}\,\bar{\check{Z}},$$

$$P = \bar{W}\,\bar{L}\,\bar{R}\,\bar{Y}\,\bar{S}\,\bar{N}\,\bar{\check{S}}\,\bar{T},$$

$$K = \bar{W}\,\bar{L}\,\bar{R}\,\bar{S}\,\bar{N}\,\bar{F}\,\bar{T}\,\bar{V},$$

$$F = \bar{W}\,\bar{L}\,\bar{R}\,\bar{Y}\,\bar{S}\,\bar{K}\,\bar{\check{S}}\,\bar{T},$$

$$M = \bar{W}\,\bar{L}\,\bar{R}\,\bar{Y}\,\bar{S}\,\bar{N}\,\bar{\check{S}}\,\bar{Z},$$

$$\check{S} = \bar{W}\,\bar{L}\,\bar{R}\,\bar{N}\,\bar{P}\,\bar{F}\,\bar{M}\,\bar{T},$$

$$T = \bar{W}\,\bar{R}\,\bar{Y}\,\bar{S}\,\bar{P}\,\bar{K}\,\bar{F}\,\bar{\check{S}},$$

$$Z = \bar{W}\,\bar{L}\,\bar{Y}\,\bar{M}\,\bar{B}\,\bar{G}\,\bar{D},$$

$$V = \bar{W}\,\bar{L}\,\bar{R}\,\bar{Y}\,\bar{K}\,\bar{\check{Z}},$$

$$B = \bar{W}\,\bar{L}\,\bar{R}\,\bar{Y}\,\bar{Z},$$

$$G = \bar{W}\,\bar{L}\,\bar{R}\,\bar{N}\,\bar{Z},$$

$$D = \bar{W}\,\bar{R}\,\bar{Y}\,\bar{Z},$$

$$H = \bar{W}\,\bar{L}\,\bar{R},$$

$$\check{Z} = \bar{W}\,\bar{N}\,\bar{V},$$

$$Ţ = \bar{W}\,\bar{Y},$$

and thus coincides with the system of Boolean equations which was solved in Example II.9. Hence the graph has 28 maximal internally stable sets, whose characteristic vectors are precisely the 28 solutions found in Example II.9.

The minimal chromatic decompositions of this graph will be determined in Example 16.2.

Theorem 4. *Let S be a set of vertices and $(x_1^*, \ldots, x_n^*)$ its characteristic vector. The set S is a maximal internally stable set if and only if the function*

$$(16)\qquad f(x_1, \ldots, x_n) = \sum_{i=1}^{n} x_i - (n+1) \sum_{i=1}^{n} \sum_{j=1}^{n} a_{ij}\, x_i\, x_j$$

reaches a local maximum in $(x_1^*, \ldots, x_n^*)$ and $f(x_1^*, \ldots, x_n^*) \geqq 0$.

Proof. Follows immediately from Theorem 1 and the theorem proved in Chapter VII, § 1, note added in proofs (the function to be minimized being here $F = -\sum_{i=1}^{n} x_i + (n+1) \sum_{i=1}^{n} \sum_{j=1}^{n} a_{ij} x_i x_j$).

Taking into account Theorem VII.1 we deduce the following

Corollary 1. All maximal internally stable sets may be obtained by solving the following system of pseudo-Boolean inequalities:

$$(17) \quad (2x_i - 1)\left[1 - 2(n+1)\, a_{ii} - (n+1) \sum_{\substack{k=1 \\ k \neq i}}^{n} (a_{ik} + a_{ki})\, x_k\right] \geqq 0 \quad (i = 1, \ldots, n),$$

$$(18) \quad \sum_{h=1}^{n} x_h - (n+1) \sum_{h=1}^{n} \sum_{k=1}^{n} a_{hk}\, x_h x_k \geqq 0.$$

Example 6. In the case of the graph of Fig. 2, the function (16) has the following expression:

$$(19) \quad f = x_a + x_b + x_c + x_d + x_e + x_f + x_g + x_h + x_i -$$
$$- 10(2x_a x_b + 2x_a x_h + 2x_b x_c + 2x_b x_d + 2x_b x_e + 2x_b x_h +$$
$$+ 2x_c x_d + 2x_d x_e + 2x_d x_f + 2x_d x_g + 2x_d x_h + 2x_d x_i + 2x_e x_f +$$
$$+ 2x_f x_g + 2x_h x_i);$$

he coefficient 2 is due to the symmetry of the graph. The inequations (17) and 18) become

$$(2x_a - 1)(1 - 20x_b - 20x_h) \geqq 0,$$
$$(2x_b - 1)(1 - 20x_a - 20x_c - 20x_d - 20x_e - 20x_h) \geqq 0,$$
$$(2x_c - 1)(1 - 20x_b - 20x_d) \geqq 0,$$
$$(2x_d - 1)(1 - 20x_b - 20x_c - 20x_e - 20x_f - 20x_g - 20x_h - 20x_i) \geqq 0,$$
$$(2x_e - 1)(1 - 20x_b - 20x_d - 20x_f) \geqq 0,$$
$$(2x_f - 1)(1 - 20x_d - 20x_e - 20x_g) \geqq 0,$$
$$(2x_g - 1)(1 - 20x_d - 20x_f) \geqq 0,$$
$$(2x_h - 1)(1 - 20x_a - 20x_b - 20x_d - 20x_i) \geqq 0,$$
$$(2x_i - 1)(1 - 20x_d - 20x_h) \geqq 0,$$
$$f \geqq 0.$$

Solving this system with the methods of Chapter IV, we find the same result as in Examples 4 and 5.1.

Y. Malgrange [1] reduces the problem of finding the maximal internally stable sets of a graph to that of determining the "prime" submatrices of the matrix $\bar{A}$; the latter problem is solved with the aid of an algorithm which uses the lattice-theoretical properties of the set of all "complete" submatrices of $\bar{A}$.

Theorem 5 (P. L. HAMMER and I. ROSENBERG [1]). *Consider the pseudo-Boolean function*

$$(16)\qquad f(x_1, \ldots, x_n) = \sum_{i=1}^{n} x_i - (n+1) \sum_{i=1}^{n} \sum_{j=1}^{n} a_{ij}\, x_i\, x_j;$$

let S be a subset of N and denote by $(x_1^*, \ldots, x_n^*)$ its characteristic vector. Then:

(i) *The number $\alpha(G)$ of internal stability of the graph is equal to the global maximum of the pseudo-Boolean function f.*

(ii) *The set $S \subseteq N$ is an absolutely maximal internally stable set if and only if f reaches its global maximum in $(x_1^*, \ldots, x_n^*)$.*

Proof. The problem of finding $\alpha(G)$ and the absolutely maximal internally stable sets of N is obviously equivalent to the following pseudo-Boolean program: maximize $\sum_{i=1}^{n} x_i$, under the constraint (7); in view of Chapter VI, § 2, the latter problem is in its turn equivalent to that of maximizing the unrestricted pseudo-Boolean function (16).

Example 7. Let us determine the absolutely maximal internally stable sets and the number of internal stability of the graph in Fig. 2.

The corresponding function (16) was already written in Example 6; thus we have to find the minimum of the function

$$(20)\qquad \begin{aligned} f_1 = -f = & -x_a - x_b - x_c - x_d - x_e - x_f - x_g - x_h - x_i + \\ & + 20(x_a x_b + x_a x_h + x_b x_c + x_b x_d + x_b x_e + x_b x_h + \\ & + x_c x_d + x_d x_e + x_d x_f + x_d x_g + x_d x_h + x_d x_i + \\ & + x_e x_f + x_f x_g + x_h x_i). \end{aligned}$$

We find

$$(21)\qquad (f_1)_{\min} = -5,$$

the minimizing point being

$$(22)\qquad \begin{aligned} & x_a = 1, \quad x_b = 0, \quad x_c = 1, \quad x_d = 0, \quad x_e = 1, \\ & x_f = 0, \quad x_g = 1, \quad x_h = 0, \quad x_i = 1. \end{aligned}$$

From (21) and (22) we see that $\alpha(G) = f_{\max} = -(f_1)_{\min} = 5$, and that the unique absolutely maximal internally stable set is $\{a, c, e, g, i\}$ (which is the set S_1 obtained in Example 4).

A natural generalization of the concept of internally stable set is the following: a set $S \subseteq N$ is called "*d-internally stable*", if for any two vertices $i, j \in S$ there is no path of length at most d joining them. Similarly, we can define the notions of (*absolutely*) *maximal d-internally stable set* and *number of d-internal stability*. Notice that a d-internally stable set is also d'-internally stable, for any integer d' with $0 < d' \leq d$. The 1-internally stable sets are simply the internally stable sets.

The above described computational methods can be adapted to the new concepts, by simply taking the matrix

$$A^{[d]} = A \cup A^2 \cup \cdots \cup A^d \tag{23}$$

instead of A. Indeed, a d-internally stable set is nothing but an internally stable set of the graph having $A^{[d]}$ as incidence matrix.

Example 8. The absolutely maximal 2-internally stable sets of the graph in Fig. 3 can be determined as follows. The incidence matrix of this graph is

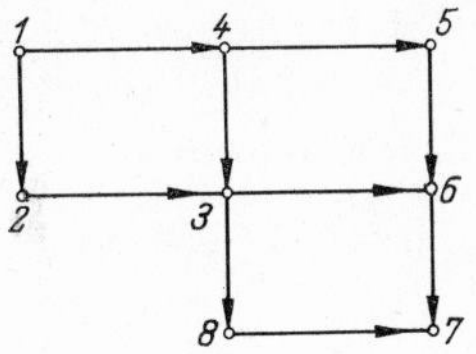

Fig. 3.

$$A = \begin{bmatrix} 0 & 1 & 0 & 1 & 0 & 0 & 0 & 0 \\ 0 & 0 & 1 & 0 & 0 & 0 & 0 & 0 \\ 0 & 0 & 0 & 0 & 0 & 1 & 0 & 1 \\ 0 & 0 & 1 & 0 & 1 & 0 & 0 & 0 \\ 0 & 0 & 0 & 0 & 0 & 1 & 0 & 0 \\ 0 & 0 & 0 & 0 & 0 & 0 & 1 & 0 \\ 0 & 0 & 0 & 0 & 0 & 0 & 0 & 0 \\ 0 & 0 & 0 & 0 & 0 & 0 & 1 & 0 \end{bmatrix},$$

hence,

$$A^2 = \begin{bmatrix} 0 & 0 & 1 & 0 & 1 & 0 & 0 & 0 \\ 0 & 0 & 0 & 0 & 0 & 1 & 0 & 1 \\ 0 & 0 & 0 & 0 & 0 & 0 & 1 & 0 \\ 0 & 0 & 0 & 0 & 0 & 1 & 0 & 1 \\ 0 & 0 & 0 & 0 & 0 & 0 & 1 & 0 \\ 0 & 0 & 0 & 0 & 0 & 0 & 0 & 0 \\ 0 & 0 & 0 & 0 & 0 & 0 & 0 & 0 \\ 0 & 0 & 0 & 0 & 0 & 0 & 1 & 0 \end{bmatrix},$$

$$A^{[2]} = A \cup A^2 = \begin{bmatrix} 0 & 1 & 1 & 1 & 1 & 0 & 0 & 0 \\ 0 & 0 & 1 & 0 & 0 & 1 & 0 & 1 \\ 0 & 0 & 0 & 0 & 0 & 1 & 1 & 1 \\ 0 & 0 & 1 & 0 & 1 & 1 & 0 & 1 \\ 0 & 0 & 0 & 0 & 0 & 1 & 1 & 0 \\ 0 & 0 & 0 & 0 & 0 & 0 & 1 & 0 \\ 0 & 0 & 0 & 0 & 0 & 0 & 0 & 0 \\ 0 & 0 & 0 & 0 & 0 & 0 & 1 & 0 \end{bmatrix}.$$

The absolutely maximal 2-internally stable sets coincide with the absolutely maximal internally stable sets of the graph $G^{[2]}$ having $A^{[2]}$ as incidence matrix. By applying any one of the methods described in this section, we find the following sets:

$$\{1, 6, 8\} \text{ and } \{2, 4, 7\}.$$

§ 3. Externally Stable Sets

It follows from Definitions 4 and 10 that a set S is externally stable if and only if its characteristic vector $(x_1, \ldots, x_n)$ satisfies the following condition: for any $i \in N$, if $x_i = 0$, then there exists a vertex $j \in N$, so that $a_{ij} = x_j = 1$. This condition is obviously equivalent to relation $x_i \cup \bigcup_{j=1}^{n} a_{ij} x_j = 1$, which may be written in the form $\bigcup_{j=1}^{n} a'_{ij} x_j = 1$, where

$$a'_{ij} = a_{ij} \cup \delta_{ij} \tag{IX.14}$$

denote the elements of the unitary incidence matrix, as defined in § 2 of the preceding chapter. Hence:

Theorem 6 (K. Maghout [1, 6], P. L. Hammer and I. Rosenberg [1]). *A set $S \subseteqq N$ is externally stable if and only if its characteristic vector satisfies the Boolean equation*

$$\prod_{i=1}^{n} \bigcup_{j=1}^{n} a'_{ij} x_j = 1, \tag{24}$$

or, equivalently, the pseudo-Boolean equation

$$\sum_{i=1}^{n} \prod_{j=1}^{n} (1 - a'_{ij} x_j) = 0. \tag{25}$$

The proof of this theorem is completed by observing that (25) is nothing but the pseudo-Boolean translation of the dual form of equation (24).

Theorem 6 provides us with a method for finding all externally stable sets of a graph, including all (absolutely) minimal externally stable sets.

Using the same idea as that employed for the determination of the maximal internally stable sets, K. Maghout suggests the following way for solving equation (24). Perform all the multiplications in the left-hand side of (24), (using the dual absorption and distributive law) and all possible absorptions, obtaining thus an equation of the form

$$\bigcup x_{h_1} x_{h_2} \ldots x_{h_{k(h)}} = 1. \tag{26}$$

We have the following

Theorem 7. *If* $x_{h_1} x_{h_2} \ldots x_{h_{k(h)}}$ *is one of the conjunctions appearing in the left-hand side of equation* (26), *then the vector* $(x_1^*, \ldots, x_n^*)$ *defined by*

$$(27) \qquad x_i^* = \begin{cases} 1, & \text{if } i = h_1, h_2, \ldots, h_{k(h)}, \\ 0, & \text{otherwise,} \end{cases}$$

is the characteristic vector of a minimal externally stable set, and all minimal externally stable sets may be obtained in this way.

Proof. Similar to that of Theorem 2.

Example 9. Find the minimal externally stable sets of the graph in Fig. 2. Equation (24) becomes

$$(28) \quad (x_a \cup x_b \cup x_h)(x_a \cup x_b \cup x_c \cup x_d \cup x_e \cup x_h)(x_b \cup x_c \cup x_d)(x_b \cup x_c \cup \\ \cup x_d \cup x_e \cup x_f \cup x_g \cup x_h \cup x_i)(x_b \cup x_d \cup x_e \cup x_f)(x_d \cup x_e \cup x_f \cup \\ \cup x_g)(x_d \cup x_f \cup x_g)(x_a \cup x_b \cup x_d \cup x_h \cup x_i)(x_d \cup x_h \cup x_i) = 1,$$

which, in view of the dual absorption law, reduces to

$$(29) \quad (x_a \cup x_b \cup x_h)(x_b \cup x_c \cup x_d)(x_b \cup x_d \cup x_e \cup x_f)(x_d \cup x_f \cup x_g)(x_d \cup x_h \cup x_i) = 1,$$

and, eventually, to

$$x_a x_d \cup x_b x_d \cup x_d x_h \cup x_b x_f x_h \cup x_b x_f x_i \cup x_b x_g x_h \cup \\ \cup x_b x_g x_i \cup x_c x_f x_h \cup x_a x_c x_f x_i \cup x_c x_e x_g x_h \cup x_a x_c x_e x_g x_i = 1,$$

so that the minimal externally stable sets are

$$\{a, d\}, \quad \{b, d\}, \quad \{d, h\}, \quad \{b, f, h\}, \quad \{b, f, i\},$$
$$\{b, g, h\}, \quad \{b, g, i\}, \quad \{c, f, h\}, \quad \{a, c, f, i\}, \quad \{c, e, g, h\}, \quad \{a, c, e, g, i\}.$$

Remark 2. The determination of the minimal externally stable sets (including in particular, the absolutely minimal ones), is, in fact, an indirect way for determining all externally stable sets, because any externally stable set is characterized by the property of including a minimal externally stable set.

Another procedure for finding the minimal externally stable sets is given by the following

Theorem 8 (S. Rudeanu [9]; see also K. Maghout [1]). *A set* $S \subseteq N$ *is minimal externally stable if and only if its characteristic vector* $(x_1, \ldots, x_n)$ *satisfies the system of Boolean equations*

$$(30) \qquad x_i = \bigcup_{h=1}^{n} \prod_{\substack{j=1 \\ j \neq i}}^{n} \overline{(a'_{hj} \cup \bar{x}_j)} \qquad (i = 1, \ldots, n).$$

Proof*. For the sake of simplicity, let us denote equation (24), which characterizes the externally stable sets, by $f(x_1, \ldots, x_n) = 1$.

* For another proof see Corollary VII.2.

The characteristic vector $(x_1, \ldots, x_n)$ of a minimal externally stable set obviously satisfies the following two conditions:

(31) for any i, if $x_i = 0$, then $f(x_1, \ldots, x_{i-1}, 0, x_{i+1}, \ldots, x_n) = 1$;

(32) for any i, if $x_i = 1$, then $f(x_1, \ldots, x_{i-1}, 0, x_{i+1}, \ldots, x_n) = 0$;

(because $x_i = 0$ implies $f(x_1, \ldots, x_{i-1}, 0, x_{i+1}, \ldots, x_n) = f(x_1, \ldots, x_n) = 1$, while property (32) expresses the minimality).

Conversely, let S be a set whose characteristic vector $(x_1, \ldots, x_n)$ satisfies conditions (31) and (32). If all $x_i = 1$, then $f(x_1, \ldots, x_n) = f(1, \ldots, 1) = 1$, because of (24) and of relation $a'_{ii} = 1$; if there is an index i for which $x_i = 0$, then $f(x_1, \ldots, x_n) = f(x_1, \ldots, x_{i-1}, 0, x_{i+1}, \ldots, x_n) = 1$, by (31). In both cases, $f(x_1, \ldots, x_n) = 1$, showing that S is an externally stable set; moreover, S is minimal, according to (32).

We have thus proved that the minimal externally stable sets are characterized by properties (31) and (32), which can simply be written in the form

$$x_i = \bar{f}(x_1, \ldots, x_{i-1}, 0, x_{i+1}, \ldots, x_n) \qquad (i = 1, \ldots, n) \tag{33}$$

and thus coincide with (30).

Computational remark. If equation (24) can be written in a simpler form $g(x_1, \ldots, x_n) = 1$, then it is convenient to replace the system (33) by

$$x_i = \bar{g}(x_1, \ldots, x_{i-1}, 0, x_{i+1}, \ldots, x_n) \qquad (i = 1, \ldots, n). \tag{34}$$

Example 10. We have seen that in the case of the graph in Fig. 2, equation (24) becomes (28) and can be written in the simpler form (29). Thus, in view of the above remark, the equations which characterize the minimal externally stable sets are the following:

$$\begin{aligned}
x_a &= \bar{x}_b \bar{x}_h \cup \bar{x}_b \bar{x}_c \bar{x}_d \cup \bar{x}_b \bar{x}_d \bar{x}_e \bar{x}_f \cup \bar{x}_d \bar{x}_f \bar{x}_g \cup \bar{x}_d \bar{x}_h \bar{x}_i, \\
x_b &= \bar{x}_a \bar{x}_h \cup \bar{x}_c \bar{x}_d \cup \bar{x}_d \bar{x}_e \bar{x}_f \cup \bar{x}_d \bar{x}_f \bar{x}_g \cup \bar{x}_d \bar{x}_h \bar{x}_i, \\
x_c &= \bar{x}_a \bar{x}_b \bar{x}_h \cup \bar{x}_b \bar{x}_d \cup \bar{x}_b \bar{x}_d \bar{x}_e \bar{x}_f \cup \bar{x}_d \bar{x}_f \bar{x}_g \cup \bar{x}_d \bar{x}_h \bar{x}_i \\
&= \bar{x}_a \bar{x}_b \bar{x}_h \cup \bar{x}_b \bar{x}_d \cup \bar{x}_d \bar{x}_f \bar{x}_g \cup \bar{x}_d \bar{x}_h \bar{x}_i, \\
x_d &= \bar{x}_a \bar{x}_b \bar{x}_h \cup \bar{x}_b \bar{x}_c \cup \bar{x}_b \bar{x}_e \bar{x}_f \cup \bar{x}_f \bar{x}_g \cup \bar{x}_h \bar{x}_i, \\
x_e &= \bar{x}_a \bar{x}_b \bar{x}_h \cup \bar{x}_b \bar{x}_c \bar{x}_d \cup \bar{x}_b \bar{x}_d \bar{x}_f \cup \bar{x}_d \bar{x}_f \bar{x}_g \cup \bar{x}_d \bar{x}_h \bar{x}_i, \\
x_f &= \bar{x}_a \bar{x}_b \bar{x}_h \cup \bar{x}_b \bar{x}_c \bar{x}_d \cup \bar{x}_b \bar{x}_d \bar{x}_e \cup \bar{x}_d \bar{x}_g \cup \bar{x}_d \bar{x}_h \bar{x}_i, \\
x_g &= \bar{x}_a \bar{x}_b \bar{x}_h \cup \bar{x}_b \bar{x}_c \bar{x}_d \cup \bar{x}_b \bar{x}_d \bar{x}_e \bar{x}_f \cup \bar{x}_d \bar{x}_f \cup \bar{x}_d \bar{x}_h \bar{x}_i \\
&= \bar{x}_a \bar{x}_b \bar{x}_h \cup \bar{x}_b \bar{x}_c \bar{x}_d \cup \bar{x}_d \bar{x}_f \cup \bar{x}_d \bar{x}_h \bar{x}_i, \\
x_h &= \bar{x}_a \bar{x}_b \cup \bar{x}_b \bar{x}_c \bar{x}_d \cup \bar{x}_b \bar{x}_d \bar{x}_e \bar{x}_f \cup \bar{x}_d \bar{x}_f \bar{x}_g \cup \bar{x}_d \bar{x}_i, \\
x_i &= \bar{x}_a \bar{x}_b \bar{x}_h \cup \bar{x}_b \bar{x}_c \bar{x}_d \cup \bar{x}_b \bar{x}_d \bar{x}_e \bar{x}_f \cup \bar{x}_d \bar{x}_f \bar{x}_g \cup \bar{x}_d \bar{x}_h.
\end{aligned}$$

This system of Boolean equations was solved in Example II.7; we see that we have, of course, obtained the same result as in the previous Example 9.

Theorem 9. *Let S be a set of vertices and $(x_1^*, \ldots, x_n^*)$ its characteristic vector. The set S is a minimal externally stable set, if and only if the pseudo-Boolean function*

$$(35)\qquad f(x_1, \ldots, x_n) = \sum_{i=1}^{n} x_i + (n+1) \sum_{i=1}^{n} \prod_{j=1}^{n} (1 - a'_{ij} x_j)$$

reaches a local minimum in $(x_1^*, \ldots, x_n^*)$ and $f(x_1^*, \ldots, x_n^*) \leqq n$.

Proof. Follows immediately from Theorem 6 and the theorem proved in Chapter VII, § 1, note added in proofs.

Computational remarks. 1) It is convenient to perform in (35) the transformations $x_i = 1 - y_i$ $(i = 1, \ldots, n)$.

2) If $g = 1$ is a simpler form of the Boolean equation (24) then it is convenient to replace the function (35) by $\sum_{i=1}^{n} x_i + (n+1) g'$, where g' is the pseudo-Boolean form of g. Hence: if π and ϱ are products of the variables $y_i = \bar{x}_i$ and both π and $\pi\,\varrho$ appear in the expression (35), then we can simply delete $\pi\,\varrho$.

Taking into account Theorem VII.1, we deduce the following

Corollary 2. All minimal externally stable sets may be obtained by solving the following system of pseudo-Boolean inequalities:

$$(36)\qquad (2x_i - 1)\,[f(x_1, \ldots, x_{i-1}, 1, x_{i+1}, \ldots, x_n) - f(x_1, \ldots, x_{i-1}, 0, x_{i+1}, \ldots, x_n)] \leqq 0 \qquad (i = 1, \ldots, n),$$

$$(37)\qquad \sum_{h=1}^{n} x_h + (n+1) \sum_{h=1}^{n} \prod_{k=1}^{n} (1 - a'_{hk} x_k) \leqq n.$$

Example 11. In case of the graph of Fig. 2, the function (35) becomes, taking into account the above computational remarks and the simplified form (29) of equation (28),

$$(38)\qquad f = 9 - y_a - y_b - y_c - y_d - y_e - y_f - y_g - y_h - y_i + 10\,(y_a y_b y_h + y_b y_c y_d + y_b y_d y_e y_f + y_d y_f y_g + y_d y_h y_i),$$

hence the inequations (36) and (37) are

$$(2y_a - 1)\,(-1 + 10 y_b y_h) \leqq 0,$$

$$(2y_b - 1)\,[-1 + 10\,(y_a y_h + y_c y_d + y_d y_e y_f)] \leqq 0,$$

$$(2y_c - 1)\,(-1 + 10 y_b y_d) \leqq 0,$$

$$(2y_d - 1)\,[-1 + 10\,(y_b y_c + y_b y_e y_f + y_f y_g + y_h y_i)] \leqq 0,$$

$$(2y_e - 1)\,(-1 + 10 y_b y_d y_f) \leqq 0,$$

$$(2y_f - 1)\,[-1 + 10\,(y_b y_d y_e + y_d y_g)] \leqq 0,$$

$$(2y_g - 1)\,(-1 + 10 y_d y_f) \leqq 0,$$

$$(2y_h - 1)\,[-1 + 10\,(y_a y_b + y_d y_i)] \leqq 0,$$

$$(2y_i - 1)\,(-1 + 10 y_d y_h) \leqq 0,$$

$$f \leqq 9,$$

which are easily solved by the methods described in Chapter IV and lead to the same results as in Examples 9 and 10.

Theorem 10 (P. L. Hammer and I. Rosenberg [1]). *Consider the pseudo-Boolean function*

$$(35)\qquad f(x_1, \ldots, x_n) = \sum_{i=1}^{n} x_i + (n+1) \sum_{i=1}^{n} \prod_{j=1}^{n} (1 - a'_{ij}\, x_j);$$

let S be a subset of N and denote by $(x_1^*, \ldots, x_n^*)$ its characteristic vector. Then:

(i) *The number $\beta(G)$ of external stability of the graph is equal to the global minimum of the pseudo-Boolean function f.*

(ii) *The set $S \subseteqq N$ is an absolutely minimal externally stable set if and only if f reaches its global minimum in $(x_1^*, \ldots, x_n^*)$.*

Proof. Similar to that of Theorem 5.

Example 12. In order to determine the absolutely minimal externally stable sets and the number $\beta(G)$ of external stability of the graph G in Fig. 2, we have to minimize the pseudo-Boolean function (38).

Using the algorithm for minimizing pseudo-Boolean functions, we find that $\beta(G) = f_{\min} = 2$ and the absolutely minimal externally stable sets are $\{a, d\}$, $\{b, d\}$ and $\{d, h\}$.

As in the case of internally stable sets, we can introduce a generalization. Namely, we shall say that a set $S \subseteqq N$ is *d-externally stable*, if for any vertex i not in S, there exists a vertex $j \in S$ so that there is a path of length at most d from i to j. Similarly, we can define the notions of (*absolutely*) *minimal d-externally stable set* and *number of d-external stability*. Notice that a d-externally stable set is also d'-externally stable, for any integer $d' \geqq d$. The 1-externally stable sets are simply the externally stable sets.

As in the preceding section, the above described computational methods can be adapted to the new concepts, by taking the matrix

$$(23)\qquad A^{[d]} = A \cup A^2 \cup \cdots \cup A^d$$

instead of A.

The notion of d-externally stable set was introduced, in a slightly different form, by S. L. Hakimi [1], who studied the following problem: what is the minimum number of policemen that can be distributed in a highway network so that no point is farther away from a policeman than a given distance d. In the terminology adopted here, this is simply the problem of finding the number of d-external stability and an absolutely minimal d-externally stable set.

Various natural generalizations of the above discussed concepts may be of importance. For instance, Hakimi [1] solves the problem of distributing a minimum number of policemen within the distance d_1 and such that each policeman is within a distance d_2 from another policeman. If $(x_1, \ldots, x_n)$ denotes the characteristic vector of the sought set and $a_{ij}^{[d]}$ are the elements of the matrix (23), then the above problem may be translated either as the problem of finding those solutions of the Boolean equation

$$\prod_{i=1}^{n}\left[x_i \bigcup_{\substack{j=1\\j\neq i}}^{n} a_{ij}^{[d_2]} x_j \cup \bar{x}_i \bigcup_{\substack{j=1\\j\neq i}}^{n} a_{ij}^{[d_1]} x_j\right] = 1 \tag{39}$$

which have the minimum number of ones, or as the problem of minimizing the pseudo-Boolean function

$$f = \sum_{i=1}^{n} x_i + (n+1)\sum_{i=1}^{n}\left[x_i \prod_{\substack{j=1\\j\neq i}}^{n}(1 - a_{ij}^{[d_2]} x_j) + (1 - x_i)\prod_{\substack{j=1\\j\neq i}}^{n}(1 - a_{ij}^{[d_1]} x_j)\right]. \tag{40}$$

§ 4. Kernels of Graphs

We recall that a subset K of N is called a *kernel*, if it is both internally and externally stable (Definition 7). In this section we shall be concerned with the Boolean treatment of a problem which was paid a great attention in graph theory, namely that of determining the kernels of a graph (if any).

Theorem 11 (S. Rudeanu [11]). *Each of the following four systems of Boolean equations characterizes the kernels of a graph:* (6) *and* (24), (6) *and* (30), (24) *and* (11), (11) *and* (30).

Proof. It is well-known (see, for instance, C. Berge [2]) and, in fact, it follows easily from the definitions, that each of the following four conditions is equivalent to the property of S to be a kernel:

S is both internally and externally stable;

S is both internally and minimal externally stable;

S is both externally and maximal internally stable;

S is both maximal internally and minimal externally stable.

Taking into account, Theorems 1, 3, 6 and 8, we see that the above four conditions are translated by the four systems of equations given in the statement of Theorem 11.

Computational remark. The system consisting of equations (24) and (11) seems to be the most convenient for computation.

Example 13. For the graph in Fig. 4 of the preceding chapter, equations (11) and (24) become

$$x_1 = \bar{x}_2, \tag{41.1}$$

$$x_2 = \bar{x}_1 \bar{x}_3 \bar{x}_5, \tag{41.2}$$

$$x_3 = 0, \tag{41.3}$$

$$x_4 = \bar{x}_3 \bar{x}_5 \bar{x}_6, \tag{41.4}$$

$$x_5 = \bar{x}_2 \bar{x}_4 \bar{x}_6, \tag{41.5}$$

$$x_6 = \bar{x}_4 \bar{x}_5 \bar{x}_7, \tag{41.6}$$

$$x_7 = \bar{x}_6, \tag{41.7}$$

and

$$(x_1 \cup x_2)(x_2 \cup x_1 \cup x_5)(x_3 \cup x_2)(x_4 \cup x_3 \cup x_6)(x_5 \cup x_4)(x_6 \cup x_7)\, x_7 = 1, \tag{42}$$

respectively.

Equation (42) shows that $x_7 = 1$, hence $x_6 = 0$, by (41.7). It also implies that $x_3 \cup x_2 = 1$, hence $x_2 = 1$, by (41.3). Therefore we deduce from (41.2) that $x_1 = x_5 = 0$, hence $x_4 = 1$ by (41.4). These values satisfy all equations (41) and (42), so that $\{2, 4, 7\}$ is the unique kernel of the graph.

Another system of equations characterizing the kernels of a graph is given in the following:

Theorem 12 (C. Berge [2], P. L. Hammer and I. Rosenberg [1]). *A set S of vertices is a kernel if and only if its characteristic vector $(x_1, \ldots, x_n)$ satisfies the system of Boolean equations*

$$x_i = \prod_{j=1}^{n} (\bar{a}_{ij} \cup \bar{x}_j) \qquad (i = 1, \ldots, n), \tag{43}$$

or, equivalently, the pseudo-Boolean equation

$$\sum_{i=1}^{n} \left[(1 - x_i) \prod_{j=1}^{n} (1 - a_{ij}\, x_j) + \sum_{j=1}^{n} a_{ij}\, x_i\, x_j \right] = 0. \tag{44}$$

Proof. According to Theorem 1, an internally stable set is characterized by equation (6), which is equivalent to the system

$$x_i \bigcup_{j=1}^{n} a_{ij}\, x_j = 0 \qquad (i = 1, \ldots, n)$$

and this may also be written in the form

$$x_i \leqq \prod_{j=1}^{n} (\bar{a}_{ij} \cup \bar{x}_j) \qquad (i = 1, \ldots, n). \tag{45}$$

Further, Theorem 6 states that an externally stable set is characterized by equation (24). Since $a'_{ii}\, x_i = x_i = x_i \cup a_{ii}\, x_i$ and $a'_{ij} = a_{ij}$ for $i \neq j$, equation (24) is equivalent to the system

$$x_i \cup \bigcup_{j=1}^{n} a_{ij}\, x_j = 1 \qquad (i = 1, \ldots, n)$$

which may also be written in the form

$$\prod_{j=1}^{n} (\bar{a}_{ij} \cup \bar{x}_j) \leqq x_i \qquad (i = 1, \ldots, n). \tag{46}$$

Thus a set S is a kernel if and only if it satisfies both (45) and (46); but these inequalities are equivalent to the system of equations (43).

Further, the system (43) may also be written in the form

$$\bigcup_{i=1}^{n} \left[\bar{x}_i \prod_{j=1}^{n} (\bar{a}_{ij} \cup \bar{x}_j) \cup x_i \bigcup_{j=1}^{n} a_{ij}\, x_j \right] = 0 \tag{47}$$

since the terms occuring in equation (47) are non-negative, we may replace (47) by (44).

Remark 3. The system (43) can also be written in the form

$$(43') \qquad x_i = \overline{a_{ii}} \prod_{\substack{j=1 \\ j \neq i}}^{n} (\overline{a_{ij}} \cup \overline{x_j}) \qquad (i = 1, \ldots, n),$$

because (43) and (43′) coincide if $a_{ii} = 0$, while for $a_{ii} = 1$ both (43) and (43′) imply $x_i = 0$.

Hence we deduce:

Remark 4. In case of symmetric graphs $a_{ij}^* = a_{ij}$, therefore Theorems 3 and 12 show that for these graphs, the concepts of maximal internally stable set and kernel coincide.

Example 14. 1) The kernels of the symmetric graph in Fig. 2 coincide with its maximal internally stable sets, which were determined, by applying various methods, in Examples 4, 5.1 and 6.

2) For the graph in Fig. 4 of the preceding chapter, equations (43) become

$$(48.1) \qquad x_1 = \bar{x}_2,$$

$$(48.2) \qquad x_2 = \bar{x}_1 \bar{x}_5,$$

$$(48.3) \qquad x_3 = \bar{x}_2 \bar{x}_3,$$

$$(48.4) \qquad x_4 = \bar{x}_3 \bar{x}_6,$$

$$(48.5) \qquad x_5 = \bar{x}_4,$$

$$(48.6) \qquad x_6 = \bar{x}_7,$$

$$(48.7) \qquad x_7 = 1.$$

Equation (48.3) shows that $x_3 \leqq \bar{x}_3$, implying thus $x_3 = 0$. Therefore (48.3) reduces to $0 = \bar{x}_2$. Thus $x_2 = 1$ and we have $x_1 = 0$ and $1 = \bar{x}_5$, by (48.1) and (48.2), respectively. Since $x_5 = 0$, equation (48.5) shows that $x_4 = 1$. Finally, $x_6 = 0$ by (48.6) and (48.7). We have, of course, obtained the same result as in Example 13.

Theorem 13 (P. L. Hammer and I. Rosenberg [1]). *Let S be a set of vertices and $(x_1^*, \ldots, x_n^*)$ its characteristic vector. The set S is a kernel if and only if the pseudo-Boolean function*

$$(49) \qquad f(x_1, \ldots, x_n) = \sum_{i=1}^{n} \left[(1 - x_i) \prod_{j=1}^{n} (1 - a_{ij} x_j) + \sum_{j=1}^{n} a_{ij} x_i x_j \right]$$

reaches its minimum in $(x_1^*, \ldots, x_n^*)$ and this minimum is zero. If $f_{\min} > 0$, then the graph has no kernel.

Proof. Immediate from Theorem 12 and the fact that $f(x_1, \ldots, x_n) \geqq 0$ for all $x_1, \ldots, x_n$.

Example 15. For the graph in Fig. 4 of the preceding chapter, the function (49) becomes

$$f = (1 - x_1)(1 - x_2) + x_1 x_2 + (1 - x_2)(1 - x_1)(1 - x_5) + x_2 x_1 + x_2 x_5 + \\ + (1 - x_3)(1 - x_2)(1 - x_3) + x_3 x_2 + x_3 x_3 + (1 - x_4)(1 - x_3)(1 - x_6) + \\ + x_4 x_3 + x_4 x_6 + (1 - x_5)(1 - x_4) + x_5 x_4 + (1 - x_6)(1 - x_7) + \\ + x_6 x_7 + (1 - x_7)$$

and reaches the minimum 0 in the point $(0, 1, 0, 1, 0, 0, 1)$.

§ 5. Chromatic Decompositions

We recall that a chromatic decomposition of a graph $G = (N, \Gamma)$ is a family of disjoint internally stable sets covering the set N (Definition 8). A chromatic decomposition with the smallest number $\gamma(G)$ of sets is called a minimal chromatic decomposition, while $\gamma(G)$ is the chromatic number.

Lemma 1. *Let γ_1 denote the smallest number of internally stable sets (not necessarily disjoint) covering N. Then:*

(i) $\gamma(G) = \gamma_1$;

(ii) *If $S_1, S_2, \ldots, S_{\gamma_1}$ are γ_1 internally stable sets covering N, then the disjoint internally stable sets*

$$C_1 = S_1, \tag{50.1}$$

$$C_2 = S_2 - S_1, \tag{50.2}$$

$$\cdots\cdots\cdots\cdots$$

$$C_i = S_i - \bigcup_{j=1}^{i-1} S_j, \tag{50. i}$$

$$\cdots\cdots\cdots\cdots$$

$$C_{\gamma_1} = S_{\gamma_1} - \bigcup_{j=1}^{\gamma_1 - 1} S_j \tag{50. γ_1}$$

cover the set N.

Proof. If $C_i = \emptyset$, then $S_i \subseteq \bigcup_{j=1}^{i-1} S_j$ and the family $\{S_1, \ldots, S_{i-1}, S_{i+1}, \ldots, S_{\gamma_1}\}$ would be a covering of N with $\gamma_1 - 1$ internally stable sets, in contradiction with the definition of γ_1. Therefore all $C_i \neq \emptyset$ and thus there are γ_1 disjoint internally stable (for $C_i \subseteq S_i$) sets C_i which, of course, cover N. This shows that $\gamma \leqq \gamma_1$; since the converse inequality does obviously hold, the proof is complete.

Lemma 2. *Let γ_2 be the smallest number of maximal internally stable sets covering N. Then $\gamma_1 = \gamma_2$.*

Proof. Any internally stable set can be imbedded in a maximal internally stable one. Hence, for any family $C_1, \ldots, C_{\gamma_1}$ of internally stable sets covering N, there exists a family $D_1, \ldots, D_k$ of maximal internally stable sets covering N with $k \leqq \gamma_1$. Hence $\gamma_2 \leqq \gamma_1$; as the converse relation does obviously hold, the lemma is proved.

Theorem 14 (J. WEISSMAN [1], P. L. HAMMER and I. ROSENBERG [1]. *The chromatic number γ of a graph $G = (N, \Gamma)$ is equal to the smallest number of maximal internally stable sets covering N. If $S_1, \ldots, S_\gamma$ are γ maximal internally stable sets covering N, and the sets $C_1, \ldots, C_\gamma$ are defined by relations* (50), *then $\{C_1, \ldots, C_\gamma\}$ is a minimal chromatic decomposition of G.*

Proof. Obvious from Lemmas 1 and 2.

This theorem reduces the problem of finding a minimal chromatic decomposition to that of finding a covering of N with the smallest number of maximal internally stable sets. These sets being determined by any of the procedures given in § 2, the latter problem will be solved using methods similar to those employed in the previous sections.

Definition 11. Let $M_1, \ldots, M_m$ be all the maximal internally stable sets of the graph $G = (N, \Gamma)$. The *covering matrix* is the $n \times m$ Boolean matrix $C = (c_{ij})$ defined by

$$c_{ij} = \begin{cases} 1, & \text{if} \quad i \in M_j, \\ 0, & \text{if} \quad i \notin M_j. \end{cases} \tag{51}$$

Definition 12. Let $M_1, \ldots, M_m$ be all the maximal internally stable sets of the graph $G = (N, \Gamma)$. To each family $\mathcal{F}$ of maximal internally stable sets we associate a *characteristic vector* $(y_1, \ldots, y_m)$, defined by

$$y_j = \begin{cases} 1, & \text{if} \quad M_j \in \mathcal{F}, \\ 0, & \text{if} \quad M_j \notin \mathcal{F}. \end{cases} \tag{52}$$

Theorem 15*. *A necessary and sufficient condition for a family $\mathcal{F}$ of maximal internally stable sets to cover N, is the fulfilment of the Boolean equation*

$$\prod_{i=1}^{n} \bigcup_{j=1}^{m} c_{ij} \, y_j = 1, \tag{53}$$

or, equivalently, of the pseudo-Boolean equation

$$\sum_{i=1}^{n} \prod_{j=1}^{m} (1 - c_{ij} \, y_j) = 0. \tag{54}$$

Proof. The element $i \in N$ is covered by at least one set M_j if and only if $\bigcup_{j=1}^{m} c_{ij} \, y_j = 1$. As this relation must hold for each i, we obtain equation (53), the dual of which may be written in the pseudo-Boolean form (54).

* This theorem (or variants of it) was independently found by several authors.

Now we are faced with the problem of solving equation (53) with the smallest number of ones.

To do this, let us notice that equation (53) is of the same type as equation (24): more precisely, in the case when $m = n$, relation (53) reduces to an equation of the form (24). Therefore the techniques employed in § 3 for determining the absolutely minimal externally stable sets are suitable for finding (all!) coverings with the smallest number of maximal externally stable sets.

We have seen in § 3 that the absolutely minimal externally stable sets may be found either by determining all minimal externally stable sets (as done in Theorem 7, or 8, or 9) and then selecting the absolutely minimal ones, or by directly applying the pseudo-Boolean method, as done in Theorem 10, which supplies us exactly with the absolutely minimal externally stable sets.

Another method for finding the minimal coverings will be suggested in Example 16.2 below.

Example 16.

1) The graph in Fig. 2 has 9 vertices $a, b, \ldots, i$, and 7 maximal internally stable sets $M_1, \ldots, M_7$, which were determined in Example 4 and also in Examples 5.1 and 6. In this case, the covering matrix is the 9×7 Boolean matrix

$$C = \begin{bmatrix} 1 & 0 & 1 & 0 & 1 & 0 & 0 \\ 0 & 0 & 0 & 0 & 0 & 1 & 1 \\ 1 & 1 & 1 & 1 & 0 & 0 & 0 \\ 0 & 0 & 0 & 0 & 1 & 0 & 0 \\ 1 & 1 & 0 & 0 & 0 & 0 & 0 \\ 0 & 0 & 1 & 1 & 0 & 1 & 0 \\ 1 & 1 & 0 & 0 & 0 & 0 & 1 \\ 0 & 1 & 0 & 1 & 0 & 0 & 0 \\ 1 & 0 & 1 & 0 & 0 & 1 & 1 \end{bmatrix}, \tag{55}$$

so that equation (53) becomes

$$(y_1 \cup y_3 \cup y_5)(y_6 \cup y_7)(y_1 \cup y_2 \cup y_3 \cup y_4)\, y_5 (y_1 \cup y_2) \cdot \tag{56}$$
$$\cdot (y_3 \cup y_4 \cup y_6)(y_1 \cup y_2 \cup y_7)(y_2 \cup y_4)(y_1 \cup y_3 \cup y_6 \cup y_7) = 1$$

and may be written in the simpler form

$$(y_6 \cup y_7)\, y_5 (y_1 \cup y_2)(y_3 \cup y_4 \cup y_6)(y_2 \cup y_4) = 1, \tag{57}$$

because of the absorptions.

By applying, for instance, the Maghout procedure, we perform the multiplications indicated in the left-hand side of (57), obtaining thus

$$y_2 y_5 y_6 \cup y_1 y_4 y_5 y_6 \cup y_2 y_3 y_5 y_7 \cup y_2 y_4 y_5 y_7 \cup y_1 y_4 y_5 y_7 = 1,$$

The shortest term in this equation is $y_2 y_5 y_6$; therefore the chromatic number is 3. Since the sets $M_2 = \{c, e, g, h\}$, $M_5 = \{a, d\}$ and $M_6 = \{b, f, i\}$ are pairwise disjoint, it follows that $\{M_2, M_5, M_6\}$ is the unique minimal chromatic decomposition.

2) We have shown that the graph studied in Example 5.2 has 28 maximal internally stable sets, whose characteristic vectors are the solutions of the system of equations solved in Example II.9. The table of solutions given there is nothing but the transpose of the covering matrix of this graph.

In order to obtain the minimal chromatic decompositions, we shall decompose equation (53) into a system of n equations

$$\bigcup_{j=1}^{m} c_{ij} y_j = 1 \qquad (i = 1, \ldots, n). \tag{58}$$

In our case, the system (58) is

(59.1) $y_1 = 1$,

(59.2) $y_2 \cup y_3 \cup y_4 \cup y_5 = 1$,

(59.3) $y_2 \cup y_4 \cup y_6 \cup y_7 \cup y_8 = 1$,

(59.4) $y_1 \cup y_6 \cup y_9 = 1$,

(59.5) $y_8 \cup y_{10} \cup y_{11} \cup y_{12} \cup y_{13} = 1$,

(59.6) $y_2 \cup y_3 \cup y_7 \cup y_{14} \cup y_{15} \cup y_{16} \cup y_{17} = 1$,

(59.7) $y_{18} \cup y_{19} \cup y_{20} \cup y_{21} \cup y_{22} \cup y_{23} = 1$,

(59.8) $y_9 \cup y_{18} \cup y_{19} \cup y_{24} \cup y_{25} = 1$,

(59.9) $y_{14} \cup y_{15} \cup y_{20} \cup y_{21} \cup y_{22} \cup y_{23} = 1$,

(59.10) $y_{19} \cup y_{22} \cup y_{23} \cup y_{26} \cup y_{27} = 1$,

(59.11) $y_9 \cup y_{10} \cup y_{11} \cup y_{12} \cup y_{13} \cup y_{24} \cup y_{25} = 1$,

(59.12) $y_3 \cup y_5 \cup y_{16} \cup y_{17} \cup y_{26} \cup y_{27} \cup y_{28} = 1$,

(59.13) $y_7 \cup y_8 \cup y_{10} \cup y_{12} \cup y_{14} \cup y_{16} \cup y_{18} \cup y_{20} \cup y_{21} \cup y_{24} \cup y_{28} = 1$,

(59.14) $y_{10} \cup y_{11} \cup y_{14} \cup y_{15} \cup y_{16} \cup y_{17} \cup y_{20} \cup y_{22} \cup y_{26} = 1$,

(59.15) $y_{11} \cup y_{13} \cup y_{15} \cup y_{17} \cup y_{19} \cup y_{22} \cup y_{23} \cup y_{25} \cup y_{26} \cup y_{27} = 1$,

(59.16) $y_9 \cup y_{11} \cup y_{13} \cup y_{19} \cup y_{22} \cup y_{23} \cup y_{25} \cup y_{26} \cup y_{27} = 1$,

(59.17) $y_3 \cup y_5 \cup y_{11} \cup y_{13} \cup y_{15} \cup y_{17} \cup y_{19} \cup y_{22} \cup y_{23} \cup y_{25} \cup y_{26} \cup y_{27} = 1$,

(59.18) $y_9 \cup \cdots \cup y_{28} = 1$,

(59.19) $y_4 \cup y_5 \cup y_6 \cup y_8 \cup y_9 \cup y_{12} \cup y_{13} \cup y_{18} \cup y_{19} \cup y_{21} \cup y_{23} \cup y_{24} \cup y_{25} \cup y_{27} \cup y_{28} = 1$,

(59.20) $y_2 \cup y_3 \cup y_4 \cup y_5 \cup y_7 \cup y_8 \cup y_{10} \cup \cdots \cup y_{28} = 1$.

But, equation (59.10) implies (59.15), (59.16) and (59.17), while equations (59.1), (59.5) and (59.8) imply (59.4), (59.20) and (59.19), respectively. Thus the system (59) is equivalent to the subsystem {(59.1), (59.2), (59.3), (59.5), (59.6), ..., (59.14)}.

Now we can apply any of the procedures indicated in this section. We suggest below another method, which seems suitable for hand computation. It runs as follows.

Notice first that equations (59.1), (59.2), (59.5), (59.8) and (59.9) correspond to disjoint sets of variables hence *the given system has no solution with less than* 5 *ones.*

Therefore, it seems natural *to seek the solutions with exactly* 5 *ones*, if any.

Remark. Suppose there exists a solution with exactly 5 ones. Then, in each of the equations (59.2), (59.5), (59.8) and (59.9), corresponding to disjoint sets of variables, one of the unknowns equals 1, the other being 0.

In other words, the 5 ones of the solution necessarily correspond to 5 unknowns appearing in (59.1), (59.2), (59.5), (59.8) and (59.9). Therefore

(60) $$y_6 = y_7 = y_{16} = y_{17} = y_{26} = y_{27} = y_{28} = 0$$

hence equations (59.12) and (59.10) reduce to

(61) $$y_3 \cup y_5 = 1,$$

(62) $$y_{19} \cup y_{22} \cup y_{23} = 1.$$

Now *we can omit equation* (59.7), since it is a consequence of (62). On the other hand, taking into account (62) and the above remark, we see that equation (59.2) reduces to

(63) $$y_2 = y_4 = 0,$$

hence (59.3) reduces to

(64) $$y_8 = 1.$$

It follows that *we can omit equations* (59.5) *and* (59.13), as being implied by (64). Moreover, equations (59.5), (64) and the above remark show that

(65) $$y_{10} = y_{11} = y_{12} = y_{13} = 0,$$

hence equation (59.11) reduces to

(66) $$y_9 \cup y_{24} \cup y_{25} = 1,$$

therefore *we can omit equation* (59.8).

Now, since we have already $y_1 = y_8 = 1$, we must solve the rest of the equations with 3 ones. But equations (62), (66) and (59.6), which reduces to

(67) $$y_3 \cup y_{14} \cup y_{15} = 1,$$

correspond to disjoint sets of variables. Therefore the rest of the variables must equal 0, i.e.

(68) $$y_5 = y_{18} = y_{20} = y_{21} = 0,$$

hence equation (61) reduces to

(69) $$y_3 = 1.$$

We notice, as before, that in each of the equations (62), (66) and (67), only one of the unknowns is equal to 1, the other being 0. Therefore, we deduce from (69) that (67) reduces to

(70) $$y_{14} = y_{15} = 0,$$

hence (59.14) implies

(71) $$y_{22} = 1,$$

which shows that (62) reduces to

(72) $$y_{19} = y_{23} = 0.$$

Now the whole system is satisfied provided that y_9, y_{24} and y_{25} satisfy equation (66). We have thus proved that the solutions of our system with the smallest number of ones, are given by the following relations

$$(73)\quad y_2 = y_4 = y_5 = y_6 = y_7 = y_{10} = y_{11} = \cdots = y_{21} = y_{23} = y_{26} = y_{27} = y_{28} = 0,$$

$$(74)\quad y_1 = y_3 = y_8 = y_{22} = 1,$$

$$(75)\quad y_9 + y_{24} + y_{25} = 1$$

(equation (75) expresses the condition that equation (66) is solved with a single one).

Hence the chromatic number is 5 and there are three minimal coverings: $\{M_1, M_3, M_8, M_9, M_{22}\}$, $\{M_1, M_3, M_8, M_{22}, M_{24}\}$ and $\{M_1, M_3, M_8, M_{22}, M_{25}\}$.

The minimal chromatic decompositions can be obtained by applying the procedure indicated in Lemma 1.

A direct procedure for finding the coverings with the smallest number of sets is given by the analogue of Theorem 10:

Theorem 16 (P. L. Hammer and I. Rosenberg [1]). *Consider the pseudo-Boolean function*

$$(76)\qquad f(y_1, \ldots, y_n) = \sum_{j=1}^{m} y_j + (m+1) \sum_{i=1}^{n} \prod_{j=1}^{m} (1 - c_{ij}\, y_j);$$

let $\mathcal{F}$ be a family of maximal internally stable sets and $(y_1^*, \ldots, y_m^*)$ its characteristic vector. Then:

(i) *The chromatic number $\gamma(G)$ is equal to the global minimum of the pseudo-Boolean function f.*

(ii) *The family $\mathcal{F}$ realizes the required covering if and only if f reaches its global minimum in $(y_1^*, \ldots, y_m^*)$.*

Proof. See Theorem 10.

Example 17. For the graph in Fig. 2, the function (76) is

$$f = 7 - z_1 - z_2 - z_3 - z_4 - z_5 - z_6 - z_7 + 8(z_1 z_3 z_5 + z_6 z_7 + z_1 z_2 z_3 z_4 + \\ + z_5 + z_1 z_2 + z_3 z_4 z_6 + z_1 z_2 z_7 + z_2 z_4 + z_1 z_3 z_6 z_7),$$

where $z_j = 1 - y_j$ $(j = 1, \ldots, 7)$, and reaches its minimum $f_{\min} = 3$ in the point $z_2 = z_5 = z_6 = 0$, $z_1 = z_3 = z_4 = z_7 = 1$; i.e. the same result as in Example 16.1.

In case of symmetric graphs, the chromatic number may be determined using the following theorem, due to L. M. Vitaver [1]. Let G be a symmetric graph; giving an arbitrary orientation to its edges, we obtain an oriented graph $\vec{G}$. Let $k(G)$ be the greatest integer m having the property that for any possible orientation, the graph $\vec{G}$ contains at least a path of length m. Then $\gamma(G) = k(G) + 1$.

J. Weissman [1] has considered the following generalization of the minimal chromatic decomposition problem. Assume the set N of vertices is divided into several pairwise disjoint classes $D_1, \ldots, D_p$. The problem is to find a family of pairwise disjoint internally stable sets $E_1, E_2, \ldots, E_q$

so that the set-theoretical union $\bigcup_{i=1}^{q} E_i$ should contain at least one vertex from each class D_j, and so that q should be minimal. (Obviously, if $p = n$ and for every i, D_i is the single-element set $\{i\}$, then the problem reduces to the former one.)

WEISSMAN gives the following interpretation of this problem: "Imagine pins sticking out of a plane surface of a box of electronic components. A list of some pairs of these pins is available. Each pair of pins in this list is to be connected electrically by etching a conducting path on a board through which all the pins protrude. Many such boards may be used, stacked one upon the other with the pins protruding through all. Each etched path must follow one of some prescribed routes between the pair of pins to be connected. Two such path-routes are said to *interfere* if and only if, when paths following these routes are etched on the same board, the paths touch and establish an unwanted electrical connection. The problem is to choose one path-route for each pair of pins to be connected and to assign each path-route to a board so that on any one board no two path-routes interfere and so that the number of boards used is the minimum".

§ 6. The Four Colouring Problem

The four-colouring problem was first stated as follows: prove that every map representing various countries can be coloured with four colours in such a way that every two neighbouring countries are not coloured identically.

Let the various countries be taken as vertices of a graph, two vertices being related by an arc if and only if the corresponding countries have common borders. Then we get a symmetric ($a_{ij} = a_{ji}$) planar graph G without loops; the term "planar" means that the graph can be drawn in plane, so that arcs are meeting only in vertices. The fact that two neighbouring countries are not coloured identically means that for any colour c, the set of all c-coloured vertices is an internally stable set of the graph G. Now the problem may be reformulated in the following way: prove that for every planar graph G without loops (symmetric or not!) there exists a decomposition of the set N of its vertices into four pairwise disjoint internally stable sets.

In view of the results in § 5, an equivalent formulation of the four-colouring conjecture is the following: *the chromatic number of a planar graph without loops is* $\leqq 4$.

Although there exists a wide literature on this famous problem, the conjecture is still unproved. However, there exist methods for obtaining a colouring of a specified graph (planar or not).

Some of these methods were described in § 5. (In particular, it was shown in Example 16.1 that the chromatic number of the planar graph in Fig. 2 is 3.)

For the specific problem of *four*-colouring a graph, R. FORTET [1] has proposed a method using Boolean equations, while G. B. DANTZIG [3] developes an idea of R. GOMORY for converting this problem into one of linear integer programming; this latter method was given a pseudo-Boolean approach in P. L. HAMMER [7].

The method of R. FORTET runs as follows. Let a, b, c, d be the four colours and let the Boolean variables x_i, y_i, be defined by the formulas

$$x_i = \begin{cases} 1, & \text{if vertex } i \text{ is either } a\text{-coloured, or } b\text{-coloured,} \\ 0, & \text{otherwise;} \end{cases} \tag{77}$$

$$y_i = \begin{cases} 1, & \text{if vertex } i \text{ is either } a\text{-coloured, or } c\text{-coloured,} \\ 0, & \text{otherwise.} \end{cases} \tag{78}$$

Then:

$$x_i y_i = 1 \quad \text{if and only if vertex } i \text{ is } a\text{-coloured,} \tag{79}$$

$$x_i \bar{y}_i = 1 \quad \text{if and only if vertex } i \text{ is } b\text{-coloured,} \tag{80}$$

$$\bar{x}_i y_i = 1 \quad \text{if and only if vertex } i \text{ is } c\text{-coloured,} \tag{81}$$

$$\bar{x}_i \bar{y}_i = 1 \quad \text{if and only if vertex } i \text{ is } d\text{-coloured.} \tag{82}$$

Now, if (a_{ij}) is the $(n \times n)$ incidence matrix of the graph G, the given condition may be written in the Boolean form

$$\bigcup_{i=1}^{n} \bigcup_{j=1}^{n} a_{ij} (x_i y_i x_j y_j \cup x_i \bar{y}_i x_j \bar{y}_j \cup \bar{x}_i y_i \bar{x}_j y_j \cup \bar{x}_i \bar{y}_i \bar{x}_j \bar{y}_j) = 0, \tag{83}$$

or, equivalently, in the pseudo-Boolean form

$$\sum_{i=1}^{n} \sum_{j=1}^{n} a_{ij} (x_i y_i x_j y_j + x_i \bar{y}_i x_j \bar{y}_j + \bar{x}_i y_i \bar{x}_j y_j + \bar{x}_i \bar{y}_i \bar{x}_j \bar{y}_j) = 0. \tag{84}$$

Every solution of the above equation yields a colouring of the graph G via formulas (79)—(82).

The GOMORY-DANTZIG approach to this problem runs as follows. Associate to each vertex i a variable t_i taking values in the set $\{0, 1, 2, 3\}$ and indicating the colouring of that vertex. For every two vertices i and j related by an arc, the condition is that $t_i - t_j \neq 0$, or, equivalently,

$$\text{either} \quad t_i - t_j \geqq 1, \quad \text{or} \quad t_j - t_i \geqq 1. \tag{85}$$

But the latter relation can be written as a system consisting of the following two inequalities:

$$t_i - t_j \geqq 1 - 4u_{ij}, \tag{86.1}$$

$$t_j - t_i \geqq -3 + 4u_{ij}, \tag{86.2}$$

where u_{ij} is a new bivalent variable. Setting $t_k = t'_k + 2t''_k$, where t'_k and t''_k are bivalent variables, the system (86) becomes

$$t'_i + 2t''_i - t'_j - 2t''_j + 4u_{ij} \geqq 1, \tag{87.1}$$

$$t'_j + 2t''_j - t'_i - 2t''_i - 4u_{ij} \geqq -3. \tag{87.2}$$

The system (87) can be solved with the methods described in Chapter III.

Chapter XI

Matchings of Bipartite Graphs

It will be shown in this chapter that numerous problems concerning the matchings of bipartite graphs (as, for instance, those on maximal matchings, minimal separation sets, deficiencies, systems of representatives, etc.) may be regarded as problems of pseudo-Boolean programming and handled with the methods described in Chapters III—VI. Outstanding examples are the assignment and transportation problems, discussed in § 2.

§ 1. Bipartite Graphs

Let $N = \{1, 2, \ldots, p\}$ be a finite set, on which a symmetric binary relation ϱ is defined (i.e. relation $i \varrho j$ implies $j \varrho i$).

Definition 1. The pair $G = (N, \varrho)$ is said to be an *undirected graph*, the elements $i \in N$ are called its *vertices* (or *nodes*), while each pair (i, j) of vertices satisfying $i \varrho j$ is termed an *edge*.

Definition 2. An undirected graph $G = (N, \varrho)$ is called *bipartite* if the set N decomposes into two disjoint sets $N' = \{n'_1, \ldots, n'_{m_1}\}$ and $N'' = \{n''_1, \ldots, n''_{m_2}\}$ so that each edge of the graph connects a node $n' \in N'$ to a node $n'' \in N''$. We shall denote the above graph by $G = (N', N''; \varrho)$.

Definition 3. A one-to-one correspondence between two subsets A' and A'' of N' and N'' respectively, is called a *matching*, if every two corresponding vertices are linked by an edge; we say also that A' is *matched* to A''.

The matching is thus characterized by the set of edges involved in it.

For instance, in the bipartite graph of Fig. 1, the edges $(1', 1'')$, $(2', 3'')$, $(3', 4'')$ define a matching.

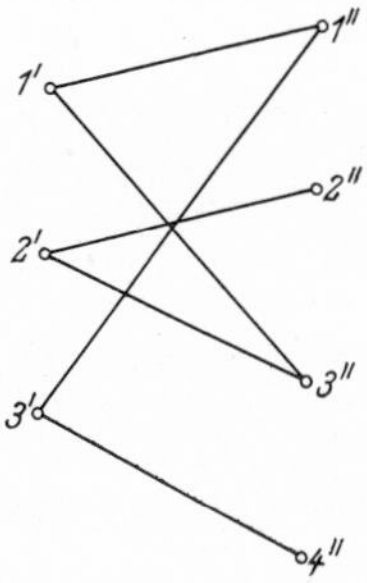

1'
2'
3'
1''
2''
3''
4''

Fig. 1.

As an important example involving matchings, let us consider the so-called *assignment problem*. Several individuals are available for certain jobs. Each individual is able to fulfil some of the existing jobs. We may be interested for instance in the following problems:

1) Which is the greatest number of people which can be assigned to jobs?

2) Is it possible to fill all the positions with qualified people?

3) Given a "qualification-matrix" (c_{ij}), which shows the productivity of the individual i in the job j, how could we make the best assignment (i.e. that one for which the global productivity is maximized)?

Considering a graph $G = (N', N''; \varrho)$ where N' corresponds to the set of individuals, N'' to the set of jobs and where $n' \in N'$ is related by an edge to $n'' \in N''$ whenever n' can perform the job n'', we see that an assignment is nothing but a matching of G, so that all the above problems may be translated as questions regarding matchings in G.

Maximal matchings

Definition 4. A matching M will be called: *maximal*, if there is no other matching M' properly including all the edges in M, and *absolutely maximal**, if no other matching involves a greater number of edges.

We shall show that the problem of determining the maximal matchings of a bipartite graph G reduces to that of determining the absolutely maximal internally stable sets of a certain (undirected) graph G^* associated to G.

An internally stable set S of an undirected graph G is defined as it was done in Chapter X for directed graphs: it is a set $S \subseteqq N$ of vertices so that for every two vertices $s_1, s_2 \subseteqq S$ there is no edge (s_1, s_2) in G.

Definition 5. The *dual* of an undirected graph $G = (N, \varrho)$ is the undirected graph $G^* = (E, \varrho^*)$, where E is the set of all edges of G, and relation $e_1 \varrho^* e_2$ holds if and only if the edges e_1 and e_2 are different and have a common endpoint $n \in N$.

For instance, the dual of the graph in Fig. 2 is the graph depicted in Fig. 3.

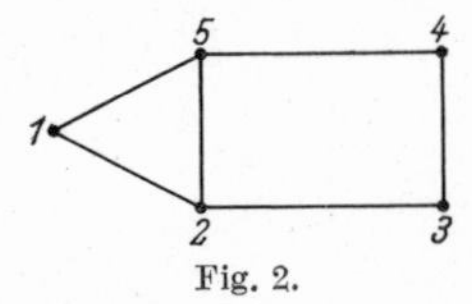

Fig. 2.

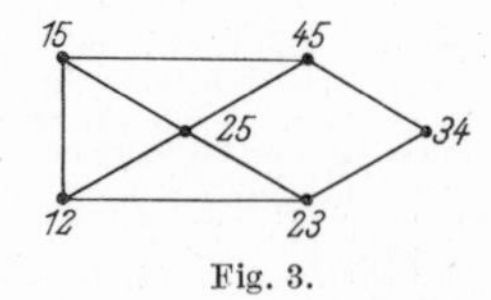

Fig. 3.

Lemma 1. *A set of edges of a bipartite graph G determines a matching of G if and only if it is an internally stable set of the dual graph G^*.*

Proof. A set E of edges of G determines a matching of G if and only if the following condition is satisfied: (α) if e_1 and e_2

* In other books the term "maximal" stands for "absolutely maximal".

have a common endpoint, then at least one of the edges e_1, e_2 is not in E^*.

But condition (α) expresses also the fact that E is an internally stable set of the graph G^*.

Observing that the results in § 2 of Chapter X apply to undirected graphs too, and taking into account Lemma 1, we deduce:

Theorem 1. *The (absolutely) maximal matchings of a bipartite graph G coincide with the (absolutely) maximal internally stable sets of the dual graph G^* and hence can be determined with the methods given in § 2 of Chapter X.*

In other words, in order to determine the maximal (absolutely maximal) matchings of G, we have to find the locally (globally) maximizing points of the pseudo-Boolean function

$$\sum_{i=1}^{n} x_i \tag{1}$$

under the conditions

$$\sum_{i=1}^{n} \sum_{j=1}^{n} a_{ij}\, x_i\, x_j = 0, \tag{2}$$

where (a_{ij}) is the incidence matrix of the dual graph G^* (i.e. $a_{ij} = a_{ji} = 1$ if and only if there is an edge (i, j) in G^*, that is if and only if the edges i and j in G have exactly one common endpoint).

Remark 1. Similar results hold for bipartite graphs with pre-assigned weights c_i for the edges i. In this case one requires the maximization of the total weight and the problem reduces to that of maximizing

$$\sum_{i=1}^{n} c_i\, x_i \tag{1'}$$

under the same constraint (2).

Example 1. Consider the bipartite graph G in Fig. 4.

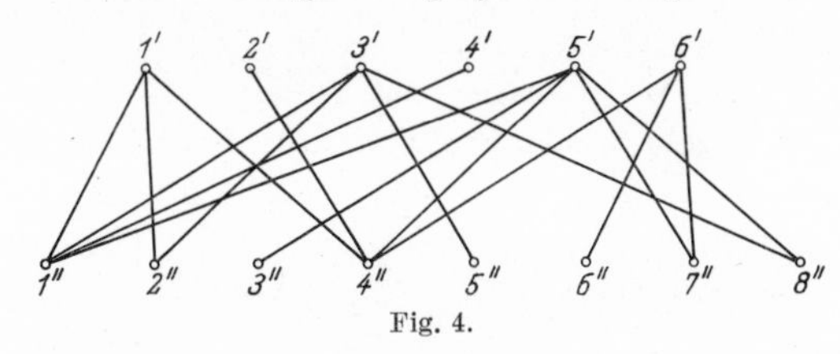

Fig. 4.

The dual graph G^* has 17 vertices, namely

$$\begin{aligned}
&1''' = (1', 1''), && 2''' = (1', 2''), && 3''' = (1', 4''),\\
&4''' = (2', 4''), && 5''' = (3', 1''), && 6''' = (3', 2''),\\
&7''' = (3', 5''), && 8''' = (3', 8''), && 9''' = (4', 1''),\\
&10''' = (5', 1''), && 11''' = (5', 3''), && 12''' = (5', 4''),\\
&13''' = (5', 7''), && 14''' = (5', 8''), && 15''' = (6', 4''),\\
&16''' = (6', 6''), && 17''' = (6', 7''),
\end{aligned} \tag{3}$$

---

* We take also into consideration the degenerate matching which involves no edges.

and the edges indicated in the incidence matrix

$$
(4) \qquad A = \begin{pmatrix}
0 & 1 & 1 & 0 & 1 & 0 & 0 & 0 & 1 & 1 & 0 & 0 & 0 & 0 & 0 & 0 & 0 \\
1 & 0 & 1 & 0 & 0 & 1 & 0 & 0 & 0 & 0 & 0 & 0 & 0 & 0 & 0 & 0 & 0 \\
1 & 1 & 0 & 1 & 0 & 0 & 0 & 0 & 0 & 0 & 0 & 1 & 0 & 0 & 1 & 0 & 0 \\
0 & 0 & 1 & 0 & 0 & 0 & 0 & 0 & 0 & 0 & 0 & 1 & 0 & 0 & 1 & 0 & 0 \\
1 & 0 & 0 & 0 & 0 & 1 & 1 & 1 & 1 & 1 & 0 & 0 & 0 & 0 & 0 & 0 & 0 \\
0 & 1 & 0 & 0 & 1 & 0 & 1 & 1 & 0 & 0 & 0 & 0 & 0 & 0 & 0 & 0 & 0 \\
0 & 0 & 0 & 0 & 1 & 1 & 0 & 1 & 0 & 0 & 0 & 0 & 0 & 0 & 0 & 0 & 0 \\
0 & 0 & 0 & 0 & 1 & 1 & 1 & 0 & 0 & 0 & 0 & 0 & 0 & 1 & 0 & 0 & 0 \\
1 & 0 & 0 & 0 & 1 & 0 & 0 & 0 & 0 & 1 & 0 & 0 & 0 & 0 & 0 & 0 & 0 \\
1 & 0 & 0 & 0 & 1 & 0 & 0 & 0 & 1 & 0 & 1 & 1 & 1 & 1 & 0 & 0 & 0 \\
0 & 0 & 0 & 0 & 0 & 0 & 0 & 0 & 0 & 1 & 0 & 1 & 1 & 1 & 0 & 0 & 0 \\
0 & 0 & 1 & 1 & 0 & 0 & 0 & 0 & 0 & 1 & 1 & 0 & 1 & 1 & 1 & 0 & 0 \\
0 & 0 & 0 & 0 & 0 & 0 & 0 & 0 & 0 & 1 & 1 & 1 & 0 & 1 & 0 & 0 & 1 \\
0 & 0 & 0 & 0 & 0 & 0 & 0 & 1 & 0 & 1 & 1 & 1 & 1 & 0 & 0 & 0 & 0 \\
0 & 0 & 1 & 1 & 0 & 0 & 0 & 0 & 0 & 0 & 0 & 1 & 0 & 0 & 0 & 1 & 1 \\
0 & 0 & 0 & 0 & 0 & 0 & 0 & 0 & 0 & 0 & 0 & 0 & 0 & 0 & 1 & 0 & 1 \\
0 & 0 & 0 & 0 & 0 & 0 & 0 & 0 & 0 & 0 & 0 & 0 & 1 & 0 & 1 & 1 & 0
\end{pmatrix} .
$$

The matchings of G correspond to the internally stable sets of G^*; hence, according to Theorem X.1, their characteristic vectors are the solution of the pseudo-Boolean equation

$$
\begin{aligned}
(5) \qquad & x_1 x_2 + x_1 x_3 + x_2 x_3 + x_5 x_6 + x_5 x_7 + x_5 x_8 + x_6 x_7 + x_6 x_8 + x_7 x_8 + \\
& + x_{10} x_{11} + x_{10} x_{12} + x_{10} x_{13} + x_{10} x_{14} + x_{11} x_{12} + x_{11} x_{13} + x_{11} x_{14} + \\
& + x_{12} x_{13} + x_{12} x_{14} + x_{13} x_{14} + x_{15} x_{16} + x_{15} x_{17} + x_{16} x_{17} + x_8 x_{14} + \\
& + x_1 x_5 + x_1 x_9 + x_1 x_{10} + x_5 x_9 + x_5 x_{10} + x_9 x_{10} + x_{13} x_{17} + \\
& + x_2 x_6 + x_3 x_4 + x_3 x_{12} + x_3 x_{15} + x_4 x_{12} + x_4 x_{15} + x_{12} x_{15} = 0 .
\end{aligned}
$$

Now, in view of Theorem 1 and § 2 of Chapter X, the absolutely maximal matchings of G correspond to the maximizing points of the pseudo-Boolean function

$$\sum_{i=1}^{17} x_i$$

under the above constraint.

Applying the method proposed in § 1 of Chapter V, we find that these points are those indicated in Table 1 below:

Table 1

x_1	x_2	x_3	x_4	x_5	x_6	x_7	x_8	x_9	x_{10}	x_{11}	x_{12}	x_{13}	x_{14}	x_{15}	x_{16}	x_{17}
0	1	0	1	0	0	1	0	1	0	0	0	1	0	0	1	0
0	1	0	1	0	0	0	1	1	0	0	0	1	0	0	1	0
0	1	0	1	0	0	1	0	1	0	0	0	0	1	0	1	0
0	1	0	1	0	0	1	0	1	0	0	0	0	1	0	0	1
0	1	0	1	0	0	1	0	1	0	1	0	0	0	0	1	0
0	1	0	1	0	0	1	0	1	0	1	0	0	0	0	0	1
0	1	0	1	0	0	0	1	1	0	1	0	0	0	0	1	0
0	1	0	1	0	0	0	1	1	0	1	0	0	0	0	0	1

From (3) and Table 1 we see that the absolutely maximal matchings are:

$$\{(1', 2''), (2', 4''), (3', 5''), (4', 1''), (5', 7''), (6', 6'')\},$$

$$\{(1', 2''), (2', 4''), (3', 8''), (4', 1''), (5', 7''), (6', 6'')\},$$

$$\{(1', 2''), (2', 4''), (3', 5''), (4', 1''), (5', 8''), (6', 6'')\},$$

$$\{(1', 2''), (2', 4''), (3', 5''), (4', 1''), (5', 8''), (6', 7'')\},$$

$$\{(1', 2''), (2', 4''), (3', 5''), (4', 1''), (5', 3''), (6', 6'')\},$$

$$\{(1', 2''), (2', 4''), (3', 5''), (4', 1''), (5', 3''), (6', 7'')\},$$

$$\{(1', 2''), (2', 4''), (3', 8''), (4', 1''), (5', 3''), (6', 6'')\},$$

$$\{(1', 2''), (2', 4''), (3', 8''), (4', 1''), (5', 3''), (6', 7'')\}.$$

Example 2. Let us consider the graph in Fig. 5.

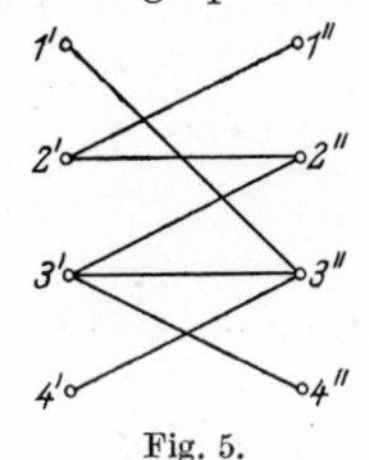

Fig. 5.

Proceeding as in Example 1, we find that the maximal matchings are

$$\{(1', 3''), (2', 1''), (3', 2'')\},$$

$$\{(1', 3''), (2', 1''), (3', 4'')\},$$

$$\{(1', 3''), (2', 2''), (3', 4'')\},$$

$$\{(2', 1''), (3', 2''), (4', 3'')\},$$

$$\{(2', 1''), (3', 4''), (4', 3'')\},$$

$$\{(2', 2''), (3', 4''), (4', 3'')\}.$$

This example shows that the number of edges involved in a maximal matching need not coincide with $\min(m_1, m_2)$*.

* For the significance of m_1 and m_2, see Definition 1.

Minimal separating sets

Definition 6. Let $G = (N', N''; \varrho)$ be a bipartite graph. A set $B = B' \cup B''$, with $B' \subseteqq N'$, $B'' \subseteqq N''$, is called a *separating set*, if every edge has at least one endpoint in B.

The problem of determining the *absolutely minimal** separating sets, i.e. the separating sets with the smallest number of nodes, is of a considerable interest.

Let us charakterize the bipartite graph $(N', N''; \varrho)$ by the Boolean $m_1 \times m_2$ matrix (a_{ij}) defined as follows:

$$a_{ij} = \begin{cases} 1 & \text{if} \quad i \varrho j, \\ 0 & \text{otherwise.} \end{cases} \tag{6}$$

If we characterize the sets B' and B'' by the Boolean vectors $(x_1, \ldots, x_{m_1})$ and $(y_1, \ldots, y_{m_2})$ we have obviously:

Lemma 2. *The set B is a separating set if and only if*

$$\sum_{i=1}^{m_1} \sum_{j=1}^{m_2} a_{ij} \bar{x}_i \bar{y}_j = 0. \tag{7}$$

Theorem 2. *Let $(x_1^*, \ldots, x_{m_1}^*)$ and $(y_1^*, \ldots, y_{m_2}^*)$ be the characteristic vectors of the sets $B' (\subseteqq N')$ and $B'' (\subseteqq N'')$, respectively, and let $B = B' \cup B''$. The set B is an absolutely minimal separating set if and only if $(x_1^*, \ldots, x_{m_1}^*)$ is a minimizing point of the pseudo-Boolean function*

$$F = \sum_{i=1}^{m_1} x_i - \sum_{j=1}^{m_2} \prod_{a_{ij}=1} x_i \tag{8}$$

and relations

$$y_j^* = 1 - \prod_{a_{ij}=1} x_i^* \qquad (j = 1, \ldots, m_2) \tag{9}$$

hold [a_{ij} *being defined by* (6)].

Proof. It follows from Lemma 2 that B is an absolutely minimal separating set if and only if its characteristic vector $(x_1^*, \ldots, x_{m_1}^*, y_1^*, \ldots, y_{m_2}^*)$ is a minimizing point of the function

$$\sum_{i=1}^{m_1} x_i + \sum_{j=1}^{m_2} y_j \tag{10}$$

under the constraint (7).

But equation (7) is equivalent to the system of Boolean equations

$$\left(\bigcup_{i=1}^{m_1} a_{ij} \bar{x}_i \right) \bar{y}_j = 0 \qquad (j = 1, \ldots, m_2), \tag{11}$$

which may also be written in the form

$$\bigcup_{i=1}^{m_1} a_{ij} \bar{x}_i \leqq y_j \qquad (j = 1, \ldots, m_2). \tag{12}$$

---

* Usually termed minimal.

Therefore the problem of determining the absolutely minimal separating sets is equivalent to that of minimizing (10) under the constraints (12).

Now, if $(x_1^*, \ldots, x_{m_1}^*, y_1^*, \ldots, y_{m_2}^*)$ is a solution to the latter problem, then relations (12) obviously become

$$\bigcup_{i=1}^{m_1} a_{ij} \overline{x_i^*} = y_j^* \qquad (j = 1, \ldots, m_2) \tag{13}$$

or, equivalently, (9), while the corresponding value of (10) is

$$\sum_{i=1}^{m_1} x_i^* + m_2 - \sum_{j=1}^{m_2} \prod_{a_{ij}=1} x_i^*, \tag{14}$$

so that $(x_1^*, \ldots, x_{m_1}^*)$ is a minimizing point of (8).

Conversely, let $(x_1^*, \ldots, x_{m_1}^*)$ be a minimizing point of (8) and let $(y_1^*, \ldots, y_{m_2}^*)$ be defined by formulas (9). Then $(x_1^*, \ldots, x_{m_1}^*, y_1^*, \ldots, y_{m_2}^*)$ satisfies (13) so that it is a solution of (12) too; moreover if $(x_1, \ldots, x_{m_1}, y_1, \ldots, y_{m_2})$ denotes an arbitrary solution of (12), then $y_j \geqq 1 - \prod_{a_{ij}=1} x_i$ and

$$\sum_{i=1}^{m_1} x_i^* + \sum_{j=1}^{m_2} y_j^* = m_2 + \sum_{i=1}^{m_1} x_i^* - \sum_{j=1}^{m_2} \prod_{a_{ij}=1} x_i^* \leqq$$

$$\leqq m_2 + \sum_{i=1}^{m_1} x_i - \sum_{j=1}^{m_2} \prod_{a_{ij}=1} x_i \leqq \sum_{i=1}^{m_1} x_i + \sum_{j=1}^{m_2} y_j,$$

completing the proof.

Remark 2. Similar results hold for bipartite graphs with pre-assigned weights c_i, d_j for the vertices i', j''. In this case one requires the minimization of the total weight and the problem reduces to that of minimizing

$$F' = \sum_{i=1}^{m_1} c_i x_i - \sum_{j=1}^{m_2} d_j \prod_{a_{ij}=1} x_i, \tag{8'}$$

the variables y_j^* being obtained afterwards via (9).

Example 3. The function (8) for the graph in Fig. 4 is

$$\begin{aligned} F = {} & x_1 + x_2 + x_3 + x_4 + x_5 + x_6 - x_1 x_3 x_4 x_5 - x_1 x_3 - x_5 - \\ & - x_1 x_2 x_5 x_6 - x_3 - x_6 - x_5 x_6 - x_3 x_5 = x_1 + x_2 + x_4 - x_1 x_3 x_4 x_5 - \\ & - x_1 x_3 - x_1 x_2 x_5 x_6 - x_5 x_6 - x_3 x_5. \end{aligned} \tag{15}$$

The minima are reached for

$$x_1^* = p, \quad x_2^* = p q, \quad x_3^* = 1, \quad x_4^* = p r, \quad x_5^* = 1, \quad x_6^* = 1, \tag{16}$$

which, introduced into (9), give:

$$y_1^* = \bar{p} \cup \bar{r}, \quad y_2^* = \bar{p}, \quad y_3^* = 0, \quad y_4^* = \bar{p} \cup \bar{q}, \quad y_5^* = y_6^* = y_7^* = y_8^* = 0. \tag{17}$$

Hence the absolutely minimal separating sets are the following:

(18.1) $$\{3', 5', 6', 1'', 2'', 4''\},$$

(18.2) $$\{1', 3', 4', 5', 6', 4''\},$$

(18.3) $$\{1', 2', 3', 4', 5', 6'\},$$

(18.4) $$\{1', 3', 5', 6', 1'', 4''\},$$

(18.5) $$\{1', 2', 3', 5', 6', 1''\}.$$

Example 4. The function (8) for the graph in Fig. 5 is

(19) $$\begin{aligned} F &= x_1 + x_2 + x_3 + x_4 - x_2 - x_2 x_3 - x_1 x_3 x_4 - x_3 \\ &= x_1 + x_4 - x_2 x_3 - x_1 x_3 x_4 . \end{aligned}$$

Its minimum is

(20) $$F_{\min} = F(0, 1, 1, 0) = -1$$

and the corresponding values of y, given by (9) are:

$$y_1 = 0, \quad y_2 = 0, \quad y_3 = 1, \quad y_4 = 0.$$

Hence, $B' = \{2', 3'\}$, $B'' = \{3''\}$ and

(21) $$B = B' \cup B'' = \{2', 3', 3''\}.$$

Deficiencies and Critical Sets

Definition 7. Let A' be an arbitrary subset of N'. Let $N''(A')$ be the subset of those elements of N'' which are connected to elements of A' by edges of the graph. If $|S|$ denotes the number of elements of a set S, then the *deficiency* $\delta(A')$ *of* A' is

(22) $$\delta(A') = |A'| - |N''(A')|.$$

If the deficiency δ_0 of a set A_0 is greater or equal to the deficiency of any other set, then A_0 is called a *set of maximal deficiency* or a *critical set.* The set N' is said to be *without deficiency* if $\delta_0 = 0$.

In order to determine δ_0 and all the critical sets of a graph, let us introduce again the Boolean variables x_i for the elements of N' (being 1 if $n'_i \in A_0$, and 0 in the other case) and y_j for the elements of N'' being 1 if $n''_j \in N''(A_0)$, and 0 in the other case]. Then, obviously

(23) $$\delta_0 = \max\left(\sum_{i=1}^{m_1} x_i - \sum_{j=1}^{m_2} y_j\right).$$

An element n''_j of N'' belongs to $N''(A_0)$ if and only if there exists an element $n'_i \in N'$ so that $a_{ij} = 1$. In other words, $y_j = 1$ if and only if there is an i so that $a_{ij} = x_i = 1$. We see that $y_j = 1$ if and only if $\bigcup_{i=1}^{m_1} a_{ij} x_i = 1$, or

(24) $$y_j = \bigcup_{i=1}^{m_1} a_{ij} x_i .$$

Relation (24) is equivalent to

$$y_j = 1 - \prod_{i=1}^{m_1} (1 - a_{ij}\, x_i). \tag{25}$$

We have thus obtained:

Theorem 3. *A set $A_0 \subseteqq N'$ is a critical set if and only if its characteristic vector $(x_1^*, \ldots, x_n^*)$ is a maximizing point of the pseudo-Boolean function*

$$G = -m_2 + \sum_{i=1}^{m_1} x_i + \sum_{j=1}^{m_2} \prod_{i=1}^{m_1} (1 - a_{ij}\, x_i). \tag{26}$$

The maximal deficiency is

$$\delta_0 = \max G. \tag{27}$$

Example 5. The function G for the graph in Fig. 4 is

$$G = -8 + x_1 + x_2 + x_3 + x_4 + x_5 + x_6 + \bar{x}_1 \bar{x}_3 \bar{x}_4 \bar{x}_5 + \tag{28}$$
$$+ \bar{x}_1 \bar{x}_3 + \bar{x}_5 + \bar{x}_1 \bar{x}_2 \bar{x}_5 \bar{x}_6 + \bar{x}_3 + \bar{x}_6 + \bar{x}_5 \bar{x}_6 + \bar{x}_3 \bar{x}_5$$

and its maximum is

$$G_{\max} = 0, \tag{29}$$

hence

$$\delta_0 = 0; \tag{30}$$

the maximizing points are

$$x_1^* = p, \quad x_2^* = p \cup q, \quad x_3^* = 0, \quad x_4^* = p \cup r, \quad x_5^* = x_6^* = 0, \tag{31}$$

hence the critical sets are

$$\{1', 2', 4'\}, \tag{32.1}$$

$$\{2', 4'\}, \tag{32.2}$$

$$\{2'\}, \tag{32.3}$$

$$\{4'\}, \tag{32.4}$$

$$\emptyset. \tag{32.5}$$

It is well-known (see O. Ore [1]) that there exists a unique (absolutely) minimal critical set A_m contained in all other critical sets, and a unique (absolutely) maximal critical set A_M containing all other critical sets.

It is obvious then, that we can obtain them by minimizing the pseudo-Boolean functions

$$G_m = (m_1 + 1)\, G + \sum_{i=1}^{m_1} x_i, \tag{33}$$

$$G_M = (m_1 + 1)\, G - \sum_{i=1}^{m_1} x_i \tag{34}$$

where G has the form given in (26).

In order to show the importance of the concept of deficiency, we point out the following three well-known theorems:
— The set N' can be matched upon N'' if and only if

$$\delta_0 = 0. \tag{35}$$

(Theorem of KÖNIG-HALL; see C. BERGE [2], p. 92).
— The number of elements of an absolutely minimal separating set is $m_1 - \delta_0$ (KÖNIG; see C. BERGE [2], p. 98).
— The number of edges in an absolutely maximal matching is $m_1 - \delta_0$ (Theorem of KÖNIG-ORE; see C. BERGE [2], p. 97).

As we can easily determine δ_0 by maximizing the pseudo-Boolean function G, we have thus a simple criterion showing whether N' may be matched upon N'' or not, determining the number of elements in a minimal separating set, and determining the maximal number of edges in a matching.

§ 2. Assignment and Transportation Problems

As it was shown at the beginning of § 1, the assignment problem may be viewed as one concerning maximal matchings in a bipartite graph G with weighted edges.

Namely, let N' be the set of individuals and N'' the set of jobs. The vertices $i' \in N'$ and $j'' \in N''$ are connected by an edge if and only if the individual i' can perform the job j''; to each edge (i', j'') corresponds a "profit" $c_{i'j''}$. The problem is to find those assignments of individuals to jobs which maximize the total profit.

It was shown in Theorem 1 and Remark 1 that this problem can be translated as the problem of finding the absolutely maximal internally stable sets of the dual graph G^* and thus can be solved by means of pseudo-Boolean programming.

Example 6. Let us consider 6 individuals and 8 jobs, the different profits being given in the following table:

Table 2

N' \ N''	1''	2''	3''	4''	5''	6''	7''	8''
1'	6	5	—	8	—	—	—	—
2'	—	—	—	3	—	—	—	—
3'	7	6	—	—	5	—	—	8
4'	8	—	—	—	—	—	—	—
5'	6	—	4	5	—	—	7	4
6'	—	—	—	6	—	5	6	—

where the dashes indicate the absence of edges.

We see that this is the graph G studied in Example 1. It was shown there that the incidence matrix of the dual graph G^* is given by formula (4), so that the characteristic equation of the internally stable sets is (5).

Hence we have to maximize the pseudo-Boolean function

$$f = 6x_1 + 5x_2 + 8x_3 + 3x_4 + 7x_5 + 6x_6 + 5x_7 + 8x_8 + 8x_9 + 6x_{10} + 4x_{11} + 5x_{12} + 7x_{13} + 4x_{14} + 6x_{15} + 5x_{16} + 6x_{17} \tag{36}$$

under the constraint (5).

Applying our algorithm, we find that there are two maximizing points

$$x_3 = x_8 = x_9 = x_{13} = x_{16} = 1, \quad \text{the other } x_i = 0; \tag{37.1}$$

$$x_2 = x_4 = x_8 = x_9 = x_{13} = x_{16} = 1, \quad \text{the other } x_i = 0; \tag{37.2}$$

the maximum of (36) being

$$f_{\max} = 36. \tag{38}$$

Taking into account formulas (3), we see that the corresponding optimal assignments are

$$1' \to 4'', \quad 3' \to 8'', \quad 4' \to 1'', \quad 5' \to 7'', \quad 6' \to 6''; \tag{39.1}$$

$$1' \to 2'', \quad 2' \to 4'', \quad 3' \to 8'', \quad 4' \to 1'', \quad 5' \to 7'', \quad 6' \to 6''. \tag{39.2}$$

Both solutions assure a total profit of 36 units. The second solution assigns each individual to a job, while the first one leaves the individual $2'$ free.

By the transportation problem we mean the problem of determining the real values z_{ij} $(i = 1, 2, \ldots, m;\ j = 1, 2, \ldots, n)$ which minimize the sum

$$E = \sum_{i=1}^{m} \sum_{j=1}^{n} c_{ij} z_{ij} \tag{40}$$

subject to the following constraints:

$$\begin{cases} \sum_{j=1}^{n} z_{ij} = a_i & (i = 1, \ldots, m), \\ \sum_{i=1}^{m} z_{ij} = b_j & (j = 1, \ldots, n), \end{cases} \tag{41}$$

$$z_{ij} \geqq 0 \quad (i = 1, \ldots, m;\ j = 1, \ldots, n) \tag{42}$$

where a_i, b_j and c_{ij} are fixed positive numbers which satisfy the condition

$$\sum_{i=1}^{m} a_i = \sum_{j=1}^{n} b_j. \tag{43}$$

The economic interpretation of the problem is the following: if $i = 1, 2, \ldots, m$ represent the production centres and $j = 1, 2, \ldots, n$ the consumption centres in which a certain item is produced or consumed in the quantities a_i and b_j respectively, and if c_{ij} represents the unit cost of transportation from the centre i to the centre j, then the values x_{ij} for which the sum E attains its minimum, represent those quantities of the given item that have to be transported from the centres i to the centres j, in order that the total cost of transportation be a minimum.

The assignment problem can be viewed as a particular case of the transportation problem (if $m = n$, $a_i = b_j = 1$ for all i, j, the quantities $-c_{ij}$ are interpreted as profits, and it is required to assign necessarily each individual to a job). However, it is well-known (see, for instance G. B. DANTZIG [3], Chapter 15) that the converse relationship also holds, so that the two problems are in fact equivalent.

One of the frequently utilized methods for solving transportation problems, the so-called *"Hungarian method"* due to J. EGERVÁRY [1, 2] and H. W. KUHN [1, 2], runs as follows:

1) For each i_0 choose one of the smallest $c_{i_0 j}$ $(j = 1, \ldots, n)$ and subtract it from all $c_{i_0 j}$.

2) For each j_0 choose one of the smallest resulted $c'_{i j_0}$ $(i = 1, \ldots, m)$ and subtract it from all $c'_{i j_0}$.

Denote the resulted elements by c^1_{ij}. Notice that in each row and in each column of the matrix (c^1_{ij}) there is at least one zero.

A set $S_1 = \{i_1, i_2, \ldots, i_h; j_1, j_2, \ldots, j_k\}$ of rows and columns is termed a *covering* of the zeros of an $m \times n$ matrix, if each zero of that matrix belongs to a row and/or a column of S_1. The *weight* of S is defined as being the sum

$$w(S_1) = a_{i_1} + a_{i_2} + \cdots + a_{i_h} + b_{j_1} + b_{j_2} + \cdots + b_{j_k}. \tag{44}$$

3) Choose a minimum covering of (c^1_{ij}), i.e. a covering of the zeros of the matrix (c^1_{ij}) with a minimum weight. Let c^{1*}_{ij} be the smallest uncovered element.

4) Construct the matrix (c^2_{ij}) with the elements

$$c^2_{ij} = \begin{cases} c^1_{ij} - c^{1*}_{ij}, & \text{for the uncovered elements } c^1_{ij}; \\ c^1_{ij} + c^{1*}_{ij}, & \text{for the element covered both by a row and by a column;} \\ c^1_{ij}, & \text{for the other elements.} \end{cases} \tag{45}$$

5) Repeat steps 3 and 4 until obtaining a matrix (c^p_{ij}) for which the minimum covering S_p has the weight

$$w(S_p) = \sum_{i=1}^{m} a_i = \sum_{j=1}^{n} b_j \tag{46}$$

(it is easy to observe that in the previous steps $q\,(q < p)$, the weight of the minimum covering is $< \sum_{i=1}^{m} a_i$).

6) Put $z_{ij} = 0$ whenever $c_{ij}^p \neq 0$ and solve, for the remaining variables, the system (41) (which will have a "triangular" form, enabling the immediate determination of its solutions).

The values obtained in this way solve the original problem.

When the method was first published, no procedure was indicated for performing the above step 3. Meanwhile, various combinatorial methods were proposed for solving this step.

The problem to be solved in step 3 may be given the following graph-theoretical formulation. Let $N' = \{1, \ldots, m\}$ be the set of the rows of the given matrix, $N'' = \{1, \ldots, n\}$ the set of all its columns and, for every $i' \in N'$, $j'' \in N''$, let relation $i' \varrho j''$ mean that the element $c_{i'j''}$ of the matrix is a zero. If the bipartite graph $G = (N', N''; \varrho)$ is introduced, then we have to find a separating set which minimizes the sum of the weights.

In view of Theorem 2 and Remark 2, the problem admits the following pseudo-Boolean formulation: minimize the pseudo-Boolean function

$$w(S) = \sum_{i=1}^{m} a_i x_i - \sum_{j=1}^{n} b_j \prod_{c_{ij}=0} x_i + \sum_{i=1}^{m} a_i; \tag{47}$$

after obtaining a minimizing point $(x_1^*, \ldots, x_n^*)$, determine the vector $(y_1^*, \ldots, y_n^*)$ by formulas

$$y_j^* = 1 - \prod_{c_{ij}=0} x_i^* \qquad (j = 1, \ldots, n). \tag{48}$$

The required minimal weighted covering consists of those rows i for which $x_i^* = 1$ and of those columns j for which $y_j^* = 1$; introducing these values into (47), we obtain the minimum weight.

This approach was first suggested by P. L. Hammer and S. Rudeanu [1, 2].

Remark 3. When performing the step 3 of the Hungarian method, we need only one minimal covering, hence only one minimizing point of the function (47). Therefore the minimizing algorithm becomes more simple, as it was shown in Remark VI.10.

Example 7. Let us consider the transportation problem with the following data:

$$(c_{ij}) = \begin{pmatrix} 12 & 27 & 61 & 49 & 83 & 35 \\ 23 & 39 & 78 & 28 & 65 & 42 \\ 67 & 56 & 92 & 24 & 53 & 54 \\ 71 & 43 & 91 & 67 & 40 & 49 \end{pmatrix}, \tag{49}$$

$$a_1 = 18, \quad a_2 = 32, \quad a_3 = 14, \quad a_4 = 9, \tag{50}$$

$$b_1 = 9, \quad b_2 = 11, \quad b_3 = 28, \quad b_4 = 6, \quad b_5 = 14, \quad b_6 = 5. \tag{51}$$

The first two steps yield the matrix

$$(52.1) \qquad (c^1_{ij}) = \begin{pmatrix} 0 & 12 & 0 & 37 & 61 & 14 \\ 0 & 13 & 6 & 5 & 42 & 10 \\ 43 & 29 & 19 & 0 & 29 & 11 \\ 31 & 0 & 2 & 27 & 0 & 0 \end{pmatrix}.$$

The corresponding function (47) is

$$(53.1) \qquad \begin{aligned} w(S_1) &= 18x_1 + 32x_2 + 14x_3 + 9x_4 - 9x_1 x_2 - 11x_4 - \\ &\qquad - 28x_1 - 6x_3 - 14x_4 - 5x_4 + 73 \\ &= -10x_1 + 32x_2 + 8x_3 - 21x_4 - 9x_1 x_2 + 73; \end{aligned}$$

the minimum of this function is 42 and it is attained for $x_1^* = 1$, $x_2^* = 0$, $x_3^* = 0$, $x_4^* = 1$, yielding, by (48), $y_1^* = 1$, $y_2^* = 0$, $y_3^* = 0$, $y_4^* = 1$, $y_5^* = 0$, $y_6^* = 0$.

The minimal covering consists of rows 1 and 4 and of columns 1 and 4; the smallest uncovered element is thus $c^1_{23} = 6$. Hence, performing step 4, we get

$$(52.2) \qquad (c^2_{ij}) = \begin{pmatrix} 6 & 12 & 0 & 43 & 61 & 14 \\ 0 & 7 & 0 & 5 & 36 & 4 \\ 43 & 23 & 13 & 0 & 23 & 5 \\ 37 & 0 & 2 & 33 & 0 & 0 \end{pmatrix}.$$

The corresponding function (47) is

$$(53.2) \qquad \begin{aligned} w(S_2) &= 18x_1 + 32x_2 + 14x_3 + 9x_4 - 9x_2 - 11x_4 - \\ &\qquad - 28x_1 x_2 - 6x_3 - 14x_4 - 5x_4 + 73 \\ &= 18x_1 + 23x_2 + 8x_3 - 21x_4 - 28x_1 x_2 + 73; \end{aligned}$$

the minimum of this function is 52 and it is attained for $x_1^* = x_2^* = x_3^* = 0$, $x_4^* = 1$, yielding $y_1^* = y_3^* = y_4^* = 1$, $y_2^* = y_5^* = y_6^* = 0$.

The minimal covering is $\{4; 1, 3, 4\}$ and the smallest uncovered element is $c^2_{26} = 4$. Hence we obtain

$$(52.3) \qquad (c^3_{ij}) = \begin{pmatrix} 6 & 8 & 0 & 43 & 57 & 10 \\ 0 & 3 & 0 & 5 & 32 & 0 \\ 43 & 19 & 13 & 0 & 19 & 1 \\ 41 & 0 & 6 & 37 & 0 & 0 \end{pmatrix}.$$

The function (47) becomes

$$(53.3) \qquad \begin{aligned} w(S_3) &= 18x_1 + 32x_2 + 14x_3 + 9x_4 - 9x_2 - 11x_4 - \\ &\qquad - 28x_1 x_2 - 6x_3 - 14x_4 - 5x_2 x_4 + 73 \\ &= 18x_1 + 23x_2 + 8x_3 - 16x_4 - 28x_1 x_2 - 5x_2 x_4 + 73; \end{aligned}$$

the minimum of this function is 57 and it is attained for $x_1^* = x_2^* = x_3^* = 0$, $x_4^* = 1$, yielding $y_1^* = y_3^* = y_4^* = y_6^* = 1$, $y_2^* = y_5^* = 0$.

Now the minimal covering is $\{4; 1, 3, 4, 6\}$ and the smallest uncovered element is $c^3_{22} = 3$. Hence

$$(52.4) \qquad (c^4_{ij}) = \begin{pmatrix} 6 & 5 & 0 & 43 & 54 & 10 \\ 0 & 0 & 0 & 5 & 29 & 0 \\ 43 & 16 & 13 & 0 & 16 & 1 \\ 44 & 0 & 9 & 40 & 0 & 3 \end{pmatrix}.$$

The corresponding function (47) is

$$(53.4) \qquad \begin{aligned} w(S_4) &= 18x_1 + 32x_2 + 14x_3 + 9x_4 - 9x_2 - 11x_2 x_4 - \\ &\qquad - 28x_1 x_2 - 6x_3 - 14x_4 - 5x_2 + 73 \\ &= 18x_1 + 18x_2 + 8x_3 - 5x_4 - 11x_2 x_4 - 28x_1 x_2 + 73; \end{aligned}$$

the minimum of this function is 65 and it is attained for $x_1^* = x_2^* = x_4^* = 1$, $x_3^* = 0$, yielding $y_4^* = 1$, $y_1^* = y_2^* = y_3^* = y_5^* = y_6^* = 0$.

The minimal covering is $\{1, 2, 4; 4\}$ and the smallest uncovered element is $c^4_{36} = 1$. Hence

$$(52.5) \qquad (c^5_{ij}) = \begin{pmatrix} 6 & 5 & 0 & 44 & 54 & 10 \\ 0 & 0 & 0 & 6 & 29 & 0 \\ 42 & 15 & 12 & 0 & 15 & 0 \\ 41 & 0 & 9 & 41 & 0 & 3 \end{pmatrix}.$$

It follows that (47) becomes

$$(53.5) \qquad \begin{aligned} w(S_5) &= 18x_1 + 32x_2 + 14x_3 + 9x_4 - 9x_2 - 11x_2 x_4 - \\ &\qquad - 28x_1 x_2 - 6x_3 - 14x_4 - 5x_2 x_3 + 73 \\ &= 18x_1 + 23x_2 + 8x_3 - 5x_4 - 11x_2 x_4 - 28x_1 x_2 - 5x_2 x_3 + 73; \end{aligned}$$

the minimum of (53.5) is 68 and it is attained for $x_1^* = x_2^* = x_3^* = 0$, $x_4^* = 1$, yielding $y_1^* = y_2^* = y_3^* = y_4^* = y_6^* = 1$, $y_5^* = 0$.

The minimal covering is now $\{4; 1, 2, 3, 4, 6\}$ and the smallest uncovered element is $c^5_{35} = 15$. Therefore

$$(52.6) \qquad (c^6_{ij}) = \begin{pmatrix} 6 & 5 & 0 & 44 & 39 & 10 \\ 0 & 0 & 0 & 6 & 14 & 0 \\ 42 & 15 & 12 & 0 & 0 & 0 \\ 56 & 15 & 24 & 56 & 0 & 18 \end{pmatrix}$$

and (47) becomes

$$(53.6) \qquad \begin{aligned} w(S_6) &= 18x_1 + 32x_2 + 14x_3 + 9x_4 - 9x_2 - 11x_2 - \\ &\qquad - 28x_1 x_2 - 6x_3 - 14x_3 x_4 - 5x_2 x_3 + 73 \\ &= 18x_1 + 12x_2 + 8x_3 + 9x_4 - 28x_1 x_2 - 14x_3 x_4 - 5x_2 x_3 + 73 \end{aligned}$$

and the minimum is now $73 = \sum a_i = \sum b_j$.

We have now to perform step 6: we put

$$(54) \qquad \begin{aligned} &z_{11} = z_{12} = z_{14} = z_{15} = z_{16} = z_{24} = z_{25} = z_{31} = z_{32} = z_{33} = z_{41} \\ &= z_{42} = z_{43} = z_{44} = z_{46} = 0 \end{aligned}$$

and solve the remaining system (41):

$$(55)\qquad \begin{cases} z_{13} = 18, \\ z_{21} + z_{22} + z_{23} + z_{26} = 32, \\ z_{34} + z_{35} + z_{36} = 14, \\ z_{45} = 9, \\ z_{21} = 9, \\ z_{22} = 11, \\ z_{13} + z_{23} = 28, \\ z_{34} = 6, \\ z_{35} + z_{45} = 14, \\ z_{26} + z_{36} = 5, \end{cases}$$

obtaining thus

$$(56)\qquad z_{13} = 18, \quad z_{21} = 9, \quad z_{22} = 11, \quad z_{23} = 10, \quad z_{26} = 2, \quad z_{34} = 6, \quad z_{35} = 5, \quad z_{36} = 3, \quad z_{45} = 9.$$

The solution of the problem is thus given by formulas (54) and (56), or by the following table:

Table 3

		18			
9	11	10			2
			6	5	3
				9	

§ 3. Systems of Representatives

Definition 8. Let $E = \{e_1, \ldots, e_m\}$ be a set and $\mathscr{S} = (S_1, \ldots, S_n)$ an ordered sequence of subsets S_j of E. An ordered sequence $R = (e_{i_1}, \ldots, e_{i_n})$ of elements of E is called a *system of representatives* for $\mathscr{S}$, if

$$(57)\qquad e_{i_j} \in S_j \qquad (j = 1, \ldots, n);$$

and e_{i_j} is said to represent S_j.

Definition 9. A system of representatives for $\mathscr{S}$ is called a *system of distinct representatives,* if the elements $e_{i_k} \in R$ are distinct.

Of course, a necessary condition for the existence of systems of distinct representatives is that $m \geqq n$.

Considering the bipartite graph $G = (E, \mathscr{S}; \in)$, we see that the following theorem holds:

Theorem 4. *An ordered sequence $R = (e_{i_1}, \ldots, e_{i_n})$ is a system of distinct representatives for $\mathscr{S}$, if and only if the set $\{(e_{i_1}, S_1), \ldots, (e_{i_n}, S_n)\}$ is an absolutely maximal matching of the bipartite graph $G = (E, \mathscr{S}; \in)$.*

Corollary 1. A system of distinct representatives exists if and only if the absolutely maximal matchings of G involve n edges each.

Corollary 2. A system of distinct representatives exists if and only if the number of internal stability of the dual G^* of the graph $G = (E, \mathcal{S}; \in)$ is equal to n.

This results immediately from Corollary 1, Theorem 1 and Definition X.3.

Corollary 3. Pseudo-Boolean procedures may be applied for establishing the existence of the systems of distinct representatives, as well as for their effective detection.

To do this, we associate to each system R of representatives for $\mathcal{S}$ a *characteristic vector* consisting of p Boolean variables x_{ij} ($p =$ the number of edges of G), defined as follows:

$$x_{ij} = \begin{cases} 1, & \text{if the element } e_i \text{ is taken as a representative of } S_j, \\ 0, & \text{in the opposite case.} \end{cases} \tag{58}$$

We see that x_{ij} must satisfy

$$\sum_i x_{ij} = 1 \qquad (j = 1, \ldots, n). \tag{59}$$

Moreover, *R is a system of distinct representatives if and only if its characteristic vector satisfies* (59) *and*

$$\sum_j x_{ij} \leqq 1 \qquad (i = 1, \ldots, m). \tag{60}$$

However, we prefer to handle (x_{ij}) as the characteristic vector of an absolutely maximal internally stable set of G^*, thus using the results of Chapter X, § 2.

Example 8. Let $E = \{1, 2, 3, 4, 5, 6, 7, 8\}$ and suppose $\mathcal{S}$ is composed of $S_1 = \{1, 2, 4\}$, $S_2 = \{4\}$, $S_3 = \{1, 2, 5, 8\}$, $S_4 = \{1\}$, $S_5 = \{1, 3, 4, 7, 8\}$, $S_6 = \{4, 6, 7\}$.

The associated graph $G = (E, \mathcal{S}; \in)$ is that studied in Example 1, where the maximal matchings of G were determined. Since the number of edges of each absolutely maximal matching is 6 (i.e. it is equal to n), it results from Example 1, Theorem 4 and Corollaries 1, 2, that there exist systems of distinct representatives for $\mathcal{S}$, and that they are those given in Table 4 below:

Table 4

	S_1	S_2	S_3	S_4	S_5	S_6
R_1	2	4	5	1	7	6
R_2	2	4	8	1	7	6
R_3	2	4	5	1	8	6
R_4	2	4	5	1	8	7
R_5	2	4	5	1	3	6
R_6	2	4	5	1	3	7
R_7	2	4	8	1	3	6
R_8	2	4	8	1	3	7

The number of systems of distinct representatives is here *incidentally* equal to the number of elements in E.

Various problems concerning systems of distinct representatives may be translated into the pseudo-Boolean language.

For instance, a theorem of P. HALL [1], states that $\mathscr{S} = (S_1, \ldots, S_n)$ has a system of distinct representatives if and only if every union of k sets of $\mathscr{S}$ contains at least k distinct elements, $k = 1, \ldots, n$.

This result yields the following

Theorem 5. *Let $A = (a_{ij})$ be the incidence matrix of the graph $G = (E, \mathscr{S}; \in)$. Then $\mathscr{S}$ has a system of distinct representatives if and only if the minimum of the pseudo-Boolean function*

$$m - \sum_{j=1}^{n} x_j - \sum_{i=1}^{m} \prod_{j=i}^{n} (1 - a_{ij} x_j) \tag{61}$$

is non-negative.

Proof. Let $\mathscr{S}'$ be a subsystem of $\mathscr{S}$ and $(x_1, \ldots, x_n)$ its characteristic vector, i.e.

$$x_j = \begin{cases} 1, & \text{if} \quad S_j \in \mathscr{S}', \\ 0, & \text{if} \quad S_j \notin \mathscr{S}'. \end{cases} \tag{62}$$

The element $e_i \in E$ belongs to the union $\bigcup_{S_j \in \mathscr{S}'} S_j$ if and only if

$$\prod_{j=1}^{n} (1 - a_{ij} x_j) = 0. \tag{63}$$

Hence the number of elements *not* in $\bigcup_{S_j \in \mathscr{S}'} S_j$ is equal to

$$\sum_{i=1}^{m} \prod_{j=1}^{n} (1 - a_{ij} x_j). \tag{64}$$

On the other hand, the number of sets in $\mathscr{S}'$ is $\sum_{j=1}^{m} x_j$.

The condition given in HALL's theorem is that the number of elements in $\bigcup_{S_j \in \mathscr{S}'} S_j$ is not less than the number of sets in $\mathscr{S}'$. In other words this conditions requires that relation

$$m - \sum_{i=1}^{m} \prod_{j=1}^{n} (1 - a_{ij} x_j) \geqq \sum_{j=1}^{n} x_j \tag{65}$$

holds for all $(x_1, \ldots, x_n) \in B_2^n$.

This completes the proof.

Definition 10. A system R of representatives for $\mathscr{S}$ will be called a *system of restricted representatives* with respect to the couples of numbers d_i, d_i', where

$$0 \leqq d_i \leqq d_i' \qquad (i = 1, \ldots, m), \tag{66}$$

if each element $e_i \in E$ occurs in the system of representatives R at least d_i times and at most d'_i times.

The following theorem is obvious:

Theorem 6. *R is a system of restricted representatives for $\mathscr{S}$ with respect to d_i, d'_i, if and only if its characteristic vector satisfies*

$$\sum_{i=1}^{m} x_{ij} = 1 \qquad (j = 1, \ldots, n) \tag{59}$$

and

$$d_i \leqq \sum_{j=1}^{n} x_{ij} \leqq d'_i \qquad (i = 1, \ldots, m). \tag{67}$$

Example 9. Let E and $\mathscr{S}$ be defined as in Example 8, and let us seek the systems of restricted representatives for $\mathscr{S}$ with respect to the numbers d_i, d_i given in Table 5 below:

Table 5

	1	2	3	4	5	6	7	8
d_i	2	0	0	2	0	0	1	1
d'_i	4	1	1	3	1	1	1	2

There are 17 variables x_{ij}, which we re-index as $x_1, \ldots, x_{17}$, in accordance with relations (3), where the indices i are $1'', \ldots, 8''$, while $1', \ldots, 6'$ are the indices j. The system (59) becomes

$$x_1 + x_2 + x_3 = 1, \tag{68.1}$$

$$x_4 = 1, \tag{68.2}$$

$$x_5 + x_6 + x_7 + x_8 = 1, \tag{68.3}$$

$$x_9 = 1, \tag{68.4}$$

$$x_{10} + x_{11} + x_{12} + x_{13} + x_{14} = 1, \tag{68.5}$$

$$x_{15} + x_{16} + x_{17} = 1, \tag{68.6}$$

while the system (67) becomes

$$2 \leqq x_1 + x_5 + x_9 + x_{10} \leqq 4, \tag{69.1}$$

$$0 \leqq x_2 + x_6 \leqq 1, \tag{69.2}$$

$$0 \leqq x_{11} \leqq 1, \tag{69.3}$$

$$2 \leqq x_3 + x_4 + x_{12} + x_{15} \leqq 3, \tag{69.4}$$

$$0 \leqq x_7 \leqq 1, \tag{69.5}$$

$$0 \leqq x_{16} \leqq 1, \tag{69.6}$$

$$1 \leqq x_{13} + x_{17} \leqq 1, \tag{69.7}$$

$$1 \leqq x_8 + x_{14} \leqq 2. \tag{69.8}$$

Using the methods given in Chapter III, we find that the above system (68), (69) has the following solutions:

Table 6

x_1	x_2	x_3	x_4	x_5	x_6	x_7	x_8	x_9	x_{10}	x_{11}	x_{12}	x_{13}	x_{14}	x_{15}	x_{16}	x_{17}
0	0	1	1	0	0	0	1	1	1	0	0	0	0	0	0	1
1	0	0	1	0	0	0	1	1	0	0	1	0	0	0	0	1
1	0	0	1	0	0	0	1	1	0	0	0	1	0	1	0	0
0	0	1	1	1	0	0	0	1	0	0	0	0	1	0	0	1

Hence we have found the following systems of restricted representatives with respect to the numbers in Table 5:

$$R_1 = (4, 4, 8, 1, 1, 7), \tag{70.1}$$

$$R_2 = (1, 4, 8, 1, 4, 7), \tag{70.2}$$

$$R_3 = (1, 4, 8, 1, 7, 4), \tag{70.3}$$

$$R_4 = (4, 4, 1, 1, 8, 7). \tag{70.4}$$

L. R. Ford Jr. and D. R. Fulkerson [3] have shown that the following condition is necessary and sufficient for the existence of a system of restricted representatives: every subset X of the set of indices $\{1, \ldots, n\}$ has no more than

$$\min\left(n - \sum_{i=1}^{m} d_i + \sum_{i \in I(X)} d_i, \sum_{i \in I(X)} d_i'\right) \tag{71}$$

elements, where $I(X) \subseteqq \{1, \ldots, m\}$ is the index set of $\bigcup_{j \in X} S_j$.

This theorem may be translated into the pseudo-Boolean language as follows:

Theorem 7. *A system of restricted representatives for $\mathcal{S}$ with respect to d_i, d_i' exists if and only if the minimum of the pseudo-Boolean function*

(72) $f(x_0, x_1, \ldots, x_n)$

$$= x_0\left[n + \sum_{i=1}^{m} d_i \prod_{j=1}^{n} (1 - a_{ij} x_j)\right] + \bar{x}_0 \sum_{i=1}^{m} d_i'\left[1 - \prod_{j=1}^{n} (1 - a_{ij} x_j)\right] - \sum_{j=1}^{n} x_j$$

is non-negative, where $a_{ij} = 1$ if $e_i \in S_j$, and $a_{ij} = 0$ if $e_i \notin S_j$.

Proof. Let $(x_1, \ldots, x_n)$ be the characteristic vector of the set X. Then obviously the following relations hold:

$$n - \sum_{i=1}^{m} d_i + \sum_{i \in I(X)} d_i = n + \sum_{i \notin I(X)} d_i = n + \sum_{i=1}^{m} d_i \prod_{j=1}^{n} (1 - a_{ij} x_j), \tag{73}$$

$$\sum_{i \in I(X)} d_i' = \sum_{i=1}^{m} d_i'\left[1 - \prod_{j=1}^{n} (1 - a_{ij} x_j)\right]. \tag{74}$$

Denoting the right-hand sides of (73) and (74) by $g(x_1, \ldots, x_n)$ and $h(x_1, \ldots, x_n)$, respectively, the condition given by FORD and FULKERSON becomes

$$\sum_{j=1}^{n} x_j \leqq \min[g(x_1, \ldots, x_n), h(x_1, \ldots, x_n)], \tag{75}$$

for all $(x_1, \ldots, x_n) \in B_2^n$. In other terms,

$$\sum_{j=1}^{n} x_j \leqq x_0 \, g(x_1, \ldots, x_n) + \bar{x}_0 \, h(x_1, \ldots, x_n) \tag{76}$$

for all $(x_0, x_1, \ldots, x_n) \in B_2^{n+1}$, i.e.

$$\min\left[x_0 \, g(x_1, \ldots, x_n) + \bar{x}_0 \, h(x_1, \ldots, x_n) - \sum_{j=1}^{n} x_j\right] \geqq 0, \tag{77}$$

which coincides with (72).

Remark 4. Taking $d_i = 0$, $d_i' = 1$ $(i = 1, \ldots, m)$, and $x_0 = 0$, Theorem 7 reduces to Theorem 5.

Several other problems concerning systems of representatives (as those on systems of common representatives, marginal elements, matrices of zeros and ones, etc.) may also be handled by means of pseudo-Boolean techniques. As a matter of fact, several proposals were made for using integer linear programming procedures (especially the GOMORY algorithms) in order to solve combinatorial problems belonging to the above discussed class. The bibliography concerning systems of representatives is rich enough. The reader is referred, for instance, to L. R. FORD JR. and D. R. FULKERSON [3, 4], D. R. FULKERSON and H. J. RYSER [1], M. HALL JR. [1], P. HALL [1], P. R. HALMOS and H. E. VAUGHAN [1], A. J. HOFFMAN and H. W. KUHN [1, 2], H. B. MANN and H. J. RYSER [1], N. S. MENDELSOHN and A. L. DULMAGE [1], O. ORE [1], H. J. RYSER [1] (Chapter 5).

Chapter XII

Flows in Networks and Chains in Partially Ordered Sets

The aim of this chapter is to apply pseudo-Boolean programming to problems concerning flows in networks and minimal decompositions of partially ordered sets into chains.

§ 1. Flows in Networks

The aim of this section is to apply (following P. L. HAMMER [6]) the method of pseudo-Boolean programming to the determination of the minimal cut and of the value of the maximal flow through a network, to extend this procedure to the case when lower bounds on the arc flows are given and to apply this method to the determination of the feasibility of some supply-demand problems and of the existence of circuits in a network*.

Throughout this section we shall use the terminology of L. R. FORD JR. and D. R. FULKERSON [4].

Maximal Flows in Networks

Let $G = (N, \Gamma)$ be a graph; let $N = \{s_1, s_2, \ldots, s_n\}$ be the set of its nodes, among which two vertices are distinguished: s_1, called the *source*, and s_n, called the *sink*.

We shall suppose that besides the incidence matrix (a_{ij}), we are also given a matrix $C = (c_{ij})$ of non-negative elements, called *arc-capacities* satisfying

(1) $$a_{ij} = 0 \Rightarrow c_{ij} = 0.$$

A graph with arc-capacities is usually termed a *network*.

A *flow* of value v from the source s_1 to the sink s_n is a matrix $F = (f_{ij})$ of non-negative elements called arc-flows satisfying the following conditions:

(2) $$\sum_{j=1}^{n} a_{ij} f_{ij} - \sum_{j=1}^{n} a_{ji} f_{ji} = \begin{cases} v & \text{if } i = 1, \\ 0 & \text{if } i \neq 1, n, \\ -v & \text{if } i = n, \end{cases}$$

(3) $$f_{ij} \leqq c_{ij}.$$

---

* After the completion of this section, we were informed about the interesting thesis of U. S. R. MURTY [1], in which pseudo-Boolean programming is applied to the solution of problems occurring in communication networks.

The *maximal flow problem* is that of finding those f_{ij} for which v is maximal, conditions (2) and (3) being satisfied.

If $M \subseteqq N$ we shall put $M' = N - M$ and $I_M = \{i \mid s_i \in M\}$. For any function

$$g : U \to R,$$

where U is the set of arcs, we put

$$g(M, M') = \sum_{i \in I_M} \sum_{j \in I_{M'}} g_{ij}. \tag{4}$$

By a *cut* $\mathscr{C}$ in (N, Γ) separating s_1 and s_n we mean a set $K \subset N$ so that $s_1 \in K$, $s_n \in K'$. The capacity of a cut is $c(K, K')$.

The *minimal cut problem* is that of determining a cut of minimal capacity.

Ford and Fulkerson [1, 2] have proved the following fundamental result in the theory of flows in networks:

For any network, the maximal flow value from s_1 to s_n is equal to the minimal cut capacity of all cuts separating s_1 and s_n.

When a minimal cut K_0 is known then

$$v_{\max} = c(K_0, K_0').$$

Any subset M of N may be characterized by the vector $(x_1, \ldots, x_n)$ where

$$x_i = \begin{cases} 1 & \text{if} \quad i \in I_M, \\ 0 & \text{if} \quad i \notin I_M. \end{cases}$$

A cut $\mathscr{C}$ separating s_1 and s_n will be characterized by a vector $(1, x_2, x_3, \ldots, x_{n-1}, 0)$.

The capacity of a cut is

$$c(K, K') = \sum_{i \in I_M} \sum_{j \in I_{M'}} c_{1j} = \sum_{i=1}^{n} \sum_{j=1}^{n} c_{ij} x_i \bar{x}_j.$$

As $x_1 = 1$, $x_n = 0$, we have

Theorem 1. *If $(1, x_2^*, x_3^*, \ldots, x_{n-1}^*, 0)$ gives a minimum of the pseudo-Boolean function*

$$G = c_{1n} + \sum_{j=2}^{n-1} c_{1j} \bar{x}_j + \sum_{i=2}^{n-1} c_{in} x_i + \sum_{i=2}^{n-1} \sum_{j=2}^{n-1} c_{ij} x_i \bar{x}_j, \tag{5}$$

then

$$K^* = \{s_i \mid x_i^* = 1\}$$

is a minimal cut separating s_1 and s_n, and conversely, any minimal cut separating s_1 and s_n may be obtained in this way.

Thus, the problem of finding a minimal cut separating the source and the sink (and implicitly that of determining the value of a maximal flow) in a network, is reduced to a problem of pseudo-Boolean programming.

Example 1. Let us consider the network discussed in FULKERSON [3]

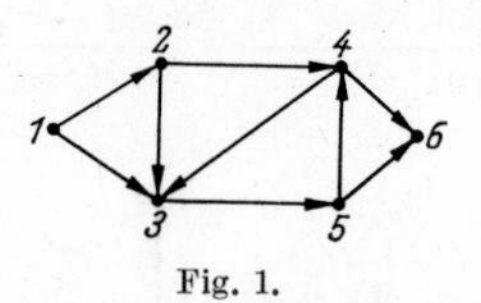

Fig. 1.

c_{ij}	1	2	3	4	5	6
1	0	7	3	0	0	0
2	0	0	1	6	0	0
3	0	0	0	0	8	0
4	0	0	3	0	0	2
5	0	0	0	2	0	8
6	0	0	0	0	0	0

Putting

$$x_1 = 1, \quad x_6 = 0, \tag{6}$$

we have

$$G = 7\bar{x}_2 + 3\bar{x}_3 + 2x_4 + 8x_5 + x_2\bar{x}_3 + 6x_2\bar{x}_4 + 8x_3\bar{x}_5 + 3x_4\bar{x}_3 + 2x_5\bar{x}_4$$
$$= 10 + 5x_3 + 5x_4 + 10x_5 - x_2x_3 - 6x_2x_4 - 3x_3x_4 - 8x_3x_5 - 2x_4x_5.$$

We find

$$G_{\min} = 9, \tag{7}$$

and the minimizing point

$$x_1^* = 1, \quad x_2^* = 1, \quad x_3^* = 0, \quad x_4^* = 1, \quad x_5^* = 0, \quad x_6^* = 0. \tag{8}$$

Hence, the minimal cut is

$$K = \{s_1, s_2, s_4\}, \quad K' = \{s_3, s_5, s_6\}. \tag{9}$$

Lower Bounds on Arc Flows

Let us now suppose that we are given a matrix $B = (b_{ij})$ of real-valued elements called *lower bounds*, satisfying

$$0 \leqq b_{ij} \leqq c_{ij} \tag{10}$$

and we are seeking the maximal v for which f_{ij} satisfy (2) and

$$b_{ij} \leqq f_{ij} \leqq c_{ij}. \tag{11}$$

FORD and FULKERSON [4] proved that:

If there is a matrix (f_{ij}) satisfying (2) and (11) for some number v, then the maximal value of v subject to these constraints is equal to the minimum of

$$H = c(M, M') - b(M', M) \tag{12}$$

taken over all $M \subseteqq N$ with $s_1 \in M$, $s_n \in M'$.

It is easy to see that

$$H = \sum_{i=1}^{n} \sum_{j=1}^{n} c_{ij}\, x_i\, \bar{x}_j - \sum_{i=1}^{n} \sum_{j=1}^{n} b_{ij}\, \bar{x}_i\, x_j \tag{13}$$

and, as $x_1 = 1$, $x_n = 0$ we have

Theorem 2. *If there is a matrix* (f_{ij}) *satisfying* (2) *and* (11) *for some* v, *then the maximal value of* v *subject to these constraints is equal to the minimum of the pseudo-Boolean function*

$$H = c_{1n} - b_{n1} + \sum_{j=2}^{n-1} (c_{1j}\, \bar{x}_j - b_{nj}\, x_j) + \sum_{i=2}^{n-1} (c_{in}\, x_i - b_{i1}\, \bar{x}_i) + \\ + \sum_{i=2}^{n-1} \sum_{j=2}^{n-1} (c_{ij}\, x_i\, \bar{x}_j - b_{ij}\, \bar{x}_i\, x_j). \tag{14}$$

Example 2. If in the previous example we take $b_{24} = 3$, $b_{35} = 4$ and $b_{54} = 2$ and all the other $b_{ij} = 0$, and if we put

$$x_1 = 1, \quad x_6 = 0, \tag{15}$$

then we have

$$H = 10 + 5x_3 + 6x_5 - x_2\, x_3 - 3x_2\, x_4 - 3x_3\, x_4 - 4x_3\, x_5.$$

We find the minimizing point

$$x_1^* = 1, \quad x_2^* = 1, \quad x_3^* = 0, \quad x_4^* = 1, \quad x_5^* = 0, \quad x_6^* = 0 \tag{16}$$

and

$$H_{\min} = 7. \tag{17}$$

Some Feasibility Theorems

The Supply-Demand Theorem.

Let (N, Γ) be a network and let N be partitioned in sources S, intermediate nodes R and sinks T. For any $s_i \in S$ a non-negative number α_i (the supply) is defined; for any $s_j \in T$ a non-negative number β_j (the demand) is defined.

GALE's theorem [1] may be now formulated as follows:

Theorem 3. *The constraints*

$$\sum_{j=1}^{n} a_{ij}\, f_{ij} - \sum_{j=1}^{n} a_{ji}\, f_{ji} \begin{cases} \leqq \alpha_i & \text{if } s_i \in S, \\ = 0 & \text{if } s_i \in R, \\ \geqq \beta_i & \text{if } s_i \in T, \end{cases} \tag{18}$$

$$0 \leqq f_{ij} \leqq c_{ij} \tag{19}$$

are feasible if and only if the minimum of the pseudo-Boolean function

$$\sum_{i=1}^{n} \sum_{j=1}^{n} c_{ij}\, x_i\, \bar{x}_j + \sum_{h=1}^{n} \alpha_h\, \bar{x}_h - \sum_{h=1}^{n} \beta_h\, \bar{x}_h \tag{20}$$

is non-negative.

A Symmetric Supply-Demand Theorem.

Let us suppose that besides α_i and β_j we are also given some non-negative numbers $\alpha'_i (s_i \in S)$ and $\beta'_j (s_j \in T)$ satisfying

$$0 \leqq \alpha_i \leqq \alpha'_i \qquad (s_i \in S), \tag{21}$$

$$0 \leqq \beta_j \leqq \beta'_j \qquad (s_j \in T) \tag{22}$$

and we are interested in the feasibility of the flow problem with the conditions:

$$\alpha_i \leqq \sum_{j=1}^{n} a_{ij} f_{ij} - \sum_{j=1}^{n} a_{ji} f_{ji} \leqq \alpha'_i \qquad (s_i \in S), \tag{23}$$

$$\sum_{j=1}^{n} a_{ij} f_{ij} - \sum_{j=1}^{n} a_{ji} f_{ji} = 0 \qquad (s_i \in R), \tag{24}$$

$$\beta_j \leqq \sum_{j=1}^{n} a_{ij} f_{ij} - \sum_{j=1}^{n} a_{ji} f_{ji} \leqq \beta'_j \qquad (s_i \in T), \tag{25}$$

$$0 \leqq f_{ij} \leqq c_{ij} \qquad (i, j = 1, \ldots, n). \tag{26}$$

FULKERSON's feasibility condition [2] now becomes:

Theorem 4. *The relations* (23), (24), (25), (26) *are feasible if and only if the minima of the pseudo-Boolean functions*

$$\sum_{i=1}^{n} \sum_{j=1}^{n} c_{ij} x_i \bar{x}_j + \sum_{h=1}^{n} \alpha'_h \bar{x}_h - \sum_{h=1}^{n} \beta_h \bar{x}_h$$

and

$$\sum_{i=1}^{n} \sum_{j=1}^{n} c_{ij} x_i \bar{x}_j - \sum_{h=1}^{n} \alpha_h x_h + \sum_{h=1}^{n} \beta'_h x_h$$

are non-negative.

Circulation Theorem.

In networks with no source and sink specifications we are interested in the existence of circulations f_{ij} satisfying

$$\sum_{j=1}^{n} a_{ij} f_{ij} - \sum_{j=1}^{n} a_{ji} f_{ji} = 0 \qquad (i = 1, \ldots, n), \tag{27}$$

$$b_{ij} \leqq f_{ij} \leqq c_{ij} \qquad (i, j = 1, \ldots, n). \tag{28}$$

HOFFMAN's feasibility condition [1] now becomes:

Theorem 5. *The conditions* (27) *and* (28) *are feasible if and only if the minimum of the pseudo-Boolean function*

$$\sum_{i=1}^{n} \sum_{j=1}^{n} c_{ij} x_i \bar{x}_j - \sum_{i=1}^{n} \sum_{j=1}^{n} b_{ij} \bar{x}_i x_j \tag{29}$$

is non-negative.

Another theorem of HOFFMAN [1] states

Theorem 6. *The constraints* (28) *and*

$$\alpha_i \leqq \sum_{j=1}^{n} a_{ij} f_{ij} - \sum_{j=1}^{n} a_{ji} f_{ji} \leqq \alpha'_i \tag{30}$$

[*where* $0 \leqq \alpha_i \leqq \alpha'_i$, $(i = 1, \ldots, n)$ *are given real numbers*] *are feasible if and only if the minimum of the pseudo-Boolean function of* $n + 1$ *variables*

$$(31) \qquad \sum_{i=1}^{n} \sum_{j=1}^{n} c_{ij}\, x_i\, \bar{x}_j - \sum_{i=1}^{n} \sum_{j=1}^{n} b_{ij}\, \bar{x}_i\, x_j - u \sum_{h=1}^{n} \alpha_h\, \bar{x}_h - \bar{u} \sum_{h=1}^{n} \alpha'_h\, x_h$$

is non-negative.

§ 2. Minimal Decomposition of Partially Ordered Sets into Chains

The aim of the present section is to show that the problem of finding the number N of chains in a minimal decomposition of a finite partially ordered set S, as well as that of actually determining those chains, may be reduced to that of minimizing pseudo-Boolean functions. DILWORTH's [1], DANTZIG and HOFFMAN's [1] and FULKERSON's [2] results on the number of chains in a minimal decomposition of S offer various means of translating this problem into the pseudo-Boolean language.

Let us consider a finite partially ordered set $S = \{s_1, \ldots, s_n\}$ with a strict ordering denoted by "$\prec$".

Let us form an $n \times n$ matrix $B = (b_{ij})$, by putting

$$(32) \qquad b_{ij} = \begin{cases} 1 & \text{if} \quad s_i \prec s_j, \\ 0 & \text{if} \quad s_i, s_j \text{ are not in the relation } \prec. \end{cases}$$

We shall suppose that $b_{ii} = 0$ $(i = 1, \ldots, n)$.

A *chain* is a subset $C = \{s_{i_1}, \ldots, s_{i_k}\}$ of S, so that for any $s_{i_m}, s_{i_h} \in C$, with $m \neq h$ either $s_{i_m} \prec s_{i_h}$, or $s_{i_h} \prec s_{i_m}$. An *antichain* is a subset $D = \{s_{i_1}, \ldots, s_{i_r}\}$ os S, so that for any $s_{j_u}, s_{j_v} \in D$, neither $s_{j_u} \prec s_{j_v}$, nor $s_{j_v} \prec s_{j_u}$. A *decomposition* of S is a family $\{C_1, \ldots, C_N\}$ of disjoint chains, the union of which is S.

A decomposition $\{C_1, \ldots, C_N\}$ is called *minimal*, if for any other decomposition $\{C'_1, \ldots, C'_{N'}\}$, $N \leqq N'$.

Economic Interpretation

The problem of finding the minimal decomposition of a finite partially ordered set into chains was shown by D. R. FULKERSON [4], to have an important economic interpretation.

Suppose there are n jobs $1, 2, \ldots, n$ with specified start and finish times $s_1, s_2, \ldots, s_n$ and $f_1, f_2, \ldots, f_n$, respectively (with $s_i \leqq f_i$). The duration times of the jobs are $t_i = f_i - s_i$. Assume that we have a number of identical machines, each of which can perform any job in the specified time. The reassignment time required for a machine to go from job i to job j is the given non-negative number r_{ij}. The problem is to find the minimum number of machines required to meet the given job schedule.

As a concrete example, FULKERSON indicates the problem of an airline which wants to meet a fixed flight schedule with the minimum number of planes, all of the same type. Start and finish times are known for each flight, and the times r_{ij} to return from the destination point of flight i to the origin point of flight j are also known.

Assuming that relations

$$r_{ij} \leqq r_{ik} + r_{kj} \tag{33}$$

hold for all i, j, k, it is obvious that the relation $i\,R\,j$ defined by

$$i\,R\,j \quad \text{if and only if} \quad f_i + r_{ij} \leqq s_j \tag{34}$$

is a relation of strict ordering among jobs.

The order relations among jobs can be depicted by means of an acyclic directed network whose *arcs* represent jobs.

For example if we are given fixed jobs with the order relations

$$1R3,\ 1R4,\ 2R4,\ 1R5,\ 2R5,\ 3R5,\ 4R5, \tag{35}$$

then we can depict them by means of the network in Fig. 2.

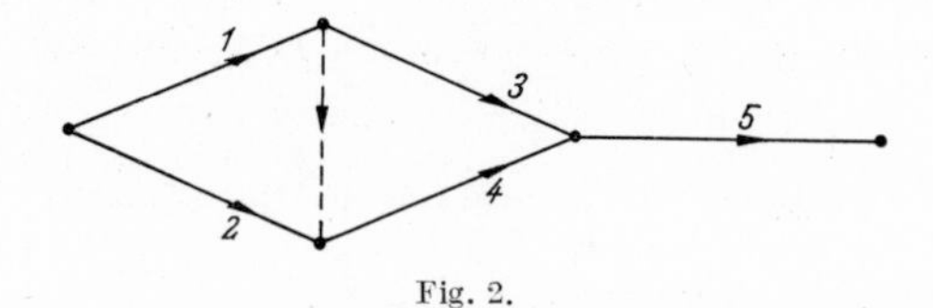

Fig. 2.

The dotted arc stands for a "dummy" job, added to maintain the proper order relations among the jobs.

It is easy to show that the use of dummies permits a network representation of this kind for any finite partially ordered set.

Since a chain of arcs in this network represents a possible assignment of jobs to one machine, *the problem is to cover all non-dummy arcs with the minimum number of chains.*

Remark 1. The above problem can be converted into one of flows in networks as follows. Add a node to the network (the source) and direct dummy arcs from this node to all nodes of the network that have only outward-pointing arcs. Similarly add a node to the network (the sink) and direct dummy arcs into this from all nodes having only inward-pointing arcs. Now place a lower bound of 1 on each non-dummy arc, a lower bound of 0 on each dummy arc, and take all arc capacities infinite. Then a minimal flow of value v picks out v chains (not necessarily distinct) covering all non-dummy arcs. We can now obtain immediately a minimal covering of the non-dummy arcs with distinct chains.

In the next pages of this section we shall describe the results given in P. L. HAMMER [2] for finding the minimal number of chains covering a partially ordered set.

Maximal anti-chains

R. P. DILWORTH [1] has proved that the number N of chains in a minimal decomposition of a finite partially ordered set, is equal to the number M of elements contained in a maximal anti-chain of S. We shall give here a way of determining all the maximal anti-chains of S.

A subset D of S is an *anti-chain* if and only if for any $s_i, s_j \in D$, we have

$$b_{ij} d_i d_j = b_{ji} d_j d_i = 0, \tag{36}$$

where

$$d_i = \begin{cases} 1, & \text{if } i \in D, \\ 0, & \text{if } i \notin D. \end{cases} \tag{37}$$

Hence:

Lemma 1. *The subset D of S is an anti-chain if and only if*

$$\sum_{i=1}^{n} \sum_{j=1}^{n} b_{ij} d_i d_j = 0. \tag{38}$$

The number of elements of D is

$$\sum_{h=1}^{n} d_h. \tag{39}$$

Let us now consider the pseudo-Boolean function

$$F = -\sum_{h=1}^{n} d_h + (n+1) \sum_{i=1}^{n} \sum_{j=1}^{n} b_{ij} d_i d_j \tag{40}$$

and let $(d_1^*, \ldots, d_n^*) \in B_2^n$ be a minimizing point of F.

It is now obvious that:

Theorem 7. *Let $(d_1^*, \ldots, d_n^*) \in B_2^n$ be a minimizing point of the pseudo-Boolean function* (40); *then*

$$D^* = \{s_i \mid d_i^* = 1, \ i = 1, \ldots, n\} \tag{41}$$

is a maximal anti-chain of S and the number of its elements is

$$M = -F(d_1^*, \ldots, d_n^*). \tag{42}$$

Conversely, given a maximal anti-chain $D^* = \{s_{i_1}, \ldots, s_{i_m}\}$ of S, we have $m = M$, and there exists a minimizing point $(d_1^*, \ldots, d_n^*)$ of the pseudo-Boolean function (40), *so that* (41) *holds.*

From DILWORTH's theorem we have the following

Corollary 1:

$$N = -F(d_1^*, \ldots, d_n^*). \tag{43}$$

The DANTZIG-HOFFMAN *Formula*

Let T now be an arbitrary subset of S and let us put

$$T^* = \{s_j \in S \mid \exists\, s_i \in T, \ s_i \prec s_j\}. \tag{44}$$

G. B. DANTZIG and A. J. HOFFMAN [1] have proved that

$$N = \max_{T \subseteq S} (|T| - |T^*|) \tag{45}$$

where $|T|$ and $|T^*|$ represent the number of elements of T and of T^*, respectively.

Obviously

$$|T| = \sum_{i=1}^{n} t_i, \tag{46}$$

$$|T^*| = \sum_{i=1}^{n} t_i^*, \tag{47}$$

where $(t_1, \ldots, t_n)$ and $(t_1^*, \ldots, t_n^*)$ are the characteristic vectors of T and T^*, respectively.

An element $s_j \in S$ does not belong to T^* if and only if for any i, $b_{ij} = 0$ or $t_i = 0$. Hence, $t_j^* = 0$ if and only if

$$\bigcup_{i=1}^{n} b_{ij} t_i = 0 \tag{48}$$

or, using the DE MORGAN formula:

$$\prod_{i=1}^{n} (\bar{b}_{ij} \cup \bar{t}_i) = 1. \tag{49}$$

If $b_{ij} = 0$, then $\bar{b}_{ij} \cup \bar{t}_i = 1$, and so (49) becomes

$$\prod_{b_{ij}=1} \bar{t}_i = 1. \tag{50}$$

Thus

$$\overline{t_j^*} = \prod_{b_{ij}=1} \bar{t}_i \tag{51}$$

or

$$t_j^* = 1 - \prod_{b_{ij}=1} \bar{t}_i \tag{52}$$

and

$$|T^*| = n - \sum_{j=1}^{n} \prod_{b_{ij}=1} \bar{t}_i \tag{53}$$

Thus, we have

Theorem 8. *If* $(t_1^*, \ldots, t_n^*) \in B_2^n$ *gives a maximum of the pseudo-Boolean function*

$$K = \sum_{h=1}^{n} t_h + \sum_{j=1}^{n} \prod_{b_{ij}=1} \bar{t}_i - n, \tag{54}$$

then

$$N = K(t_1^*, \ldots, t_n^*). \tag{55}$$

Cuts and Joins

FULKERSON [1] has pointed out the strong relationship between the problem of finding the minimal decomposition of S and the problems of finding a minimal cut or a maximal join of a certain graph G. G is defined as consisting of $2n$ vertices $U = \{v_1, \ldots, v_n, v_{n+1}, \ldots, v_{2n}\}$ and arcs defined by the rule: if $s_i \prec s_j$ then $v_i v_{n+j}$ is an arc of G, and these are all the arcs of G. Let $V' = \{v_1, \ldots, v_n\}$, $V'' = \{v_{n+1}, \ldots, v_{2n}\}$. $A(V', V'')$-cut C of G is a subset of V, so that every arc joining a vertex of V' to one of V'' has at least one endpoint in C. A (V', V'')-join J of G is a set of arcs, each of them joining a vertex of V' to one of V'', and no two of them having a vertex in common. A maximal (V', V'')-join is one having the greatest number of elements. The cuts and joins below are relative to V', V''. The results of FULKERSON based on a theorem of KÖNIG ([1], p. 232) permit to construct a minimal decomposition of S if we know a maximal join of G, and to construct a maximal anti-chain of S, if we know a minimal cut of G.

A necessary and sufficient condition that a subset C of U be a cut is that for any $i, j = 1, \ldots, n$, $b_{ij} = 1$ should imply

$$c_i \cup c_{n+j} = 1, \tag{56}$$

where $(c_1, \ldots, c_n)$ is the characteristic vector of C. Or,

$$b_{ij} \leqq c_i \cup c_{n+j}. \tag{57}$$

Hence,

$$b_{ij}\, \bar{c}_i\, \bar{c}_{n+j} = 0 \tag{58}$$

or,

$$\sum_{i=1}^{n} b_{ij}\, \bar{c}_i\, \bar{c}_{n+j} = 0. \tag{59}$$

We have

Lemma 2. *The subset C of S is a cut in G if and only if*

$$\sum_{i=1}^{n} \sum_{j=1}^{n} b_{ij}\, \bar{c}_i\, \bar{c}_{n+j} = 0. \tag{60}$$

Following the same line as above, we can prove

Theorem 9. *Let $(c_1^*, \ldots, c_n^*, c_{n+1}^*, \ldots, c_{2n}^*) \in B_2^n$ be a minimizing point of the pseudo-Boolean function*

$$H = -\sum_{h=1}^{2n} \bar{c}_h + (2n+1) \sum_{i=1}^{2n} \sum_{j=1}^{2n} b_{ij}\, \bar{c}_i\, \bar{c}_j. \tag{61}$$

Then

$$C^* = \{s_i \mid c_i^* = 1,\ i = 1, \ldots, 2n\} \tag{62}$$

is a minimal cut of G and the number of its elements is

$$L = 2n + H(c_1^*, \ldots, c_{2n}^*). \tag{63}$$

Conversely, given a minimal cut $C^* = \{v_{i_1}, \ldots, v_{i_l}\}$, of G, we have $l = L$, and there exists a minimizing point $(c_1^*, \ldots, c_{2n}^*)$ of the pseudo-Boolean function H, so that (62) *holds.*

From FULKERSON's results we have

Corollary 2.

$$N = 2n + H(c_1^*, \ldots, c_{2n}^*). \tag{64}$$

Chapter XIII

Various Applications

In this chapter we have collected some typical examples of practical applications of Boolean methods. These applications include a sequencing problem due to R. FORTET, a problem of timetable scheduling (written by I. ROSENBERG and including also results of G. MIHOC and E. BALAS), a problem in coding theory, a problem of plant location, and various other applications.

§ 1. A Sequencing Problem

R. FORTET [1] studied the following sequencing problem.

Consider m goods $G_1, \ldots, G_m$ to be produced by n machines $M_1, \ldots, M_n$. Each good G_i can be produced on any one of the machines $M_{i_1}, \ldots, M_{i_{k(i)}}$, the family of which will be denoted by E_i. The time is divided into indivisible intervals, labelled $0, 1, 2, \ldots$ The production of the good G_i lasts d_i time intervals and it cannot be fractioned; the starting of the G_i's production must take place in one of the intervals $b_i, b_i + 1, \ldots, c_i$. Each good is to be assigned to a single machine, while a machine can produce in different time intervals different goods; however, at a specified time interval, a machine can produce at most one good.

The problem is to find all possible good-machine-time assignments. Of course, one could select an efficiency criterion and choose the corresponding best sequencing.

Let us introduce, with FORTET, the following bivalent variables:

$$(1) \qquad x_i^h = \begin{cases} 1, & \text{if the production of } G_i \text{ begins in interval } h, \\ 0, & \text{if not,} \end{cases}$$

where $i = 1, \ldots, m$; $h = b_i, b_i + 1, \ldots, c_i$,

$$(2) \qquad y_{ij} = \begin{cases} 1, & \text{if } G_i \text{ is assigned to } M_j, \\ 0, & \text{if not,} \end{cases}$$

where $i = 1, \ldots, m$; $j \in E_i$;

$$(3) \qquad t_{ij}^k = \begin{cases} 1, & \text{if } G_i \text{ is being produced by } M_j \text{ in interval } k, \\ 0, & \text{if not,} \end{cases}$$

where $i = 1, \ldots, m$; $j \in E_i$; $k = b_i, b_i + 1, \ldots, c_i + d_i - 1$.

The values x_i^h and y_{ij} characterize the solutions (if any). They are related to the variables t_{ij}^k via formulas

$$(4) \qquad t_{ij}^k = y_{ij} \bigcup_{h \in N_{ik}} x_i^h,$$

where N_{ik} is the segment $[\max(b_i, k - d_i + 1), \min(k, c_i)]$. Further x_i^h, y_{ij} and t_{ij}^k must satisfy

$$(5) \qquad \sum_{h=b_i}^{c_i} x_i^h = 1 \qquad (i = 1, \ldots, m),$$

$$(6) \qquad \sum_{j \in E_i} y_{ij} = 1 \qquad (i = 1, \ldots, m),$$

$$(7) \qquad \sum_{i \in P_{jk}} t_{ij}^k \leqq 1 \qquad (j = 1, \ldots, n;\ k = t, \ldots, T);$$

where $t = \min_i b_i$ and $T = \max(c_i + d_i - 1)$, whereas P_{jk} is the set of those indices i satisfying both $j \in E_i$ and $b_i \leqq k \leqq c_i + d_i - 1$.

Remarks 1 and 2 below show that some of the equations (7) may be dropped.

Remark 1. If, for a certain index k, there exists at most one index i_0 satisfying $b_i \leqq k \leqq c_i + d_i - 1$, then for all $j = 1, \ldots, n$, we have either $P_{jk} = \varnothing$, or $\sum_{i \in P_{jk}} t_{ij}^k = t_{i_0 j}^k \leqq 1$, so that the corresponding equations (7) are identically satisfied.

Remark 2. If, for a certain index k, there exist exactly two indices i_1, i_2 satisfying $b_i \leqq k \leqq c_i + d_i - 1$, but $E_{i_1} \cap E_{i_2} = \varnothing$, then for every $j = 1, \ldots, n$, we have $P_{jk} = \{i_1, i_2\}$ and $t_{i_1 j}^k + t_{i_2 j}^k \leqq 1$, so that the corresponding equations (7) are identically satisfied. Obviously, these conditions characterize the solutions.

Notice now that a parametric solution of a pseudo-Boolean inequality of the form

$$(8) \qquad z_1 + \cdots + z_p \leqq 1$$

can easily be found. Namely, let q denote the smallest integer satisfying $2^q \geqq p$, let $u_1, u_2, \ldots, u_q$ be q bivalent parameters, and let $c_0(u_1, \ldots, u_q)$, $c_1(u_1, \ldots, u_q), \ldots, c_{2^q-1}(u_1, \ldots, u_q)$ be the 2^q complete elementary conjunctions in $u_1, \ldots, u_q$. Then formulas

$$(9.r) \qquad z_r = c_r(u_1, \ldots, u_q) \qquad (r = 1, \ldots, p)$$

give a parametric solution of (8).

Notice also that a parametric solution of the pseudo-Boolean equation

(10) $$z_1 + \cdots + z_p = 1$$

is obviously given by formulas (9.1), (9.2), ..., (9. $p - 1$) and

(11) $$z_p = c_p(u_1, \ldots, u_q) \cup \cdots \cup c_{2^q-1}(u_1, \ldots, u_q).$$

Using the above formulas, R. FORTET proposes the following procedure for the solution of the system (4)—(7). Solve separately the system (5) and (6), and introduce their parametric solutions $x_i^h = x_i^h(v_1, v_2, \ldots)$ and $y_{ij} = y_{ij}(w_1, w_2, \ldots)$ into (4). Introduce the resulting expressions $t_{ij}^{k'} = t_{ij}^k(v_1, v_2, \ldots; w_1, w_2, \ldots)$ into (7) and solve parametrically the resulting system in v_r and $w_s : v_r = v_r(u_1, u_2, \ldots)$, $w_s = w_s(u_1, u_2, \ldots)$. Introduce these expressions into the relations $x_i^h = x_i^h(v_1, v_2, \ldots)$ and $y_{ij} = y_{ij}(w_1, w_2, \ldots)$, obtaining thus the parametric solution $x_i^h = X_i^h(u_1, u_2, \ldots)$, $y_{ij} = Y_{ij}(u_1, u_2, \ldots)$ of the given problem.

Example 1 (R. FORTET [1]). Let the data of the problem be the following: $m = 6$, $n = 8$ and

Table 1

Good	b_i	c_i	d_i	E_i
G_1	5	7	2	$\{M_1, M_7\}$
G_2	2	5	3	$\{M_3, M_7\}$
G_3	1	2	4	$\{M_2, M_5\}$
G_4	13	14	3	$\{M_1, M_4, M_7, M_8\}$
G_5	11	12	4	$\{M_2, M_4, M_6, M_7\}$
G_6	11	12	3	$\{M_3, M_6\}$

The system (5) becomes

(12.1) $$x_1^5 + x_1^6 + x_1^7 = 1,$$

(12.2) $$x_2^2 + x_2^3 + x_2^4 + x_2^5 = 1,$$

(12.3) $$x_3^1 + x_3^2 = 1,$$

(12.4) $$x_4^{13} + x_4^{14} = 1,$$

(12.5) $$x_5^{11} + x_5^{12} = 1,$$

(12.6) $$x_6^{11} + x_6^{12} = 1.$$

The system (6) becomes

(13.1) $$y_{11} + y_{17} = 1,$$

(13.2) $$x_{23} + y_{27} = 1,$$

(13.3) $$y_{32} + y_{35} = 1,$$

(13.4) $$y_{41} + y_{44} + y_{47} + y_{48} = 1,$$

(13.5) $$y_{52} + y_{54} + y_{56} + y_{57} = 1,$$

(13.6) $$y_{63} + y_{66} = 1.$$

In view of Remarks 1 and 2, the system (7) reduces here to

(14.1) $$t_{17}^{5} + t_{27}^{5} \leqq 1,$$

(14.2) $$t_{17}^{6} + t_{27}^{6} \leqq 1,$$

(14.3) $$t_{17}^{7} + t_{27}^{7} \leqq 1,$$

(14.4) $$t_{56}^{11} + t_{66}^{11} \leqq 1,$$

(14.5) $$t_{56}^{12} + t_{66}^{12} \leqq 1,$$

(14.6) $$t_{44}^{13} + t_{54}^{13} \leqq 1,$$

(14.7) $$t_{56}^{13} + t_{66}^{13} \leqq 1,$$

(14.8) $$t_{47}^{13} + t_{57}^{13} \leqq 1,$$

(14.9) $$t_{44}^{14} + t_{54}^{14} \leqq 1,$$

(14.10) $$t_{56}^{14} + t_{66}^{14} \leqq 1,$$

(14.11) $$t_{47}^{14} + t_{57}^{14} \leqq 1,$$

(14.12) $$t_{44}^{15} + t_{54}^{15} \leqq 1,$$

(14.13) $$t_{47}^{15} + t_{57}^{15} \leqq 1.$$

A parametric solution of the system (12) is given by formulas

(15.1) $$x_1^5 = v_1 v_2, \quad x_1^6 = v_1 \bar{v}_2, \quad x_1^7 = \bar{v}_1,$$

(15.2) $$x_2^2 = v_3 v_4, \quad x_2^3 = v_3 \bar{v}_4, \quad x_2^4 = \bar{v}_3 v_4, \quad x_2^5 = \bar{v}_3 \bar{v}_4,$$

(15.3) $$x_3^1 = v_5, \quad x_3^2 = \bar{v}_5,$$

(15.4) $$x_4^{13} = v_6, \quad x_4^{14} = \bar{v}_6,$$

(15.5) $$x_5^{11} = v_7, \quad x_5^{12} = \bar{v}_7,$$

(15.6) $$x_6^{11} = v_8, \quad x_6^{12} = \bar{v}_8,$$

while the system (13) is solved by

(16.1) $$y_{11} = w_1, \quad y_{17} = \bar{w}_1,$$

(16.2) $$y_{23} = w_2, \quad y_{27} = \bar{w}_2,$$

(16.3) $$y_{32} = w_3, \quad y_{35} = \bar{w}_3,$$

(16.4) $$y_{41} = w_4 w_5, \quad y_{44} = w_4 \bar{w}_5, \quad y_{47} = \bar{w}_4 w_5, \quad y_{48} = \bar{w}_4 \bar{w}_5,$$

(16.5) $$y_{52} = w_6 w_7, \quad y_{54} = w_6 \bar{w}_7, \quad y_{56} = \bar{w}_6 w_7, \quad y_{57} = \bar{w}_6 \bar{w}_7,$$

(16.6) $$y_{63} = w_8, \quad y_{66} = \bar{w}_8.$$

Now we have to construct the system (4), which, in view of (15) and (16), becomes

$$t_{17}^{5} = y_{17}\,x_1^5 = \overline{w}_1\,v_1\,v_2, \tag{17.1}$$
$$t_{27}^{5} = y_{27}(x_2^3 \cup x_2^4 \cup x_2^5) = \overline{w}_2(\overline{v}_3 \cup \overline{v}_4), \tag{17.2}$$
$$t_{17}^{6} = y_{17}(x_1^5 \cup x_1^6) = \overline{w}_1\,v_1, \tag{17.3}$$
$$t_{27}^{6} = y_{27}(x_2^4 \cup x_2^5) = \overline{w}_2\,\overline{v}_3, \tag{17.4}$$
$$t_{17}^{7} = y_{17}(x_1^6 \cup x_1^7) = \overline{w}_1(\overline{v}_1 \cup \overline{v}_2), \tag{17.5}$$
$$t_{27}^{7} = y_{27}\,x_2^5 = \overline{w}_2\,\overline{v}_3\,\overline{v}_4, \tag{17.6}$$
$$t_{56}^{11} = y_{56}\,x_5^{11} = \overline{w}_6\,w_7\,v_7, \tag{17.7}$$
$$t_{66}^{11} = y_{66}\,x_6^{11} = \overline{w}_8\,v_8, \tag{17.8}$$
$$t_{56}^{12} = y_{56}(x_5^{11} \cup x_5^{12}) = \overline{w}_6\,w_7, \tag{17.9}$$
$$t_{66}^{12} = y_{66}(x_6^{11} \cup x_6^{12}) = \overline{w}_8, \tag{17.10}$$
$$t_{44}^{13} = y_{44}\,x_4^{13} = w_4\,\overline{w}_5\,v_6, \tag{17.11}$$
$$t_{54}^{13} = y_{54}(x_5^{11} \cup x_5^{12}) = w_6\,\overline{w}_7, \tag{17.12}$$
$$t_{56}^{13} = y_{56}(x_5^{11} \cup x_5^{12}) = \overline{w}_6\,w_7, \tag{17.13}$$
$$t_{66}^{13} = y_{66}(x_6^{11} \cup x_6^{12}) = \overline{w}_8, \tag{17.14}$$
$$t_{47}^{13} = y_{47}\,x_4^{13} = \overline{w}_4\,w_5\,v_6, \tag{17.15}$$
$$t_{57}^{13} = y_{57}(x_5^{11} \cup x_5^{12}) = \overline{w}_6\,\overline{w}_7, \tag{17.16}$$
$$t_{44}^{14} = y_{44}(x_4^{13} \cup x_4^{14}) = w_4\,\overline{w}_5, \tag{17.17}$$
$$t_{54}^{14} = y_{54}(x_5^{11} \cup x_5^{12}) = w_6\,\overline{w}_7, \tag{17.18}$$
$$t_{56}^{14} = y_{56}(x_5^{11} \cup x_5^{12}) = \overline{w}_6\,w_7, \tag{17.19}$$
$$t_{66}^{14} = y_{66}\,x_6^{12} = \overline{w}_8\,\overline{v}_8, \tag{17.20}$$
$$t_{47}^{14} = y_{47}(x_4^{13} \cup x_4^{14}) = \overline{w}_4\,w_5, \tag{17.21}$$
$$t_{57}^{14} = y_{57}(x_5^{11} \cup x_5^{12}) = \overline{w}_6\,\overline{w}_7, \tag{17.22}$$
$$t_{44}^{15} = y_{44}(x_4^{13} \cup x_4^{14}) = w_4\,\overline{w}_5, \tag{17.23}$$
$$t_{54}^{15} = y_{54}\,x_5^{12} = w_6\,\overline{w}_7\,\overline{v}_7, \tag{17.24}$$
$$t_{47}^{15} = y_{47}(x_4^{13} \cup x_4^{14}) = \overline{w}_4\,w_5, \tag{17.25}$$
$$t_{57}^{15} = y_{57}\,x_5^{12} = \overline{w}_6\,\overline{w}_7\,\overline{v}_7. \tag{17.26}$$

We introduce now (17) into (14), which becomes

$$\overline{w}_1\,v_1\,v_2 + \overline{w}_2(\overline{v}_3 \cup \overline{v}_4) \leqq 1, \tag{18.1}$$
$$\overline{w}_1\,v_1 + \overline{w}_2\,\overline{v}_3 \leqq 1, \tag{18.2}$$
$$\overline{w}_1(\overline{v}_1 \cup \overline{v}_2) + \overline{w}_2\,\overline{v}_3\,\overline{v}_4 \leqq 1, \tag{18.3}$$
$$\overline{w}_6\,w_7\,v_7 + \overline{w}_8\,v_8 \leqq 1, \tag{18.4}$$
$$\overline{w}_6\,w_7 + \overline{w}_8 \leqq 1, \tag{18.5}$$
$$w_4\,\overline{w}_5\,v_6 + w_6\,\overline{w}_7 \leqq 1, \tag{18.6}$$
$$\overline{w}_6\,w_7 + \overline{w}_8 \leqq 1, \tag{18.7}$$
$$\overline{w}_4\,w_5\,v_6 + \overline{w}_6\,\overline{w}_7 \leqq 1, \tag{18.8}$$
$$w_4\,\overline{w}_5 + w_6\,\overline{w}_7 \leqq 1, \tag{18.9}$$
$$\overline{w}_6\,w_7 + \overline{w}_8\,\overline{v}_8 \leqq 1, \tag{18.10}$$
$$\overline{w}_4\,w_5 + \overline{w}_6\,\overline{w}_7 \leqq 1, \tag{18.11}$$
$$w_4\,\overline{w}_5 + w_6\,\overline{w}_7\,\overline{v}_7 \leqq 1, \tag{18.12}$$
$$\overline{w}_4\,w_5 + \overline{w}_6\,\overline{w}_7\,\overline{v}_7 \leqq 1. \tag{18.13}$$

Notice that the inequality (5) coincides with (7) and implies (4) and (10); also, (9) implies (6) and (12), whereas (11) implies (8) and (13). Hence the whole system (18) reduces to the subsystem (18.1), (18.2), (18.3), (18.5), (18.9), (18.11). Taking into account that the pseudo-Boolean relation $x + y \leqq 1$ is equivalent to $x\,y = 0$, we deduce that the subsystem may be written as a Boolean equation, namely

$$\bar{w}_1 \bar{w}_2 v_1 v_2 (\bar{v}_3 \cup \bar{v}_4) \cup \bar{w}_1 \bar{w}_2 v_1 \bar{v}_3 \cup \bar{w}_1 \bar{w}_2 (\bar{v}_1 \cup \bar{v}_2) \bar{v}_3 \bar{v}_4 \cup \bar{w}_6 w_7 \bar{w}_8 \cup \cup w_4 \bar{w}_5 w_6 \bar{w}_7 \cup \bar{w}_4 w_5 \bar{w}_6 \bar{w}_7 = 0. \tag{19}$$

Equation (19) is equivalent to the system

$$\bar{w}_1 \bar{w}_2 (v_1 v_2 \bar{v}_4 \cup v_1 \bar{v}_3 \cup \bar{v}_3 \bar{v}_4) = 0, \tag{20.1}$$

$$(w_4 \bar{w}_5 w_6 \cup \bar{w}_4 w_5 \bar{w}_6) \bar{w}_7 = 0, \tag{20.2}$$

$$\bar{w}_6 w_7 \bar{w}_8 = 0, \tag{20.3}$$

yielding the parametric solution

$$\begin{cases} v_1 = u_1, \quad v_2 = u_2, \quad v_3 = u_3, \quad v_4 = u_4, \quad w_1 = u_5, \\ w_2 = \bar{u}_5 (u_1 u_2 \bar{u}_4 \cup u_1 \bar{u}_3 \cup \bar{u}_3 \bar{u}_4) \cup u_6, \end{cases} \tag{21.1}$$

$$\begin{cases} w_4 = u_7, \quad w_5 = u_8, \quad w_6 = u_9, \\ w_7 = u_7 \bar{u}_8 u_9 \cup \bar{u}_7 u_8 \bar{u}_9 \cup u_{10}, \end{cases} \tag{21.2}$$

$$w_8 = \bar{u}_9 (\bar{u}_7 u_8 \cup u_{10}) \cup u_{11}, \tag{21.3}$$

while the remaining v's and w's are arbitrary:

$$v_5 = u_{12}, \quad v_6 = u_{13}, \quad v_7 = u_{14}, \quad v_8 = u_{15}, \quad w_3 = u_{16}. \tag{21.4}$$

Introducing the solution (21) into (15) and (16) we get

$$x_1^5 = u_1 u_2, \quad x_1^6 = u_1 \bar{u}_2, \quad x_1^7 = \bar{u}_1, \tag{22.1}$$

$$x_2^2 = u_3 u_4, \quad x_2^3 = u_3 \bar{u}_4, \quad x_2^4 = \bar{u}_3 u_4, \quad x_2^5 = \bar{u}_3 \bar{u}_4, \tag{22.2}$$

$$x_3^1 = u_{12}, \quad x_3^2 = \bar{u}_{12}, \tag{22.3}$$

$$x_4^{13} = u_{13}, \quad x_4^{14} = \bar{u}_{13}, \tag{22.4}$$

$$x_5^{11} = u_{14}, \quad x_5^{12} = \bar{u}_{14}, \tag{22.5}$$

$$x_6^{11} = u_{15}, \quad x_6^{12} = \bar{u}_{15}, \tag{22.6}$$

and

$$y_{11} = u_5, \quad y_{17} = \bar{u}_5, \tag{23.1}$$

$$\begin{cases} y_{23} = \bar{u}_5 (u_1 u_2 \bar{u}_4 \cup u_1 \bar{u}_3 \cup \bar{u}_3 \bar{u}_4) \cup u_6, \\ y_{27} = [u_5 \cup (\bar{u}_1 \cup \bar{u}_2 \cup u_4) (u_3 \cup \bar{u}_1 u_4)] \bar{u}_6, \end{cases} \tag{23.2}$$

$$y_{32} = u_{16}, \quad y_{35} = \bar{u}_{16}, \tag{23.3}$$

$$y_{41} = u_7 u_8, \quad y_{44} = u_7 \bar{u}_8, \quad y_{47} = \bar{u}_7 u_8, \quad y_{48} = \bar{u}_7 \bar{u}_8, \tag{23.4}$$

$$\begin{cases} y_{52} = u_9 (u_7 \bar{u}_8 \cup u_{10}), \quad y_{54} = u_9 \bar{u}_{10} (\bar{u}_7 \cup u_8), \\ y_{56} = \bar{u}_9 (\bar{u}_7 u_8 \cup u_{10}), \quad y_{57} = \bar{u}_9 \bar{u}_{10} (u_7 \cup \bar{u}_8), \end{cases} \tag{23.5}$$

$$y_{63} = \bar{u}_9 (\bar{u}_7 u_8 \cup u_{10}) \cup u_{11}, \quad y_{66} = [u_9 \cup (u_7 \cup \bar{u}_8) \bar{u}_{10}] \bar{u}_{11}. \tag{23.6}$$

§ 2. Time-Table Scheduling*

Consider a school having t classes and u teachers. Let us suppose that we know which classes are taught by each teacher and also that the total number of hours taught by the teacher ν in the class τ is $k_{\nu\tau}$. The class-room hours are indexed from 1 to h (when, for instance, in each working day, teaching is given from 7 to 13, then $h = 36$ and we can index the class-room hours given Monday from 1 to 6, given Tuesday from 7 to 12 etc., until Saturday indexed 31—36).

We are looking for a system of elements

$$\{y_{\tau\chi\nu} \mid 1 \leqq \tau \leqq t,\ 1 \leqq \chi \leqq h,\ 1 \leqq \nu \leqq u\} \tag{24}$$

in B_2 such that each time-table schedule given by the relationship "the teacher ν teaches in class τ in hour $\chi \Longleftrightarrow y_{\tau\chi\nu} = 1$" should satisfy all the conditions imposed on a time-table schedule and should be optimal in a certain sense.

It is known *a priori* that certain $y_{\tau\chi\nu} = 0$. This situation occurs in cases when teacher ν does not teach class τ, as well as when the teacher has no free time at hour χ (for instance early in the morning, etc.), or when special class-rooms are not free at hour χ, etc.

The first condition a time-schedule must satisfy is that in a certain class, in a certain hour, no two teachers have to teach simultaneously. This condition is expressed by the pseudo-Boolean equation

$$\varphi_1 = \sum_{\tau=1}^{t} \sum_{\chi=1}^{h} \sum_{\substack{\nu_1,\nu_2=1 \\ \nu_1 \neq \nu_2}}^{u} y_{\tau\chi\nu_1} y_{\tau\chi\nu_2} = 0. \tag{25}$$

Further, we have the condition that teacher ν has to teach in class τ exactly $k_{\tau\nu}$ hours, i.e.:

$$\sum_{\chi=1}^{h} y_{\tau\chi\nu} = k_{\tau\nu}. \tag{26}$$

We obtain in this way the pseudo-Boolean condition

$$\varphi_2 = \left(\sum_{\chi=1}^{h} y_{\tau\chi\nu} - k_{\tau\nu}\right)^2 = 0. \tag{27}$$

Further conditions arise when taking into account special class-rooms (laboratories, gymnasia, etc.) in which the lesson of teacher ν in class τ is to take place. An insufficient number of class-rooms can impose further constraints. All these conditions may be handled as above, using pseudo-Boolean equations $\varphi_i = 0$ $(i = 1, 2, \ldots, m)$.

* By Ivo Rosenberg.

After formulating these constraints, the time-table schedule is feasible if and only if the system (24) satisfies the equation

$$\varphi \equiv \sum_{i=1}^{m} \varphi_i = 0. \tag{28}$$

The adequacy of a time-table schedule can be judged in several ways, both by teachers and pupils. One possible point of view could be, for instance, the compactness of the time-table schedule (there are no gaps). So, for instance, class τ has no gap between the hours $\chi - 1$ and $\chi + 1$ if relation

$$\psi_1 \equiv \sum_{\nu_1, \nu_2, \nu_3 = 1}^{u} y_{\tau, \chi - 1, \nu_1} (1 - y_{\tau \chi \nu_2})\, y_{\tau, \chi + 1, \nu_3} = 0 \tag{29}$$

holds.

Other standpoints could also be considered. So, for instance, teacher ν_0 does not like to teach in the first hour. Let us put

$$\psi_2 \equiv \sum_{\tau=1}^{t} y_{\tau 1 \nu_0}. \tag{30}$$

Now again $\psi_2 = 0$ exactly when the teacher ν_0 does not teach in the first hour.

Further advantages can also be formulated using pseudo-Boolean conditions $\psi_j = 0$ $(j = 1, \ldots, v)$ which are to be fulfilled exactly in the favorable cases.

The importance of each standpoint can be weighted using real numbers α_j $(j = 1, \ldots, v)$; the greater is α_j, the worse is the non-fulfillment of the j-th condition. We set

$$\psi = \sum_{j=1}^{v} \alpha_j \psi_j. \tag{31}$$

Thus the problem is converted into that of minimizing the pseudo-Boolean function (31) under the pseudo-Boolean condition (28).

Similar formulations may be given to problems of rather different natures, as: air traffic control, scheduling factory work, etc.

The number of variables can be reduced by using a generalized pseudo-Boolean programming, described in the Appendix.

Authors' remarks: The following more general time-table problem was studied by G. Mihoc and E. Balas [1]:

Consider m groups of students and p instructors. Each group i has t_i hours $(i = 1, \ldots, m)$, while each instructor k has t^k hours $(k = 1, \ldots, p)$ weekly. Let t_i^k be the number of hours of group i with instructor k; it follows that $\sum_{k=1}^{p} t_i^k = t_i$ $(i = 1, \ldots, m)$ and $\sum_{i=1}^{m} t_i^k = t^k$ $(k = 1, \ldots, p)$. The t working hours of a week are grouped into n

phases (days or half days) of f hours each, so that $t = n f$. Further, let s_j be the number of available class-rooms at hour j ($j = 1, \ldots, t$).

A time-table consistent with the above conditions is characterized by a three-dimensional matrix $X = (x_{ij}^k)$ defined by

$$(32) \qquad x_{ij}^k = \begin{cases} 1, & \text{if the group } i \text{ is assigned to instructor } k \text{ in the hour } j, \\ 0, & \text{in the opposite case,} \end{cases}$$

and satisfying the conditions

$$(33) \qquad \sum_{j=1}^{t} x_{ij}^k = t_i^k \qquad (i = 1, \ldots, m;\ k = 1, \ldots, p),$$

$$(34) \qquad \sum_{i=1}^{m} x_{ij}^k \leqq 1 \qquad (k = 1, \ldots, p;\ j = 1, \ldots, t),$$

$$(35) \qquad \sum_{k=1}^{p} x_{ij}^k \leqq 1 \qquad (i = 1, \ldots, m;\ j = 1, \ldots, t),$$

$$(36) \qquad \sum_{i=1}^{m} \sum_{k=1}^{p} x_{ij}^k \leqq s_j \qquad (j = 1, \ldots, t),$$

the significance of which is obvious.

A certain hour j ($j = 1, \ldots, t$) will be called a start for group i in time-table X, if $\sum_{k=1}^{p} x_{ij}^k = 1$ and at least one of the conditions $j = r f + 1\,(0 \leqq r \leqq n - 1)$ and $\sum_{k=1}^{p} x_{i,j-1}^k = 0$ holds; let $S_i(X)$ be the total number of starts for the group i in time-table X.

A time-table $\tilde{X}$ is said to be *minimal*, if

$$(37) \qquad \sum_{i=1}^{m} S_i(\tilde{X}) = \min_{X \in T} \left[\sum_{i=1}^{m} S_i(X) \right],$$

where T denotes the set of all time-tables. A minimal time-table $\tilde{X}_0$ is said to be *optimal*, if

$$(38) \qquad \max_i S_i(\tilde{X}^0) = \min_{\tilde{X} \in M} \max_i S_i(\tilde{X}),$$

where M denotes the set of all minimal time-tables.

It was proved that the problem of finding an optimal time-table is equivalent to an integer linear programming problem with $m\,t\,(p + 2)$ bivalent $(0, 1)$ variables and one integer variable. The latter problem may be solved, for instance, with the algorithm given by E. Balas [1] or with the methods described in Chapter VIII.

Other researches on this problem were carried out by J. S. Appleby, D. V. Blake and E. A. Newman [1], J. Berghius [1], W. H. Bossert and H. B. Harmon [1], J. Csima and C. C. Gotlieb [1], C. C. Gotlieb [1], K. Kirchgässner [1], G. R. Sherman [1], and P. Stahlknecht [1].

§ 3. Applications to Coding Theory

Problems of integer or bivalent programming were pointed out in coding theory. First, E. J. McCluskey Jr. [1] has shown that the problem of constructing a minimal error-correcting code can be given an approach based on integer linear programming.

A. Deleanu and P. L. Hammer [1] studied the problem of transmitting one of 2^n possible integers, $1, 2, \ldots, 2^n$, through a binary channel. To do this, we have to assign to each one of the 2^n integers an n-tuple of zeros and ones. Suppose that (for instance for technical reasons), the only type of errors likely to occur in a transmitted word are those modifying at most one binary component. If this is the case, we are faced with the problem of how to make the assignment so that the average absolute error in transmission is minimized.

From a geometrical point of view, the problem may be stated as follows: Assign the integers $1, 2, \ldots, 2^n$ to the vertices of the unit cube in n-dimensional Euclidean space, so that $\sum \Delta_{ij}$ is a minimum, where the sum runs over all neighbouring pairs (i, j) of vertices (i.e. vertices differing in only one component) and Δ_{ij} is the absolute value of the difference of the integers assigned to them:

$$\Delta_{ij} = |i - j|. \tag{39}$$

L. H. Harper [1] has given the following algorithm for solving the above problem: Assign 1 to an arbitrary vertex; having assigned $1, \ldots, h$, assign $h + 1$ to an unnumbered vertex (not necessarily unique) which has the most numbered neighbours. Harper has also shown that this algorithm produces all the assignments which minimize $\sum \Delta_{ij}$.

Harper's algorithm involves at each stage the selection of a certain vertex among the as yet unnumbered ones. This can be done by pseudo-Boolean programming.

Let us denote a vertex of the unit n-cube by $(x_1, \ldots, x_n)$, where x_i is 0 or 1 $(i = 1, \ldots, n)$. After completing the h-th step in Harper's algorithm, we introduce a function $V_h(x_1, \ldots, x_n)$ being 0 on the as yet unnumbered vertices and 1 on the numbered vertices.

Theorem 1. *The integer $h + 1$ is to be assigned to any vertex $(x_1, \ldots, x_n)$ minimizing the function*

$$W_h(x_1, \ldots, x_n) = -\sum_{i=1}^{n} V_h(x_1, \ldots, x_{i-1}, \bar{x}_i, x_{i+1}, \ldots, x_n) + \\ + (n + 1)\, V_h(x_1, \ldots, x_n). \tag{40}$$

Proof. Assume the vertex $(x_1, \ldots, x_n)$ has been numbered in one of the stages $1, 2, \ldots, h$. Then $V_h(x_1, \ldots, x_n) = 1$ and therefore $W_h(x_1, \ldots, x_n) > 0$. As $h \leqq 2^n - 1$, there exists at least one as yet unnumbered vertex and the value of W_h on that vertex is non-positive. We conclude that the vertices minimizing W_h are among the unnumbered ones.

It is easy to see that for any vertex $(x_1, \ldots, x_n)$

$$\sum_{i=1}^{n} V_h(x_1, \ldots, x_{i-1}, \bar{x}_i, x_{i+1}, \ldots, x_n) \tag{41}$$

equals the number of its numbered neighbours. Hence for every unnumbered vertex, the number of its numbered neighbours is precisely

$$- W_h(x_1, \ldots, x_n).$$

It follows that the integer $h + 1$ is to be assigned in HARPER's algorithm to any vertex minimizing the function W_h, q.e.d.

For a large n, this offers a means of avoiding the cumbersome and voluminous computations required when seeking the $(h + 1)$-th vertex without a systematic procedure. It also permits us to obtain *all* the possible optimal assignments.

HARPER's method was generalized by J. H. LINDSEY [1] and his results seem to be translatable too, into a pseudo-Boolean language.

§ 4. Miscellanies

Let us consider the following problem, variants of which are quoted in R. E. BELLMAN and S. E. DREYFUS [1] and in G. B. DANTZIG [3] (as the *knapsack problem*).

Suppose we have to load a vessel with a cargo consisting of different types of items, choosen among n possible goods $G_1, \ldots, G_n$. Every good G_j has a weight w_j and a value v_j. The problem is to determine the most valuable cargo not exceeding the maximum capacity w of the vessel.

Putting

$$x_j = \begin{cases} 1, & \text{if the good } G_j \text{ is loaded,} \\ 0, & \text{otherwise} \end{cases} \tag{42}$$

the problem is to maximize the pseudo-Boolean function

$$\sum_{j=1}^{n} v_j x_j \tag{43}$$

subject to the constraint

$$\sum_{j=1}^{n} w_j x_j \leqq w_j. \tag{44}$$

The more general case when each good G_j consists of g_j indivisible unities may be treated in a similar way.

G. Desbazeille [1] proposes the following exercise. A constructor of electronic devices can choose among several ways of assembling a radio set. For each radio set, the conditions are the following.

1) Any one of the three types T_1, T_2, T_3 of tubes may be utilized, but only one.

2) The box may be made either of wood (W), or of plastic material (M). But, when using M, dimensionality requirements impose the choice of T_2, and as there is no place for a transformer F, a special power supply S is needed.

3) T_1 needs F.

4) T_2 and T_3 need S (and not F).

The prices of the above mentioned components are:

one series of tubes T_1	28 units
one series of tubes T_2	30 units
one series of tubes T_3	31 units
one transformer F	25 units
one special power supply S	23 units
one wood box W	9 units
one plastic material box M	6 units

The other necessary components of the radio set cost:

27 units, if the tubes T_1 are utilized,
28 units, if the tubes T_2 are utilized,
25 units, if the tubes T_3 are utilized.

Which model is to be constructed in order to maximize the profit, knowing that the price of the make is 10 units for each set in all the cases and that a set is sold at 110 units when it is enclosed in a plastic material box, and at 105 units in the other case.

For each utilizable component C, we denote by c the Boolean variable

$$c = \begin{cases} 1, & \text{if } C \text{ is used,} \\ 0, & \text{in the opposite case.} \end{cases} \tag{45}$$

Then, the conditions become:

$$t_1 + t_2 + t_3 = 1, \tag{46.1}$$

$$w + m = 1, \tag{46.2}$$

$$M = 1 \text{ implies } t_2 = s = 1, \tag{46.3}$$

$$t_1 = 1 \text{ implies } f = 1, \tag{46.4}$$

$$t_2 = 1 \text{ implies } s = 1, \tag{46.5}$$

$$t_3 = 1 \text{ implies } s = 1, \tag{46.6}$$

$$f + s = 1, \tag{46.7}$$

while the function to be maximized is

$$(47)\quad 110w + 105m - (28t_1 + 30t_2 + 31t_3 + 25f + 23s + 9w + \\ + 6m + 27t_1 + 28t_2 + 25t_3 + 10) = -10 + 101w + 99m - \\ - 55t_1 - 58t_2 - 56t_3 - 25f - 23s.$$

Since conditions (46.3), (46.4), (46.5) and (46.6) are obviously equivalent to

$$(46.3')\quad m \leqq t_2 s,$$

$$(46.4')\quad t_1 \leqq f,$$

$$(46.5')\quad t_2 \leqq s,$$

$$(46.6')\quad t_3 \leqq s,$$

we are faced with a pseudo-Boolean program with a linear objective function which, solved as in Chapter V, § 1, yields immediately the single maximizing point

$$(48)\quad t_1 = 0,\quad t_2 = 0,\quad t_3 = 1,\quad f = 0,\quad s = 1,\quad w = 1,\quad m = 0,$$

which means that we have to choose the T_3 tubes, the special power supply and the wood box, assuring thus a profit of 12 units.

D. Anghel, E. Burlacu, T. Gaspar and M. Lupulescu have studied the following problem.

An agricultural area S is divided into m lots $S_1, \ldots, S_m$, to which m cultures $C_1, \ldots, C_n$ are to be assigned $(m \geqq n)$. For each lot S_i $(i = 1, \ldots, m)$, we may choose any one of the n cultures C_j $(j = 1, \ldots, n)$, but not more than one culture. When cultivating C_j on the lot S_i we have to perform some basic work which costs d_{ij} units, and we may or may not perform some supplementary work which costs c_{ij} units, we may or may not use one (not more!) of the fertilizers $F_1, \ldots, F_p$ — to each fertilizer F_k corresponding an expense of b_{ijk} units — and, finally, we may irrigate or not the lot S_i, implying an expense of a_i units (whatever the culture C_j may be).

Let p_{ijh}^{rs} $(i = 1, \ldots, m;\ j = 1, \ldots, n;\ h = 0, 1, \ldots, p;\ r, s = 0, 1)$ denote the average crop on the lot S_i when cultivating C_j and using the fertilizer F_h (the value $h = 0$ corresponding to the case when no fertilizer is used), with or without supplimentary work (corresponding to $r = 1$ and $r = 0$, respectively), with or without irrigation (corresponding to $s = 1$ and $s = 0$, respectively). Let P_j denote the C_j average crop; it is required that relations $P_j \geqq p_j$ hold for $j = 1, \ldots, p$, where p_j are prescribed quantities.

Let us introduce the following Boolean variables:

$$(49)\qquad x_{ij} = \begin{cases} 1, & \text{if the culture } C_j \text{ is assigned to the lot } S_i, \\ 0, & \text{otherwise;} \end{cases}$$

$$(50)\qquad y_{ik} = \begin{cases} 1, & \text{if the fertilizer } F_k \text{ is used on the lot } S_i, \\ 0, & \text{otherwise;} \end{cases}$$

$$(51)\qquad z_i = \begin{cases} 1, & \text{if supplementary work* is performed on the lot } S_i, \\ 0, & \text{otherwise;} \end{cases}$$

$$(52)\qquad t_i = \begin{cases} 1, & \text{if the lot } S_i \text{ is irrigated,} \\ 0, & \text{otherwise.} \end{cases}$$

The above requirements are now translated by the following pseudo-Boolean equations and inequalities:

$$(53)\qquad \sum_{\substack{j,j'=1 \\ j \neq j'}}^{n} x_{ij}\, x_{ij'} = 0 \qquad (i = 1, \ldots, m),$$

$$(54)\qquad \sum_{\substack{k,k'=1 \\ k \neq k'}}^{p} y_{ik}\, y_{ik'} = 0 \qquad (i = 1, \ldots, m),$$

$$(55)\qquad P_j = \sum_{i=1}^{m} x_{ij} \sum_{h=0}^{p} u_{ih} (p_{ijh}^{11} z_i t_i + p_{ijh}^{10} z_i \bar{t}_i + p_{ijh}^{01} \bar{z}_i t_i + p_{ijh}^{00} \bar{z}_i \bar{t}_i) \geqq p_j \qquad (j = 1, \ldots, n),$$

where

$$(56)\qquad u_{ih} = \begin{cases} y_{ih}, & \text{for } h = 1, \ldots, p; \\ \prod_{k=1}^{p} \bar{y}_{ik}, & \text{for } h = 0. \end{cases}$$

The global cost of the production will be

$$(57)\qquad W = \sum_{i=1}^{m} a_i t_i + \sum_{i=1}^{m} \sum_{j=1}^{n} x_{ij} \left(\sum_{k=1}^{p} b_{ijk} y_{ik} + c_{ij} z_i + d_{ij} \right).$$

Let v_j denote the "value" of each unit of the product C_j (v_j may have various meanings: selling price, caloric value, the quantity of proteins, etc.). Then, the global value of the average crop will be

$$(58)\qquad V = \sum_{j=1}^{n} v_j P_j.$$

Now, the following three problems may be taken into consideration and solved by means of pseudo-Boolean programming:

1. Maximize V under the restrictions (53)—(55) and $W \leqq W_0$ where W_0 is a prescribed value.

* We mean supplementary work different from irrigation and use of fertilizers.

2. Maximize $\frac{V}{W}$ under the restrictions (53)—(55).

3. Minimize W under the restrictions (53)—(55) and $V \geqq V_0$, where V_0 is a prescribed value.

§ 5. Plant Location*

The problem of plant location, as stated for instance by A. S. MANNE [1], has the following formulation:

Let $I = \{1, 2, \ldots, m\}$ be the set of places where the plants can be located, let $J = \{1, 2, \ldots, n\}$ be the set of consumers, let C_j be the anual rate of market requirements for location j, let a_i be the anual fixed cost of construction and of operation at plant i, let b_i be the manufacturing cost per unit at plant i, let c_{ij} be the transportation cost per unit from i to j.

The problem consists in finding that subset of the set I, which assures a minimum for the total anual cost of construction, manufacturing and transportation.

Let us put $y_i = 1$ if a plant is to be located in i, and $y_i = 0$ otherwise. Let x_{ij} denote the amount shiped from i to j, and let $d_{ij} = b_i + c_{ij}$. With these notations we come to the following

Problem P. Minimize

$$F(x_{11}, \ldots, x_{mn}; y_1, \ldots, y_m) = \sum_{i=1}^{m} a_i y_i + \sum_{i=1}^{m} \sum_{j=1}^{n} d_{ij} x_{ij}$$

under the constraints

$$\text{(59)} \qquad \text{for any } i, \quad y_i = 0 \text{ implies } x_{ij} = 0 \quad \text{for all } j = 1, 2, \ldots, n;$$

$$\text{(60)} \qquad \sum_{i=1}^{m} x_{ij} = C_j \qquad (j = 1, 2, \ldots, n);$$

$$\text{(61)} \qquad x_{ij} \geqq 0, \quad y_i \in \{0, 1\} \qquad (i = 1, \ldots, m; \ j = 1, \ldots, n).$$

Let us put for any $i_1, i_2 \in I$, and for any $j \in J$,

$$u_{i_1 i_2 j} = \begin{cases} 1 & \text{if} \quad d_{i_1 j} < d_{i_2 j}, \\ 1 & \text{if} \quad d_{i_1 j} = d_{i_2 j} \quad \text{and} \quad i_1 \leqq i_2, \\ 0 & \text{if} \quad d_{i_1 j} = d_{i_2 j} \quad \text{and} \quad i_1 > i_2, \\ 0 & \text{if} \quad d_{i_1 j} > d_{i_2 j}, \end{cases}$$

and let us introduce a bivalent variable t_{ij} defined by

$$\text{(62)} \qquad t_{ij} = y_i \prod_{h=1}^{m} (1 - y_h \bar{u}_{ihj}).$$

Let us now consider the following

Problem P'. Minimize the real-valued function

$$\text{(63)} \quad G(y_1, \ldots, y_m) = \sum_{i=1}^{m} a_i y_i + \sum_{j=1}^{n} \sum_{i=1}^{m} C_j d_{ij} y_i \prod_{h=1}^{m} (1 - y_h \bar{u}_{ihj}) + \\ + (M + 1) \bar{y}_1 \ldots \bar{y}_m,$$

---

* See P. L. HAMMER [10]

with bivalent variables $y_1, \ldots, y_m$, where

$$M = \sum_{i=1}^{m} \sum_{j=1}^{m} (a_i + C_j d_{ij}). \tag{64}$$

The following main result holds:

Theorem 2. *The minimum in problem P is equal to the minimum in problem P'. If $(y_1^*, \ldots, y_m^*)$ is an optimal solution of problem P', if t_{ij}^* are the corresponding values of t_{ij}, and if $x_{ij}^* = C_j t_{ij}^*$, then $(x_{11}^*, \ldots, x_{mn}^*; y_1^*, \ldots, y_m^*)$ is an optimal solution of problem P.*

Thus, the original problem P is reduced to that of minimizing the unrestricted pseudo-Boolean function G and hence it can be solved by means of the procedures developed in this book.

The proof will be given after the following

Example 2. A. S. Manne [1] discusses the following numerical example: $I = J = \{1, 2, 3, 4\}$, $a_1 = a_2 = a_3 = a_4 = 3.5$, $b_1 = b_2 = b_3 = b_4 = 0$, $C_1 = C_2 = C_3 = C_4 = 10$, and the c_{ij}'s are given in Table 2.

Table 2

$i \setminus j$	1	2	3	4
1	0	0.6	1	0.9
2	0.6	0	0.4	0.3
3	1	0.4	0	0.3
4	0.9	0.3	0.3	0

We first determine the u_{ihj}'s:

Table 3

$ih \setminus j$	1	2	3	4
11	1	1	1	1
12	1	0	0	0
13	1	0	0	0
14	1	0	0	0
21	0	1	1	1
22	1	1	1	1
23	1	1	0	1
24	0	1	0	0
31	0	1	1	1
32	0	0	1	0
33	1	1	1	1
34	0	0	1	0
41	0	1	1	1
42	1	0	1	1
43	1	1	0	1
44	1	1	1	1

Now we see that the function $G(y_1, y_2, y_3, y_4)$ is:

$$G = 3.5y_1 + 3.5y_2 + 3.5y_3 + 3.5y_4 + \quad (65)$$

$$+ 25y_1 \bar{y}_2 \bar{y}_3 \bar{y}_4 + 6\bar{y}_1 y_2 \bar{y}_4 + 4y_2 \bar{y}_3 \bar{y}_4 + 3y_2 \bar{y}_4 + 10\bar{y}_1 \bar{y}_2 y_3 \bar{y}_4 +$$

$$+ 7\bar{y}_2 y_3 \bar{y}_4 + 9\bar{y}_1 y_4 + 3\bar{y}_2 y_4 + 3\bar{y}_3 y_4 + 85\bar{y}_1 \bar{y}_2 \bar{y}_3 \bar{y}_4,$$

because $M = 84$.

We find

$$G_{\min} = G(1, 0, 0, 1) = 13. \quad (66)$$

Using formula (62) and the relation $x_{ij} = C_j t_{ij}$, we find $x_{11} = x_{42} = x_{43} = x_{44} = 10$, the other x_{ij}'s being equal to 0.

From (66) and from the last results, we see that the plants are to be located in the points 1 and 4, that the plant located in point 1 fulfils the necessities of consumer 1, and that the plant located in point 4 fulfils the necessities of the consumers 2, 3, 4.

Proofs

Let us suppose $y_1, \ldots, y_m$ fixed, not all equal to zero and let us associate to each j an index $i(j)$, defined as follows:

1⁰ $y_{i(j)} = 1$,

2⁰ for any i, if $y_i = 1$, then $d_{i(j)j} \leqq d_{ij}$,

3⁰ for any i, if $y_i = 1$, and $d_{i(j)j} = d_{ij}$, then $i(j) \leqq i$.

It is easy to notice that the mapping $j \to i(j)$ is univoque.

Lemma 1. *For any j, the bivalent variable t_{ij} is equal to 1, if and only if $i = i(j)$.*

Proof. If $t_{ij} = 1$, formula (62) shows that $y_i = 1$, and that for each $h = 1, \ldots, m$, we have $y_h \bar{u}_{ihj} = 0$. The last relation shows that $y_h = 1$ implies $u_{ihj} = 1$. The definition of u_{ihj} shows now that $y_h = 1$ implies either $d_{ij} < d_{hj}$, or $d_{ij} = d_{hj}$ and $i \leqq h$. It follows that $i = i(h)$. Conversely, it is easy to notice that $t_{i(j)j} = 1$.

Lemma 2. *Problem P is feasible if and only if at least one of the y_i's ($i = 1, \ldots, m$) is equal to 1, i.e. if and only if*

$$\bar{y}_1 \bar{y}_2 \ldots \bar{y}_m = 1. \quad (67)$$

Proof. If all $y_i = 0$ $(i = 1, \ldots, m)$, it results from (59) that all $x_{ij} = 0$ $(i = 1, \ldots, m;\ j = 1, \ldots, n)$ and hence conditions (60) cannot be fulfilled. Conversely, if at least one y_i, say y_{i_0}, is equal to one, putting

$$x_{ij} = \begin{cases} C_j & \text{if} \quad i = i_0, \\ 0 & \text{if} \quad i \neq i_0, \end{cases}$$

we see that the point $(x_{11}, \ldots, x_{mn};\ y_1, \ldots, y_m)$ fulfils conditions (59), (60), (61), and hence the problem P is feasible.

Lemma 3. *If $y_1, \ldots, y_m$ fulfil condition* (67), *and if*

$$x_{ij} = C_j t_{ij} \qquad (i = 1, \ldots, m;\ j = 1, \ldots, n), \quad (68)$$

then $(x_{11}, \ldots, x_{mn};\ y_1, \ldots, y_m)$ is a feasible solution of problem P.

Proof. If $y_i = 0$, then $t_{ij} = 0$, and hence $x_{ij} = 0$. According to Lemma 1, $t_{ij} = 1$ if and only if $i = i(j)$, hence,

$$\sum_{i=1}^{m} x_{ij} = \sum_{i=1}^{m} C_j t_{ij} = C_j.$$

Since condition (61) holds obviously, the Lemma is proved.

Let $(x_{11}^*, \ldots, x_{mn}^*; y_1^*, \ldots, y_m^*)$ be a feasible solution of problem P, and let v^* denote the value of F on this point.

Let us put

$$x_{ij}^{**} = C_j y_i^* \prod_{h=1}^{m} (1 - y_h^* \bar{u}_{ihj}); \qquad (i = 1, \ldots, m;\ j = 1, \ldots, n);$$

then $(x_{11}^{**}, \ldots, x_{mn}^{**}; y_1^*, \ldots, y_m^*)$ is a feasible solution of problem P. Let v^{**} denote the value of F on this point.

Lemma 4. $v^{**} \leqq v^*$.

Proof. $v^* = \sum_{i=1}^{m} a_i y_i^* + \sum_{i=1}^{m} \sum_{j=1}^{n} d_{ij} x_{ij}^*$

$$= \sum_{i=1}^{m} a_i y_i^* + \sum_{j=1}^{n} d_{i(j)j} C_j + \sum_{i=1}^{m} \sum_{j=1}^{n} (d_{ij} - d_{i(j)j}) x_{ij}^*$$

$$\geqq \sum_{i=1}^{m} a_i y_i^* + \sum_{j=1}^{n} d_{i(j)j} C_j = v^{**} \qquad \text{(by (60), (59) and } 2^0\text{)}.$$

It follows directly from Lemmas 2, 3 and 4:

Lemma 5. *There exist optimal solutions of problem P for which the relations* (67) *and* (68) *hold.*

Lemma 6. *Any minimizing point of the function F, under the constraints* (67) *and* (68), *is an optimal solution of problem P, and the minimum of F under constraints* (59), (60) *and* (61), *is equal to the minimum of F under constraints* (67) *and* (68).

Proof. From Lemma 3 we see that each point satisfying (67) and (68), satisfies also (59), (60) and (61), i.e. it is a feasible solution of problem P. From Lemma 5 it follows that the minimum of F under the constraints (59), (60), (61) is equal to the minimum of F under constraints (67) and (68).

Lemma 7. *The y-part of the set of minimizing points of F under the constraints* (67) *and* (68), *coincides with the set of minimizing points of the unconstrained pseudo-Boolean function* (63), *where M is given by* (64).

The proof follows directly from Theorem VII.2.

Proof of Theorem 2. Follows directly from Lemmas 6 and 7.

§ 6. Other Types of Applications

Boolean and lattice-theoretical methods can be employed to solve various types of practical problems. In this section, we shall point out several such applications which are separated from the body of our book by their specific structures.

One of these applications was given by J. R. SLAGLE [1]. A procedure for making true-false decisions which depend on the outcome of a sequence of elementary binary tests is, roughly speaking, an algorithm prescribing the order in which the tests are to be applied (this order is usually not linear and may be represented by a tree). SLAGLE considered a class of procedures which may be represented by Boolean expressions, the variables of which are associated to the elementary tests. It is assumed that the results of the elementary tests are independent and that the cost of applying each elementary test as well as the probability of its outcome may be computed; it follows that the average cost of a procedure may also be computed. The general problem solved by SLAGLE is that of finding efficiently a minimum-cost procedure equivalent to a given procedure (i.e., yielding necessarily the same decisions as the original procedure). Illustrative problems are given in the fields of computer programming, personnel selection, medical diagnosis, etc.

A Boolean method for analysing questionaries was proposed by C. FLAMENT [1].

C. CARDOT [1, 2] (see also M. DENIS-PAPIN, A. KAUFMANN and R. FAURE [1]) studied several technical problems necessitating the determination of a minimal system of generators for a specified subset of a distributive lattice.

After the completion of the manuscript, we became aware of H. GRENIEWSKI's communication [1], concerning the possibility of applying Boolean algebras to LEONTIEF's input-output analysis.

Chapter XIV

Minimization Problems in Automata Theory

As we have pointed out in the Introduction, the first practical application of Boolean algebras was given to the theory of switching circuits. Since there exists a wide literature on switching algebra, we have omitted this field in our book, with two exceptions. The first one is the description, in Chapter IX, § 8, of LUNC's methods for solving the analysis and synthesis problems in case of multi-terminal switching circuits. The second exception is the present chapter, where we shall describe a bivalent mathematical programming approach to certain minimization problems occuring in switching theory.

The chapter consists of three sections. In § 1 we shall show that the classical problem of minimizing a Boolean function in disjunctive form may be given a pseudo-Boolean treatment; various generalizations of this problem may be treated in the same way. § 2, written by I. ROSENBERG, deals with the minimization of the number of states in input-restricted machines. The last section mentions an application of pseudo-Boolean programming to the state assignment of sequential circuits, given by Y. INAGAKI and K. SUGINO.

§ 1. Minimization of Normal Forms of Boolean Functions; Generalizations

As it was shown in Chapter I, § 2, a Boolean function of p variables

$$f: B_2^p = \underbrace{B_2 \times \cdots \times B_2}_{p \text{ times}} \to B_2 \tag{1}$$

may be written in disjunctive form, i.e.

$$f = c_1 \cup \cdots \cup c_r, \tag{2}$$

where each c_i is an elementary conjunction

$$c_i = x_{i_1}^{\alpha_{i_1}} \ldots x_{i_{p(i)}}^{\alpha_{i_{p(i)}}}, \tag{3}$$

where $\{x_{i_1}, \ldots, x_{i_{p(i)}}\} \subseteqq \{x_1, \ldots, x_p\}$, each α_{i_j} is 0 or 1, and

$$x^\alpha = \begin{cases} x, & \text{if } \alpha = 1, \\ \bar{x}, & \text{if } \alpha = 0. \end{cases} \tag{4}$$

The expression (2) of a given function f is not unique; for instance, the function

$$f_0 = x\,y \cup \bar{y}\,\bar{z} \cup \bar{x}\,y\,\bar{z} \tag{5}$$

may also be written in the forms

$$f_0 = x\,y\,z \cup x\,y\,\bar{z} \cup x\,\bar{y}\,\bar{z} \cup \bar{x}\,y\,\bar{z} \cup \bar{x}\,\bar{y}\,\bar{z}, \tag{5'}$$

$$f_0 = x\,\bar{z} \cup x\,y \cup \bar{x}\,\bar{z}, \tag{5''}$$

$$f_0 = x\,y \cup \bar{z}, \tag{5'''}$$

etc.

An elementary conjunction c_i appearing in a disjunctive expression (2) of a Boolean function f is called an *implicant* of f. An equivalent definition is the following: the conjunction (3) is an implicant of the function f if $f(x_1, \ldots, x_p) = 1$ for all $(x_1, \ldots, x_p) \in B_2^p$ which satisfy $x_{i_1} = \alpha_{i_1}, \ldots, x_{i_{p(i)}} = \alpha_{i_{p(i)}}$. If (3) is an implicant of f and the deletion of any letter $x_{i_j}^{\alpha_{i_j}}$ $[j = 1, \ldots, p(i)]$ transforms (3) into a conjunction which is no longer an implicant of f, we say that (3) is a *prime implicant* of the function f.

For instance, all elementary conjunctions appearing in formulas (5)—(5'') are implicants of the function f_0; the conjunctions $x\,y$ and $\bar{z}$ are prime implicants.

Now the problem is to choose, among the various disjunctive forms to which a Boolean function f may be brought, the minimal one(s), i.e. that (those) for which the total number of occurrences of the variables $x_1, \ldots, x_p$ is minimal. From the practical point of view, this problem corresponds to that of minimizing the number of contacts of a switching circuit.

It is well-known* that a minimal disjunctive form of a Boolean function f is made up of prime implicants, so that the above minimization problem is to be solved in two steps:

1) Determination of the prime implicants.

2) Determination of a minimal disjunctive form made up of prime implicants.

There are numerous methods for solving the above problems 1) and 2); this situation is due to the fact that, for large values of the number n of variables, the computations are cumbersome, even when using an electronic computer. We shall give below a pseudo-Boolean approach to step 2).

Several authors (see I. B. PYNE and E. J. MCCLUSKEY JR. [1], T. L. MAˇSTROVA [1], M. A. BREUER [1], P. L. HAMMER, I. ROSENBERG

* See any standard textbook on switching algebra, for instance those quoted in Bibliography B.

and S. RUDEANU [2]) have observed that problem 2) may be formulated as one of bivalent linear programming*, as follows:

Let

$$f = C_1 \cup \cdots \cup C_m \tag{6}$$

be the disjunctive canonical form of the given function f; this means that $\{C_1, \ldots, C_m\} = C$ is the set of those complete elementary conjunctions $x_1^{\alpha_1} \ldots x_p^{\alpha_p}$ which are implicants of f, i.e. $f(\alpha_1, \ldots, \alpha_p) = 1$ (see Chapter I, § 2).

Let $\{P_1, \ldots, P_n\} = P$ be the set of all prime implicants of the function f; this set was determined at step 1). Each prime implicant P_j is a disjunction of $2^{p-p(j)}$ conjunctions from the set C, $p(j)$ being the number of variables which occur in P_j:

$$P_j = x_{j_1}^{\beta_{j_1}} \ldots x_{j_{p(j)}}^{\beta_{j_{p(j)}}} = \bigcup_{\alpha_{j_{p(j)+1}}, \ldots, \alpha_{j_p}} x_{j_1}^{\beta_{j_1}} \ldots x_{j_{p(j)}}^{\beta_{j_{p(j)}}} x_{j_{p(j)+1}}^{\alpha_{j_{p(j)+1}}} \ldots x_{j_p}^{\alpha_{j(p)}} \tag{7.j}$$

$$(j = 1, \ldots, n).$$

We say that a conjunction $C_i \in C$ *is covered by* P_j, if C_i appears in the right-hand side of (7. j). We say that a subset $P^* \subseteqq P$ *covers the set* C, if every conjunction $C_i \in C$ is covered by at least a prime implicant $P_j \in P^*$. If this is the case, then the function f may also be written in the disjunctive form

$$f = \bigcup_{P_j \in P^*} P_j \tag{8}$$

(for, $\bigcup_{P_j \in P^*} P_j = \bigcup_{k=1}^{m} C_k = f$).

Now problem 2) may be reformulated as follows: among all the expressions (8) of the function f, choose that (those) for which

$$\sum_{P_j \in P^*} p(j) \tag{9}$$

is minimal.

Further, let (a_{ij}) be a Boolean matrix defined as follows:

$$a_{ij} = \begin{cases} 1, & \text{if } C_i \text{ is covered by } P_j, \\ 0, & \text{otherwise.} \end{cases} \tag{10}$$

A subset $P^* \subseteqq P$ is characterized by a Boolean vector $(y_1, \ldots, y_n)$, where

$$y_j = \begin{cases} 1, & \text{if } P_j \in P^*, \\ 0, & \text{otherwise.} \end{cases} \tag{11}$$

---

* M. L. JUNCOSA has pointed out that a similar approach is given in A. COBBHAM, R. FRIDSHALL and J. H. NORTH [1].

Then the sum (9) is obviously equal to

(12) $$\sum_{j=1}^{n} p(j)\, y_j,$$

while the condition that the set P^* covers C is simply translated by the system of pseudo-Boolean inequalities

(13) $$\sum_{j=1}^{n} a_{ij}\, y_j \geqq 1 \qquad (i = 1, \ldots, m),$$

or, if preferred, by the equivalent system of Boolean equations

(13′) $$\bigcup_{j=1}^{n} a_{ij}\, y_j = 1 \qquad (i = 1, \ldots, m).$$

We have thus proved

Theorem 1. *The above problem* 2) *is equivalent to that of minimizing the linear pseudo-Boolean function* (12) *under the linear pseudo-Boolean conditions* (13).

Example 1. Let us find the minimal disjunctive forms of the Boolean function

(14) $$f = x_1 x_2 x_3 \cup x_1 x_2 \bar{x}_3 \cup \bar{x}_1 x_2 x_3 \cup x_1 \bar{x}_2 \bar{x}_3 \cup \bar{x}_1 \bar{x}_2 x_3 \cup \bar{x}_1 \bar{x}_2 \bar{x}_3 .$$

By applying the classical QUINE-MCCLUSKEY procedure, we find that there exist 6 prime implicants:

(15) $P_1 = x_1 x_2$, $P_2 = x_2 x_3$, $P_3 = x_1 \bar{x}_3$, $P_4 = \bar{x}_1 x_3$, $P_5 = \bar{x}_2 \bar{x}_3$, $P_6 = \bar{x}_1 \bar{x}_2$.

Hence the matrix (10) becomes*:

Table 1

C_i \ P_j	$x_1 x_2$	$x_2 x_3$	$x_1 \bar{x}_3$	$\bar{x}_1 x_3$	$\bar{x}_2 \bar{x}_3$	$\bar{x}_1 \bar{x}_2$
$x_1 x_2 x_3$	1	1	0	0	0	0
$x_1 x_2 \bar{x}_3$	1	0	1	0	0	0
$\bar{x}_1 x_2 x_3$	0	1	0	1	0	0
$x_1 \bar{x}_2 \bar{x}_3$	0	0	1	0	1	0
$\bar{x}_1 \bar{x}_2 x_3$	0	0	0	1	0	1
$\bar{x}_1 \bar{x}_2 \bar{x}_3$	0	0	0	0	1	1

We have $p(j) = 2$ for all $j = 1, \ldots, 6$, so that the linear pseudo-Boolean program to be solved is the following: minimize

(16) $$2y_1 + 2y_2 + 2y_3 + 2y_4 + 2y_5 + 2y_6$$

---

* Incidentally, here $m = n$.

under the restrictions

(17.1) $$y_1 + y_2 \geqq 1,$$

(17.2) $$y_1 + y_3 \geqq 1,$$

(17.3) $$y_2 + y_4 \geqq 1,$$

(17.4) $$y_3 + y_5 \geqq 1,$$

(17.5) $$y_4 + y_6 \geqq 1,$$

(17.6) $$y_5 + y_6 \geqq 1.$$

We find easily that there are two minimizing points:

(18.1) $$y_1 = 1, \quad y_2 = 0, \quad y_3 = 0, \quad y_4 = 1, \quad y_5 = 1, \quad y_6 = 0,$$

(18.2) $$y_1 = 0, \quad y_2 = 1, \quad y_3 = 1, \quad y_4 = 0, \quad y_5 = 0, \quad y_6 = 1.$$

Hence there are two minimal disjunctive forms:

(19.1) $$f = x_1 x_2 \cup \bar{x}_1 x_3 \cup \bar{x}_2 \bar{x}_3,$$

(19.2) $$f = x_2 x_3 \cup x_1 \bar{x}_3 \cup \bar{x}_1 \bar{x}_2,$$

each of them involving 6 letters*.

A similar procedure may be applied for the minimization of the so-called "partially determined" Boolean functions.

A partially determined Boolean function g is, in fact, a family of Boolean functions which have prescribed values in certain points of B_2^p and take arbitrary values in the remaining points. More precisely, the Boolean algebra B_2^p is decomposed into three pairwise disjoint sets U, Z and D, such that

(20) $$g(x_1, \ldots, x_p) = \begin{cases} 1, & \text{if } (x_1, \ldots, x_p) \in U, \\ 0, & \text{if } (x_1, \ldots, x_p) \in Z, \\ \text{arbitrary}, & \text{if } (x_1, \ldots, x_p) \in D. \end{cases}$$

Thus g has the disjunctive canonical form

(21) $$g = C_1 \cup \cdots \cup C_h \cup \lambda_1 C_{h+1} \cup \cdots \cup \lambda_k C_{h+k},$$

where $C_1, \ldots, C_{h+k}$ are complete elementary conjunctions, while $\lambda_1, \ldots, \lambda_k$ are arbitrary bivalent parameters: namely, the conjunctions $C_1, \ldots, C_h$ correspond to the h points of the set U, the conjunctions $C_{h+1}, \ldots, C_{h+k}$ correspond to the k points of the set D, while $\lambda_1, \ldots, \lambda_k$ are the values of the function g on this set.

For each fixed system of values given to the parameters $\lambda_1, \ldots, \lambda_k$, we obtain a Boolean function $g_{\lambda_1 \ldots \lambda_k}$ which has one or several minimal disjunctive forms involving $p_{\lambda_1 \ldots \lambda_k}$ letters. The problem now is to find $(\lambda_1^*, \ldots, \lambda_k^*) \in B_2^p$ for which

$$p_{\lambda_1^* \ldots \lambda_k^*} = \min_{\lambda_1 \ldots \lambda_k} p_{\lambda_1 \ldots \lambda_k}$$

* Incidentally, here min(16) $= m (= n)$.

as well as the disjunctive forms for which this minimum is attained.

It is well-known that this new problem can be solved in the following two steps:

1′) Find the set P' of all prime implicants of the Boolean function

$$g_{1\ldots 1} = C_1 \cup \cdots \cup C_h \cup C_{h+1} \cup \cdots \cup C_{h+k}. \tag{22}$$

2′) Find a subset $Q^* \subseteq P'$ covering the set $\{C_1, \ldots, C_h\}$, such that

$$\sum_{P_j \in Q^*} p(j) = \min_{Q \in \mathfrak{Q}} \sum_{P_j \in Q} p(j), \tag{23}$$

where $\mathfrak{Q}$ is the set of all subsets $Q \subseteq P'$ covering $\{C_1, \ldots, C_h\}$.

We see, as before, that step 2′) may be solved by pseudo-Boolean programming.

Example 2. Let us minimize the partially determined Boolean function

$$g = x_1 x_2 x_3 \cup x_1 x_2 \bar{x}_3 \cup \bar{x}_1 x_2 x_3 \cup x_1 \bar{x}_2 \bar{x}_3 \cup \lambda_1 \bar{x}_1 \bar{x}_2 x_3 \cup \lambda_2 \bar{x}_1 \bar{x}_2 \bar{x}_3. \tag{24}$$

In this case, the corresponding function (22) is the function f studied in Example 1. Hence the prime implicants are those given in formulas (15).

The set $\{C_1, \ldots, C_h\}$ is now $\{x_1 x_2 x_3,\ x_1 x_2 \bar{x}_3,\ \bar{x}_1 x_2 x_3,\ x_1 \bar{x}_2 \bar{x}_3\}$, so that the corresponding covering matrix is given in Table 2 below.

Table 2

C_i \ P_j	$x_1 x_2$	$x_2 x_3$	$x_1 \bar{x}_3$	$\bar{x}_1 x_3$	$\bar{x}_2 \bar{x}_3$	$\bar{x}_1 \bar{x}_2$
$x_1 x_2 x_3$	1	1	0	0	0	0
$x_1 x_2 \bar{x}_3$	1	0	1	0	0	0
$\bar{x}_1 x_2 x_3$	0	1	0	1	0	0
$x_1 \bar{x}_2 \bar{x}_3$	0	0	1	0	1	0

We have now to minimize the function

$$2y_1 + 2y_2 + 2y_3 + 2y_4 + 2y_5 + 2y_6 \tag{16}$$

under the restrictions

$$y_1 + y_2 \geqq 1, \tag{17.1}$$

$$y_1 + y_3 \geqq 1, \tag{17.2}$$

$$y_2 + y_4 \geqq 1, \tag{17.3}$$

$$y_3 + y_5 \geqq 1. \tag{17.4}$$

We see easily that the single minimizing point is

$$y_1 = 0, \quad y_2 = 1, \quad y_3 = 1, \quad y_4 = 0, \quad y_5 = 0, \quad y_6 = 0, \tag{25}$$

so that the required disjunctive form is

$$g_{00} = x_2 x_3 \cup x_1 \bar{x}_3, \tag{26}$$

corresponding to $\lambda_1 = \lambda_2 = 0$.

Other minimization problems can also be solved as above. For instance, it may be desired to minimize the number of conjunctions occurring in a disjunctive form. More generally, let us assume that to each disjunctive form D, a "cost" or "value" $v(D)$ is associated, satisfying the following properties:

(α) $v(D) > 0$;

(β) if c and c' are elementary conjunctions such that each letter $x_r^{\alpha_r}$ occurring in c occurs also in c' with the same "exponent" α_r, then $v(c) \leqq v(c')$;

(γ) if $c_1, \ldots, c_t$ are elementary conjunctions, then $v(c_1 \cup \cdots \cup c_t) = v(c_1) + \cdots + v(c_t)$.

The problem of obtaining those disjunctive forms of a (partially determined) Boolean function for which the "cost" is minimum, is now translated as the problem of minimizing the pseudo-Boolean function

$$\sum_{j=1}^{n} v_j y_j \tag{27}$$

[instead of (12)], where v_j is the "cost" of the prime implicant P_j ($j = 1, \ldots, n$), under the same restrictions as before.

Of course, dual results hold for the problem of minimizing the conjunctive form of a (partially determined) Boolean function.

Gr. C. Moisil [1, 5] has found eight interpolation formulas for Boolean functions (expressed in terms of conjunction, disjunction or Sheffer functions), including in particular the classical interpolation formulas (I.37) and (I.39). To each of these eight formulas it corresponds a minimization problem, which can be roughly stated as follows:

Let f be the given Boolean function and E its expression obtained by application of the specified interpolation formula. Find the "simplest" (in a certain sense) expression(s) E' equivalent to E; here the term "equivalent" means that E' is obtained from E by applying certain prescribed transformations. [For instance, if: (i) the interpolation formula is (I.37); (ii) the prescribed transformations are: $a \cup b = b \cup a$, $a \cup a = a$ and $a\,b \cup a\,\bar{b} = a$; (iii) the criterion of simplicity is the number of letters; then we obtain the above mentioned classical problem of minimizing the disjunctive form.] The practical significance of these problems is the simplifications of circuits with various deviees (relays, electronic tubes, etc.).

Gr. C. Moisil has shown that, for each of the eight minimization problems corresponding to the eight interpolation formulas, a concept of prime implicant may be defined so that the specified problem may be solved in two steps:

1) determination of the prime implicants by an extension of the Quine-McCluskey procedure;

2) determination of a least-cost expression made up of prime implicants.

The latter result was obtained using an axiomatic model for the eight problems; the model was given in terms of the so-called *equality algebras*. Roughly speaking, in such an algebra Boolean expressions are replaced by purely formal expressions and Boolean functions by equivalence classes of these expressions with respect to a certain congruence relation. If ω and Θ are the signs which take the places of conjunction and disjunction respectively, then the analogues of elementary conjunctions are expressions like $\Theta\left(x_{i_1}^{\alpha_{i_1}}\omega\ldots\omega\, x_{i_r}^{\alpha_{i_r}}\right)$, with $\alpha_{i_1},\ldots,\alpha_{i_r}=0$ or 1, while the analogues of disjunctive forms are expressions like $a_1\,\Theta\ldots\Theta\, a_s$, where each a_h is of the form $x_{i_1}^{\alpha_{i_1}}\omega\ldots\omega\, x_{i_r}^{\alpha_{i_r}}$.

Gr. C. Moisil [1, 5] has defined an even more general axiomatic model, termed by him the "*n-valued Quine algebra*" for which a minimization problem can also be stated and solved in the above indicated two steps. As it was remarked by S. Rudeanu [10], pseudo-Boolean programming applies to step 2 of this general case as well.

On the other hand, the actual operation of a switching circuit may be described with the aid of two-valued functions depending on three-valued arguments (Gr. C. Moisil [2]). The corresponding minimization problems may also be solved in the above discussed two steps; P. L. Hammer and S. Rudeanu [3] have solved step 2) by pseudo-Boolean programming. The case of two-valued functions with five-valued arguments, also occurring in practice, was studied by S. Median [1].

§ 2. Minimization of the Number of States in Input-Restricted Machines*

We shall consider input-restricted machines (Mealy's automata), i.e. machines where input is subject to some known restraints; we shall follow A. Gill [1].

1.

A machine is given by the following elements: three finite sets of states S, X, Z, an input alphabet, an output alphabet and two characterizing functions f_z and f_s which are defined for every pair (x, s) of a certain subset A of $X\times S$ with values in Z and S respectively. We shall denote such a machine by $M=\langle S, X, Z, A, f_z, f_s\rangle$. We can represent M by a transition diagram. This diagram describing M contains vertices (each vertex representing a different state) and oriented labelled branches [each branch goes from a state s to a state

* By Ivo Rosenberg.

$f_s(x, s)$ $[(x, s) \in A]$ and is labelled by $x/f_z(x, s)$ or possibly by the disjunction of more such expressions]. An example is given in Fig. 1.

An input sequence (i.e. a finite sequence $x_1, \ldots, x_n$ of elements of X) is said to be *acceptable* to the state $s \in S$, if, when applied to M in the state s, it does not violate the input restrictions in any of the states of M [i.e., if we define $s_0 = s$, $s_i = f_s(x_i, s_{i-1})$ for $i = 1, \ldots, n$, then $x_{i+1}, s_i \in A$ for every $i = 2, \ldots, n-1$]. Let $M' = \langle S', X, Z, A', f'_z, f'_s \rangle$ be another machine. The state s' of M' is said to be *quasi-equivalent* to the state s of M if any input sequence acceptable to s is acceptable to s' and yields identical responses when applied to M' in state s' and to M in state s. The machine M' is *quasi-equivalent* to M if to every state s' of M' there corresponds at least one state s of M such that s' is quasi-equivalent to s. A *minimal-state* machine to M is a machine $\check{M}$ quasi-equivalent to M, such that $\check{M}$ has a minimal number of states among all machines M' quasi-equivalent to M.

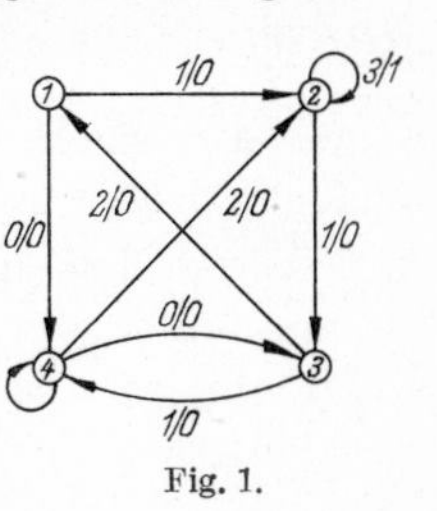

Fig. 1.

2.

The states s_1 of M and s_2 of M are *k-distinguishable* if there is an input sequence $x_1, \ldots, x_k$ acceptable to both states s_1 and s_2 such that this sequence yields different output sequences. If s_1 and s_2 are not distinguishable for any non-negative integer k, then they are said to be *compatible*. s_1 and s_2 are *incompatible*, if there is at least one input sequence acceptable to both s_1 and s_2 which yields different output sequences when applied to s_1 and s_2. We can define an unoriented graph (S, Γ) on S as follows: $y \in \Gamma x \Leftrightarrow x, y$ are incompatible. We can construct the graph (S, Γ) using the following algorithm:

1) We find directly all the pairs of 1-distinguishable states [$s_1, s_2 \in S$ such that there is an $x \in X$, $(x, s_1) \in A$, $(x, s_2) \in A$ with $f_z(x, s_1) \neq f_z(x, s_2)$].

2) If all the pairs of k-distinguishable states are established, we can proceed as follows: $s_1, s_2 \in S$ are $(k+1)$-distinguishable if and only if there is an $x \in X$ acceptable to s_1 and s_2, such that $f_s(x, s_1)$ and $f_s(x, s_2)$ are k-distinguishable elements.

3) If for some k' every pair of $(k'+1)$-distinguishable elements is a pair of k'-distinguishable states, then this construction is finished, i.e. s_1 and s_2 are incompatible if and only if s_1, s_2 are k'-distinguishable.

The internally stable sets of (S, Γ) are said to be *compatible sets* (invariant classes in E. J. McCluskey Jr. [2]). $N \subseteqq S$ is a compatible set if $N \neq \varnothing$ and for every $n_1, n_2 \in N$, n_1 and n_2 are compatible.

We shall denote by $\mathscr{N} = \{N_1, \ldots, N_u\}$ the set of all compatible sets of (S, Γ). A system $\mathscr{G} \subseteqq \mathscr{N}$ is called a *closed collection* if for every

$G \in \mathscr{G}$ and $x \in X$ there is a $G' \in \mathscr{G}$, such that $f_s(x, s) \in G'$ for every $s \in G$, $(x, s) \in A$. A closed collection G is called a *proper grouping*, if G covers S, i.e. if $\bigcup_{G \in \mathscr{G}} G = S$.

As it is known, to every machine M' which is quasi-equivalent to a machine M, there corresponds a proper grouping of S and conversely, to every proper grouping of S there corresponds a machine M' which is quasi-equivalent to M.

In this way, the minimization of a given machine is reduced to the problem of finding a proper grouping of S with minimal size (number of sets).

3.

Let $\mathscr{N} = \{N_1, N_2, \ldots, N_u\}$ be the set of all compatible sets of M. Let $S = \{s_1, s_2, \ldots, s_v\}$. We put $C_j = \{i \mid 1 \leqq i \leqq u,\ s_j = N_i\}$ $(j = 1, 2, \ldots, v)$. For $C_j = \varnothing$ we put $\prod_{i \in C_j} y_i = 0$.

Lemma 1. *Let* $y_i \in B_2 = \{0, 1\}$ $(i = 1, 2, \ldots, u)$ *and*

$$\varphi(y_1, \ldots, y_u) = \sum_{i=1}^{v} \prod_{j \in C_i} y_j . \tag{28}$$

Then $\varphi(y_1, \ldots, y_u) = 0$ *if and only if* $\mathscr{G} = \{N_k \mid 1 \leqq k \leqq u,\ y_k = 0\}$ *covers* S.

Proof. Let $\varphi(y_1, \ldots, y_u) = 0$ and assume that $1 \leqq l \leqq v$. The sum (28) contains only non-negative terms and therefore

$$\prod_{j \in C_l} y_j = 0 . \tag{29}$$

Then there exists $1 \leqq j_l \leqq u$ such that $j_l \in C_l$ and $y_{j_l} = 0$. Thus $s_l \in N_{j_l} \in \mathscr{G}$.

Let $\mathscr{G}$ cover S. We put $y_k = 0 \Leftrightarrow N_j \in \mathscr{G}$, $y_k = 1 \Leftrightarrow N_j \notin \mathscr{G}$ $(j = 1, 2, \ldots, u)$. For every $1 \leqq i \leqq v$ there exists an index j_i such that $1 \leqq j_i \leqq u$ and $s_i \in N_{j_i} \in \mathscr{G}$, i.e. $j_i \in C_i$ and $y_{j_i} = 0$. Consequently (29) holds for every $i = 1, 2, \ldots, v$ and $\varphi(y_1, \ldots, y_u) = 0$.

Let $1 \leqq i \leqq u$ and $x \in X$. We denote by L_{ix} the set of all $s' \in S$, such that there is an $s \in N_i$ with $(x, s) \in A$ and $f_s(x, s) = s'$. If L_{ix} has at least two elements, then we put $K_{ix} = \{j \mid 1 \leqq j \leqq u,\ N_j \supseteq L_{ix}\}$. If $|L_{ix}| \leqq 1$, then we put $K_{ix} = \varnothing$. We put $D = \{j \mid 1 \leqq j \leqq u,\ |N_j| \geqq 2\}$.

Lemma 2. *Let* $y_j \in B_2$ $(j = 1, 2, \ldots, u)$ *and let*

$$\psi(y_1, \ldots, y_u) = \sum_{i \in D} (1 - y_i) \sum_{x \in X} \prod_{j \in K_{ix}} y_j + \varphi(y_1, \ldots, y_u) \tag{30}$$

where for $K_{ix} = \varnothing$ *we put* $\prod_{j \in \varnothing} y_j = 0$. *Then* $\psi(y_1, \ldots, y_u) = 0$ *if and only if* $\mathscr{G} = \{N_j \mid 1 \leqq j \leqq u,\ y_j = 0\}$ *is a proper grouping.*

Proof. 1) Let $\psi(y_1, \ldots, y_u) = 0$. Every term in (30) is non-negative and therefore $\varphi(y_1, \ldots, y_u) = 0$, i.e. by Lemma 1, $\mathscr{G} = \{N_j \mid 1 \leqq j \leqq u,\ y_j = 0\}$ covers S. From $\psi(y_1, \ldots, y_u) = 0$ it follows, that

$$(1 - y_i) \sum_{x \in X} \prod_{j \in K_{ix}} y_j = 0 \tag{31}$$

for every $i \in D$. If $y_i = 0$, then $\sum_{x \in X} \prod_{j \in K_{ix}} y_j = 0$ and $\prod_{j \in K_{ix}} y_j = 0$ for every $x \in X$. Thus, either $K_{ix} = \varnothing$ or there is a $k \in K_{ix}$, such that $y_k = 0$. In the first case, for every $x \in X$ there is an $s' \in S$, such that $f_s(x, s) = s'$ for every $s \in N_i$, $(x, s) \in A$ and because $\mathscr{G}$ covers S, $f_s(x, s)$ belongs to at least one set of $\mathscr{G}$ [for every $s \in N_i$, $(x, s) \in A$]. In the second case we have obviously $f_s(x, s) \in N_k$ for every $s \in N_i$, $(x, s) \in A$. Thus $\mathscr{G}$ is a closed collection and therefore $\mathscr{G}$ is a proper grouping.

2) Let $\mathscr{G}$ be a proper grouping. We put $y_j = 0$ if $N_j \in \mathscr{G}$ and $y_j = 1$ if $N_j \notin \mathscr{G}$ $(j = 1, 2, \ldots, u)$. By Lemma 1, $\varphi(y_1, \ldots, y_u) = 0$. If $y_j = 0$ for some $1 \leqq j \leqq u$, then $N_j \in \mathscr{G}$ and by definition, for every $x \in X$ there is an $N_k \in \mathscr{G}$, such that $f_s(x, s) \in N_k$ for every $s \in N_j$, $(x, s) \in A$. Thus, if N_k has at least two elements, then $k \in K_{jx}$ and $\prod_{i \in K_{jx}} y_i = 0$. In the other cases $K_{jx} = \varnothing$ by definition and therefore $\prod_{i \in K_{jx}} y_i = 0$ too. Hence (31) holds. If $y_j = 1$, (31) holds evidently too. Thus $\psi(y_1, \ldots, y_u) = 0$.

Theorem 2. *Let* $(y_1^*, \ldots, y_u^*) \in B_2^u$. *The following conditions are equivalent:*

(I) $(y_1^*, \ldots, y_u^*)$ *is a minimizing point of the pseudo-Boolean function*

$$\Theta(y_1, \ldots, y_u) = 2u\, \psi(y_1, \ldots, y_u) - (y_1 + \cdots + y_u). \tag{32}$$

(II) $\mathscr{G} = \{N_j \mid 1 \leqq j \leqq u,\ y_j^* = 0\}$ *is a proper grouping with minimal size.*

Proof: (I) $\Rightarrow$ (II). Suppose that $\mathscr{G}$ is not a proper grouping. Then $\psi(y_1^*, \ldots, y_u^*) \neq 0$ (Lemma 2) and therefore $\psi(y_1^*, \ldots, y_u^*) \geqq 1$ and $\Theta(y_1^*, \ldots, y_u^*) \geqq 2u - u = u$. Since $(y_1^*, \ldots, y_u^*)$ is a minimizing point it follows that $\Theta(y_1, \ldots, y_u) \geqq u$ for every $y_j \in B_2$ $(j = 1, 2, \ldots, u)$. But $\Theta(0, 0, \ldots, 0) = 0$ and this is obviously a contradiction. Thus $\mathscr{G}$ is a proper grouping. From (32) it follows, that $\mathscr{G}$ is a proper grouping with minimal size.

(II) $\Rightarrow$ (I). Suppose that there is a point $(y_1^{**}, \ldots, y_u^{**}) \in B_2^u$ such that $\Theta(y_1^{**}, \ldots, y_u^{**}) < \Theta(y_1^*, \ldots, y_u^*)$. By Lemma 2, $\Theta(y_1^*, \ldots, y_u^*) = -(y_1^* + \cdots + y_u^*)$. Hence $\mathscr{G}' = \{N_j \mid 1 \leqq j \leqq u,\ y_j^{**} = 0\}$ is also a proper grouping; if not, then $\Theta(y_1^{**}, \ldots, y_u^{**}) \geqq u$ [and $\Theta(y_1^{**},\ldots,y_u^{**}) = -(y_1^{**} + \cdots + y_u^{**})$]. Thus $-(y_1^{**} + \cdots + y_u^{**}) < -(y_1^* + \cdots + y_u^*)$. Let r and s be the sizes of $\mathcal{G}$ and $\mathcal{G}'$, respectively. Obviously $y_1^* + \cdots + y_u^* = u - r$ and $y_1^{**} + \cdots + y_u^{**} = u - s$. Hence $s < r$ and this contradicts (II). Thus (I) holds.

Example 3. Let M be the machine given by the transition diagram in Fig. 1 (input alphabet $X = \{0, 1, 2, 3\}$, output alphabet $Z = \{0, 1\}$ and the set of states is $\{1, 2, 3, 4\}$).

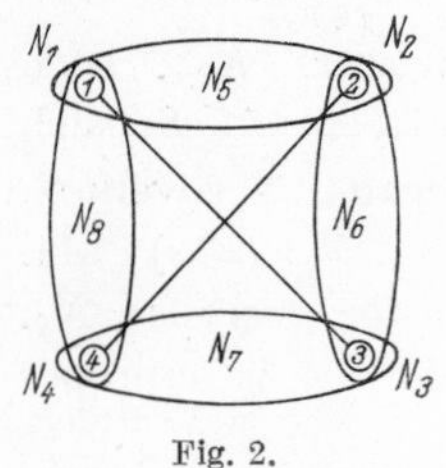

Fig. 2.

Only the states 2, 4 are 1-distinguishable. States 1, 3 are 2-distinguishable. We can easily derive, that the pairs 2, 4 and 1, 3 are incompatible, the other pairs are compatible (Fig. 2).

The compatible sets are: $N_i = \{i\}$ $(i = 1, 2, 3, 4)$ and $N_5 = \{1, 2\}, N_6 = \{2, 3\}, N_7 = \{3, 4\}, N_8 = \{4, 1\}$, hence $\varphi(y_1, \ldots, y_8) = y_1 y_5 y_8 + y_2 y_5 y_6 + y_3 y_6 y_7 + y_4 y_7 y_8$.

Evidently, $D = \{5, 6, 7, 8\}$, $L_{51} = \{N_6\}$ and consequently $K_{51} = \{6\}$. In an entirely analogous way we obtain $K_{61} = \{7\}$, $K_{72} = \{5\}$, $K_{80} = \{7\}$. The other K_{ix} are void. Therefore

$$\begin{aligned}\Theta(y_1, \ldots, y_8) &= (1 - y_5) y_6 + (1 - y_6) y_7 + (1 - y_7) y_5 + (1 - y_8) y_7 + \varphi(y_1, \ldots, y_8) \\ &= 16(y_5 + y_6 + 2y_7 - y_5 y_6 - y_5 y_7 - y_6 y_7 - y_7 y_8 + y_1 y_5 y_8 + y_2 y_5 y_6 + \\ &\quad + y_3 y_6 y_7 + y_4 y_7 y_8) - y_1 - y_2 - y_3 - y_4 - y_5 - y_6 - y_7 - y_8 \\ &= y_1(16 y_5 y_8 - 1) + y_2(16 y_5 y_6 - 1) + y_3(16 y_6 y_7 - 1) + y_4(16 y_7 y_8 - 1) + \\ &\quad + 15 y_5 + 15 y_6 + 31 y_7 - y_8 - 16 y_5 y_6 - 16 y_5 y_7 - 16 y_6 y_7 - 16 y_7 y_8.\end{aligned}$$

By applying the basic algorithm, we find that the single minimizing point of $\Theta(y_1, \ldots, y_8)$ is

$$y_1^* = 1,\ y_2^* = 1,\ y_3^* = 1,\ y_4^* = 1,\ y_5^* = 0,\ y_6^* = 0,\ y_7^* = 0,\ y_8^* = 1, \tag{33}$$

so that $\{N_5, N_6, N_7\}$ is the grouping with minimal size.

§ 3. The State Assignment of Sequential Circuits with High Reliability

Y. Inagaki and K. Sugino [1], using results obtained by K. Udagawa and Y. Inagaki [1], and by K. Udagawa, Y. Inagaki and M. Gotō [1], have shown that the problem of state assignment of sequential circuits with high reliability can be solved using pseudo-Boolean programming. We quote the following from the introduction of Inagaki and Sugino's paper:

"The state assignment is an important problem in the practice of synthesizing sequential circuits, because it has a marked influence on the complexity and reliability of the circuits. It has been already shown (K. UDAGAWA and Y. INAGAKI [1]) that the state assignment problems of reliable sequential circuits can be solved by the use of the assignment-iteration method. In this paper it is first demonstrated that the assignment improving routine of this method can be solved by pseudo-Boolean programming, ... As shown in the succeeding section, the estimation of the elements of the fundamental matrix allows us to reach the optimal state assignment after only one iteration of the assignment-iteration method, when the error probabilities of memory cells are small enough. Then, in these cases, it follows that the optimal state assignment becomes practically attainable even by the hand computations of the pseudo-Boolean programming for the problems of moderate size".

§ 4. Program Segmentation

J. KRÁL [1] has studied the following problem:

"Let us have a program P for a digital computer. Let its length be n, i.e. the program is in the machine language expressed by n instructions. If n is large, a part of P must be placed on a backing store. We shall suppose that an information from a backing store can be called only in tracks of length k and that the bounds of tracks on the backing store are given by the hardware of the computer. We shall assume that the backing store has practically unlimited capacity."

"Now, in the run time, the segment with the executed instruction must be previously transferred into the main store. If the control of the computer after the execution of the instruction on one segment needs the instruction from another segment, an administrative program must be called verifying whether the new track is in the main memory. If it is not the case, it calls the new track from the backing store. The problem of segmentation is the collecting of all the instructions of the program into the tracks so as to minimize the number of calls of the administrative program."

It is shown that this problem can be given the following graph-theoretical model:

A directed graph $G = (N, \Gamma)$ is given, where $N = \{1, 2, \ldots, n\}$ and to each edge (i, j) corresponds a flow S_{ij}. Find a partition of the set N of vertices into q subsets $B_1, B_2, \ldots, B_q$, such that:

1) the number of elements in each B_i is not greater than k;

2) the value of

$$\sum_{s=1}^{q} \sum_{i \in B_s} \sum_{j \notin B_s} S_{ij}$$

is minimal.

(The number q is not given a priori, but we can take $q \leqq [n/[k/2]]$, where $[\alpha]$ denotes the integer part of α.)

J. Král has given the following pseudo-Boolean formulation to the above problem:

Maximize

$$\sum_{i=1}^{n} \sum_{h=1}^{n} S_{ih} \sum_{j=1}^{q} x_i^j x_h^j$$

subject to the constraints

$$x_i^j \in \{0, 1\} \qquad (i = 1, \ldots, n;\ j = 1, \ldots, q),$$

$$\sum_{i=1}^{n} x_i^j \leqq k \qquad (j = 1, \ldots, q),$$

$$\sum_{j=1}^{q} x_i^j = 1 \qquad (i = 1, \ldots, n).$$

Appendix

Generalized Pseudo-Boolean Programming*

By Ivo Rosenberg

1. Let us put, for any integer $k > 1$,

$$B_k = \{0, 1, \ldots, k-1\} \tag{1}$$

and let

$$K = B_{k_1} \times B_{k_2} \times \cdots \times B_{k_n} \tag{2}$$

be the Cartesian product of the sets B_{k_j} $(j = 1, \ldots, n)$. In the sequel we shall be concerned only with functions which map the set K into the field R of reals.

For any integer $k > 1$ and any $i \in B_k$, we define, as in Gr. C. Moisil [2], the so-called *"Lagrangean functions"*

$$x^{k\,i} = \frac{x(x-1)\ldots(x-i+1)(x-i-1)\ldots(x-k+1)}{i(i-1)\ldots 1\cdot(-1)\ldots(i-k+1)} \tag{3}$$

mapping B_k into B_2. Obviously, $i^{k\,i} = 1$, $x^{k\,i} = 0$ for $x \in B_k$, $x \neq i$. For instance, for $k = 3$ we have

$$x^{30} = \tfrac{1}{2}x^2 - \tfrac{3}{2}x + 1, \quad x^{31} = -x^2 + 2x, \quad x^{32} = \tfrac{1}{2}x^2 - \tfrac{1}{2}x. \tag{4}$$

Any function $f(x_1, \ldots, x_n)$ has a Lagrangean development

$$f(x_1, \ldots, x_n) = \sum_{(\sigma_1, \ldots, \sigma_n) \in K} f(\sigma_1, \ldots, \sigma_n)\, x_1^{k_1 \sigma_1} \ldots x_n^{k_n \sigma_n}. \tag{5}$$

We see, from (3), that each $x_j^{k_j \sigma_j}$ is a polynomial with real coefficients of degree $k_r - 1$. Therefore, formula (5) shows that each function f: $K \to R$ is a polynomial with real coefficients, having as degree of the j-th variable at most $k_j - 1$ $(j = 1, \ldots, n)$.

A point $\xi = (\xi_1, \ldots, \xi_n) \in K$ will be called a *minimizing point* of the function f, if for each $(\sigma_1, \ldots, \sigma_n) \in K$, relation

$$f(\xi_1, \ldots, \xi_n) \leqq f(\sigma_1, \ldots, \sigma_n) \tag{6}$$

holds.

* See I. Rosenberg [1].

In the sequel we shall be concerned with the determination of one (all) minimizing point(s) of a function $f: K \to R$. When $k_1 = \cdots = k_n = 2$, then the problem becomes one of pseudo-Boolean programming.

2. For any $j = 1, \ldots, n$, we put

$$K_j = B_{k_j} \times B_{k_j+1} \times \cdots \times B_{k_n}. \tag{7}$$

We also put $f_1 = f$. Let us assume that for $j (1 \leq j \leq n)$, we have already determined a function $f_j(x_j, x_{j+1}, \ldots, x_n)$, mapping K_j into R. We define an auxiliary function $\varphi_j(x_{j+1}, \ldots, x_n)$ mapping K_{j+1} into B_{k_j}, as follows:

To each $(\sigma_{j+1}, \ldots, \sigma_n) \in K_{j+1}$ we associate an $\alpha \in B_{k_j}$, defined by the relation

$$f_j(\alpha, \sigma_{j+1}, \ldots, \sigma_n) = \min_{\beta_j \in B_{k_j}} f_j(\beta_j, \sigma_{j+1}, \ldots, \sigma_n). \tag{8. j}$$

In other terms, α is one of the values in B_{k_j} for which the function $f_j(x, \sigma_{j+1}, \ldots, \sigma_n)$, considered as a function of the single variable x, reaches its minimum (as x belongs to the finite set B_{k_j}, it is obvious that one or several such α do exist). This α will be chosen as a value of $\varphi_j(\sigma_{j+1}, \ldots, \sigma_n)$. In 3 we shall deal with the practical determination of the function φ_j.

Using φ_j, we define the function $f_{j+1}(x_{j+1}, \ldots, x_n)$, mapping K_{j+1} into R, as follows:

$$f_{j+1}(x_{j+1}, \ldots, x_n) = f_j(\varphi_j(x_{j+1}, \ldots, x_n), x_{j+1}, \ldots, x_n). \tag{9. j}$$

It follows from (8. j) that

$$f_{j+1}(x_{j+1}, \ldots, x_n) = \min_{\beta_j \in B_{k_j}} f_j(\beta_j, x_{j+1}, \ldots, x_n). \tag{10. j}$$

Relation (10. j) shows that the value $f_{j+1}(\sigma_{j+1}, \ldots, \sigma_n)$ does not depend on the choice of α, in case that the function $f_j(x, \sigma_{j+1}, \ldots, \sigma_n)$ reaches its minimum for several values $x \in B_{k_j}$.

We obtain thus a sequence $f_1, f_2, \ldots, f_n$. According to the definitions, φ_n is a constant for which $f_n(x_n)$ reaches its minimum. The repeated application of (10. j) for $j = n, n-1, \ldots, 1$, shows that

$$f_{n+1} = f_n(\varphi_n) = \min_{\beta_n \in B_{k_n}} \left(\min_{\beta_{n-1} \in B_{k_{n-1}}} \left(\cdots \left(\min_{\beta_1 \in B_{k_1}} f(\beta_1, \ldots, \beta_{n-1}, \beta_n) \right) \cdots \right) \right). \tag{11}$$

Therefore,

$$f_{n+1} = \min_{(\beta_1, \ldots, \beta_n) \in K} f(\beta_1, \ldots, \beta_n), \tag{12}$$

i.e., f_{n+1} is the sought minimum of the function f.

Now, it remains to determine

$$x'_n = \varphi_n, x'_{n-1} = \varphi_{n-1}(x'_n), \ldots, x'_1 = \varphi_1(x'_2, x'_3, \ldots, x'_n). \tag{13}$$

Obviously, if $(x'_1, \ldots, x'_n)$ is a minimizing point of the function f, then $f_{n+1} = f_n(\varphi_n) = f_n(x'_n) = f_{n-1}(\varphi_{n-1}(x'_n), x'_n) = f_{n-1}(x'_{n-1}, x'_n) = \cdots = f_1(x'_1, \ldots, x'_{n-1}, x'_n) = f(x'_1, \ldots, x'_n)$.

In certain cases it is necessary to determine the set M_1 of all the minimizing points. We can proceed as follows: First we determine

$$(14.\,n) \qquad M_n = \{x_n \in K_n = B_{k_n} \mid f_n(x_n) = f_{n+1}\},$$

then

$$(14.\,n-1) \quad M_{n-1} = \{(x_{n-1}, x_n) \in K_{n-1} \mid x_n \in M_n, f_{n-1}(x_{n-1}, x_n) = f_{n+1}\},$$

. . . .

finally

$$(14.1) \quad M_1 = \{(x_1, \ldots, x_n) \in K_1 = K \mid (x_2, \ldots, x_n) \in M_2, f_1(x_1, \ldots, x_n) = f_{n+1}\}.$$

3. Let us now examine the problem of the determination of the functions $\varphi_j (1 \leqq j \leqq n)$. For $\nu \in B_{k_j}$ we define the sets

$$(15) \quad P_j = \{(x_{j+1}, \ldots, x_n) \in K_{j+1} \mid f_j(\nu, x_{j+1}, \ldots, x_n) \leqq \leqq f_j(x_j, x_{j+1}, \ldots, x_n) \text{ for each } x_j \in B_{k_j}\}.$$

Thus P^j_ν is the set of those $(\sigma_{j+1}, \ldots, \sigma_n) \in K_{j+1}$ for which $\alpha = \nu$ satisfies $(8.\,j)$. Obviously,

$$(16) \qquad \bigcup_{\nu \in B_{k_j}} P^j_\nu = K_{j+1}.$$

We choose a partition $\{S_1, \ldots, S_t\}$ of K_{j+1}, so that for each $r = 1, 2, \ldots, t$, there exists an $m_r \in B_{k_j}$ so that $S_r \subseteq P^j_{m_r}$. Further, we define the characteristic functions $\Psi_r(x_{j+1}, \ldots, x_n)$ of the sets S_r $(r = 1, \ldots, t)$, that is the functions which take on S_r the value 1, and on $K_{j+1} - S_r$ the value 0. It is easy to see, that we can choose as functions $\varphi_j(x_{j+1}, \ldots, x_n)$, the functions

$$(17) \qquad \sum_{r=1}^{t} m_r \Psi_r(x_{j+1}, \ldots, x_n).$$

The sets P^j_ν may be determined as follows. We express the function $f_j(x_j, x_{j+1}, \ldots, x_n)$ as a polynomial; namely, we put

$$(18) \qquad f_j(x_j, x_{j+1}, \ldots, x_n) = \sum_{p=0}^{u} x_j^p h_p,$$

where x_j^p are powers of x_j; $h_0, h_1, \ldots, h_u$ are polynomials in the variables $x_{j+1}, \ldots, x_n$. It is easy to see that P^j is the set of all $(x_{j+1}, \ldots, x_n) \in K_{j+1}$ for which the system of $k_j - 1$ inequalities

$$(19) \qquad \sum_{p=1}^{u} (\mu^p - \nu^p)\, h_p(x_{j+1}, \ldots, x_n) \geqq 0$$

$(\mu = 0, 1, \ldots, \nu - 1, \nu + 1, \ldots, k_j - 1)$ has a solution.

4. In order to illustrate the above described methods, let us consider the following simple example. Let $n = 3$, $k_1 = k_2 = k_3 = 3$,

$$(20) \quad f(x_1, x_2, x_3) = 2x_1^2 x_2 x_3 - 3x_1 x_2^2 x_3^2 + 7x_1^2 x_2^2 x_3^2 - 3x_1^2 + \\ + 4x_2 + 6x_3 - 3.$$

The values of this function are shown in Fig. 1.

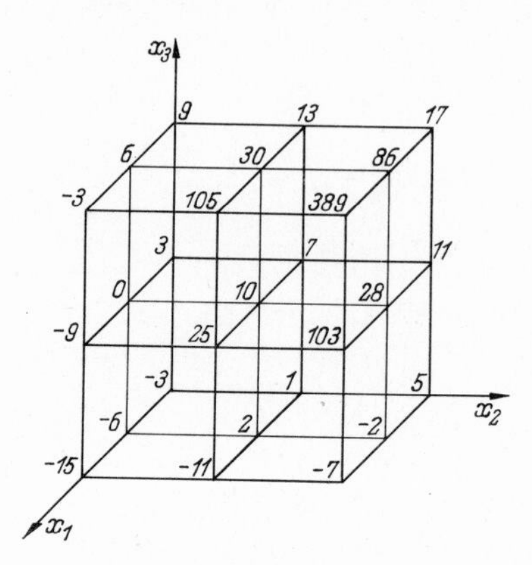

Fig. 1.

We put $f_1 = f = x_1^2(2x_1^2 x_3 + 7x_2^2 x_3^2 - 3) + x_1(-3x_2^2 x_3^2) + 4x_2 + 6x_3 - 3$.

P_1^1 is the set of all the solutions (x_2, x_3) of the inequalities

$$(21) \quad (0^2 - 1^2)(2x_2 x_3 + 7x_2^2 x_3^2 - 3) + (0 - 1)(-3x_2^2 x_3^2) \geqq 0,$$

$$(22) \quad (2^2 - 1^2)(2x_2 x_3 + 7x_2^2 x_3^3 - 3) + (2 - 1)(-3x_2^2 x_3^2) \geqq 0.$$

Relation (21) may also be written in the form $-2x_2^2 x_3^2 - 4x_2 x_3 + 3 \geqq 0$. Hence $x_2 x_3 = 0$; but for $x_2 x_3 = 0$, the inequality (22) has no solutions. Therefore $P_1^1 = \varnothing$.

P_2^1 is the set of all the solutions (x_2, x_3) of the inequalities

$$(23) \quad (0^2 - 2^2)(2x_2 x_3 + 7x_2^2 x_3^2 - 3) + (0 - 2)(-3x_2^2 x_3^2) \geqq 0,$$

$$(24) \quad (1^2 - 2^2)(2x_2 x_3 + 7x_2^2 x_3^2 - 3) + (1 - 2)(-3x_2^2 x_3^2) \geqq 0.$$

It follows from (23) that $x_2 x_3 = 0$. If $x_2 x_3 = 0$, then (24) is fulfilled too. Hence P_2^1 is the set of all $(x_2 x_3) \in B_3^2$ for which $x_2 x_3 = 0$.

If we consider the partition $\{L_3^2 - P_2^1, P_2^1\}$, then $\Psi_2 = (x_2 x_3)^{30}$ is obviously the characteristic function of the set P_2^1. According to (4), we have $\Psi_2 = \frac{1}{2}(x_2^2 x_3^2 - 3x_2 x_3 + 2)$ and by (17), $\varphi_1 = 2\Psi_2$. Since Ψ_2 takes on only the values 0 and 1, we have $\Psi_2^2 = \Psi_2$ and therefore $\varphi_1^2 = 4\Psi_2$.

We replace x_1 by φ_1 in f_1 and obtain

$$f_2 = \Psi_1[4(2x_2 x_3 + 7x_2^2 x_3^2 - 3) - 3x_2^2 x_3^2] + 4x_2 + 6x_3 - 3$$

$$= \frac{1}{2}(x_2^2 x_3^2 - 3x_2 x_3 + 2)[25x_2^2 x_3^2 + 8x_2 x_3 - 12] + 4x_2 + 6x_3 - 3$$

$$= \frac{1}{2}(25x_2^4 x_3^4 - 67x_2^3 x_3^3 + 14x_2^2 x_3^2 + 52x_2 x_3 - 24) + 4x_2 + 6x_3 - 3$$

$$= x_2^4\left(\frac{25}{2}x_3^4\right) + x_2^3\left(-\frac{67}{2}x_3^3\right) + x_2^2(7x_3^2) + x_2(26x_3 + 4) + 6x_3 - 15.$$

P_1^2 is the set of all x_3 for which

$$(0^4 - 1^4)\frac{25}{2}x_3^4 + (0^3 - 1^3)\left(-\frac{67}{2}x_3^3\right) + (0^2 - 1^2)7x_3^2 + + (0-1)(26x_3 + 4) \geqq 0, \tag{25}$$

$$(2^4 - 1^4)\frac{25}{2}x_3^4 + (2^3 - 1^3)\left(-\frac{67}{2}x_3^3\right) + (2^2 - 1^2)7x_3^2 + + (2-1)(26x_3 + 4) \geqq 0. \tag{26}$$

These inequalities may be written in the simpler forms

$$-\frac{25}{2}x_3^4 + \frac{67}{2}x_3^3 - 7x_3^2 - 26x_3 - 4 \geqq 0, \tag{27}$$

$$\frac{375}{2}x_3^4 - \frac{469}{2}x_3^3 + 21x_3^2 + 26x_3 + 4 \geqq 0. \tag{28}$$

It is easy to check that (27) has no solutions for $x_3 \in B_3$. Therefore, $P_1^2 = \varnothing$.

P_2^2 is the set of all x_3 for which

$$-200x_3^4 + 268x_3^3 - 28x_3^2 - 52x_3 - 8 \geqq 0, \tag{29}$$

$$-\frac{375}{2}x_3^4 + \frac{469}{2}x_3^2 - 21x_3^2 - 26x_3 - 4 \geqq 0. \tag{30}$$

It is easy to verify here too, that (29) has no solutions for $x_3 \in B_3$.

We can therefore choose $\varphi_2(x_3)$ identically equal to 0;

$$f_3 = 6x_3 - 15.$$

The minimum of f_3 is reached for $x_3 = 0$.

Therefore $\varphi_3 = 0$ and $f_4 = f_3(\varphi_3) = f_3(0) = -15$.

The minimizing point has $x_3' = \varphi_3 = 0$, $x_2' = \varphi_2(x_3') = 0$, $x_1' = \varphi_1(x_2', x_3') = x_2'^2 x_3'^2 - 3x_2' x_3' + 2 = 2$. Thus the minimizing point is $(2, 0, 0)$. According to the concluding remark in 3, this is the unique minimizing point.

5. Our methods are also valid for polynomials of higher degrees. According to (3) and (5), they can be represented as polynomials of degree at most $k_j - 1$ in each variable x_j. We translate the problem of finding the minimizing points of real functions of several variables, into the problem of solving a system of inequalities. In the general case, for problems involving many unknowns, this is a rather difficult task. In certain cases, however, the special properties of the function f can lead to inequalities of a peculiar type, and can thus facilitate their solution. Our aim here was only to point out the possibility of generalizing pseudo-Boolean programming.

Conclusions

The first papers dealing with applications of Boolean methods to operations research were published recently enough to assert that the investigations in this field are at their very beginning. But the field is wide, the amount of problems yet unsolved being overwhelming.

The broad range of applicability of Boolean procedures makes obvious the necessity for further researches which have to include both applications and methodology.

Integer programming, graph theory and other domains supply the researcher with many attractive problems which are easy to translate into a Boolean language. The goal of investigations could be the discovery of new applications of Boolean techniques. But investigations can also be aimed to unify classes of problems which can be handled in a similar way. Let us consider the following typical example.

In Chapter X, § 3, we have studied the problem of finding the absolutely minimal externally stable sets of a graph. In Chapter X, § 5, we have dealt with the problem of finding the minimal coverings of a graph with maximal internally stable sets (leading immediately to the minimal chromatic decompositions of the given graph). In Chapter XIV, § 1 we have studied the problem of finding the minimal disjunctive normal forms of a Boolean function. In each of the above three cases, we were led to the following problem of pseudo-Boolean programming: minimize the pseudo-Boolean function $\sum_{i=1}^{n} v_i x_i$ under the restriction $\sum_{i=1}^{n} \prod_{j=1}^{n} (1 - a_{ij} x_j) = 0$ (for the first two problems, $v_i = 1$ for $i = 1, \ldots, n$). Thus the three problems may be viewed *mutatis mutandis*, as a single one. It results, in particular, that any method (Boolean or not!) given in the literature for solving one of these problems is applicable to the other two problems as well.

Thus an axiomatic unifying approach to various optimization problems may be useful not only in systematizing the existing material, but also in drawing practical conclusions for computations. Such a study was begun by S. Rudeanu [10].

Further investigations are also necessary in developping improved techniques. For instance, in § 2 of Chapter V we have improved the method described in § 1 of the same chapter for solving linear pseudo-Boolean programs. A similar improvement should be useful for nonlinear pseudo-Boolean programs. At the same time certain extensions of the existing techniques could prove useful. For instance, the research begun by I. Rosenberg in the Appendix should be continued. Also, it would be very useful to have a pseudo-Boolean procedure for solving mixed continuous-bivalent programs.

It will be a real satisfaction for the authors to have their belief in the future of the field confirmed.

Bibliography

Appleby, J. S., D. V. Blake and E. A. Newman: [1] Techniques for Producing School Time-Tables on a Computer and Their Application to Other Scheduling Problems. Comput. J. 3 (1961) 237—245.

Balas, E.: [1] Un algorithme additif pour la résolution des programmes linéaires en variables bivalentes. C. R. Acad. Sci. Paris 258 (1964) 3817—3820.

—: [2] Extension de l'algorithme additif à la programmation en nombres entiers et à la programmation linéaire. C. R. Acad. Sci. Paris 258 (1964) 5136—5139.

—: [3] An Additive Algorithm for Solving Linear Programs with Zero-One Variables. Operations Res. 13 (1965) 517—546.

Balinski, M. L.: [1] Integer Programming: Methods, Uses, Computation. Management Sci. 12 (1965) 253—313.

Beale, E. M. L.: [1] Survey of Integer Programming. Operational Research Quarterly 16 (1965) 219—228.

Bellman, R.: [1] Maximization over Discrete Sets. Naval Research Logist. Quarterly 3 (1965) 67—70.

—: [2] Dynamic Programming. Princeton/NJ: Princeton Univ. Press 1957.

Bellman, R., and S. E. Dreyfus: [1] Applied Dynamic Programming. Princeton/NJ: Princeton Univ. Press 1962.

Ben-Israel, A., and A. Charnes: [1] On Some Problems of Diophantine Programming. Cahiers Centre Etudes Recherche Opér. 4 (1962) 215—280.

Benders, G. F., A. R. Catchpole and C. Kuiken: [1] Discrete Variables Optimization Problems. Rand. Symp. on Math. Programming, March 16—20, 1959.

Berge, C.: [1] Théorie générale des jeux à n personnes. Paris: Gauthier-Villars 1957.

—: [2] Théorie des graphes et ses applications. Paris: Dunod 1958; sec. ed., 1963.

Berge, C., A. Ghouila-Houri: [1] Programmes, jeux et réseaux de transports. Paris: Dunod 1962.

Berghius, J.: [1] La composition des horaires scolaires. Communication au 23-ième Congrès des Utilisateurs des Calculateurs Gamma-Tambour, Bologne 1961.

Bertier, P., and Ph. T. Nghiem: [1] Résolution de problèmes en variables bivalentes (Algorithme de Balas et procédure SEP). SEMA, Note de travail No. 33, janvier 1965.

Bertier, P., Ph. T. Nghiem and B. Roy: [1] Procédure SEP. Trois exemples numériques. SEMA, Note de travail No. 32, janvier 1965.

Bertier, P., and B. Roy: [1] Une procédure de résolution pour une classe de problèmes pouvant avoir un caractère combinatoire. SEMA, Note de travail No. 30 bis, décembre 1964.

Bossert, W. H., and H. B. Harmon: [1] Student Sectioning on the IBM-7090. IBM Corp., Cambridge 39 Mass. 1963.

Breuer, M. A.: [1] The Minimization of Boolean Functions Containing Unequal and Nonlinear Cost Functions. Comm. ACM 5 (1962) 374.

CAMION, P.: [1] Quelques propriétés des chemins et circuits hamiltoniens dans la théorie des graphes. Cahiers Centre Etudes Recherche Opér. 2 (1960) 5—36.

—: [2] Une méthode de résolution par l'algèbre de Boole des problèmes combinatoires ou interviennent des entiers. Cahiers Centre Etudes Recherche Opér. 2 (1960) 234—289.

CARDOT, C.: [1] Application de la théorie des treillis distributifs à l'étude de la distribution de programmes par télécommande. Revue Générale de l'Electricité Janvier 1957, 27—39.

—: [2] Application de l'algèbre des treillis distributifs aux transmetteurs automatiques à programmes. Automatisme Janvier 1959, 3—13.

CARVALLO, M.: [1] Monographie des treillis et algèbre de Boole. Paris: Gauthier-Villars 1962.

—: [2] Principes et applications de l'analyse booléenne. Paris: Gauthier-Villars 1965.

CHARNES, A., and W. W. COOPER: [1] Programming with Linear Fractional Functionals. Naval Res. Logist. Quart. 9 (1962) 181—186.

COBBHAM, A., R. FRIDSHALL and J. H. NORTH: [1] Switching Circuit Theory and Logical Design. Proc. Second Annual Sympos. on Switching Theory and Logical Design. Detroit/Mich., pp. 3—9. AIEE, New York 1961.

CONSTANTINESCU, P.: [1] On Reducing the Number of Contacts by the Introduction of Bridge Circuits (in Romanian). An. Univ. C. I. Parhon Ser. Şti. Natur. Mat.-Fiz. No. 11 (1956) 45—69.

—: [2] On Reducing the Number of Contacts by the Introduction of Bridge Circuits. Conjugate Direct Conductibilities (in Romanian). Stud. Cerc. Mat. 7 (1956) 399—419.

—: [3] On the Synthesis of Multi-Terminal Networks with Relay Contacts. (1-k)-Multi-Terminals (in Russian). Bull. Math. Soc. Sci. Math. Phys. R. P. Roumaine 1 (49) (1957) 377—392.

—: [4] On the Analysis of Series-Parallel and Bridge Circuits with Rectifiers. Stud. Cerc. Mat. 9 (1958) 165—172 (in Romanian). — Rev. Math. Pures Appl. 5 (1960) 403—410 (in Russian).

—: [5] Synthesis of the Multi-Terminal Networks with Relay Contacts and Rectifiers. Bull. Math. Soc. Sci. Math. Phys. R. P. Roumaine 2 (50) (1958) 367—396.

CSIMA, J., and C. C. GOTLIEB: [1] A Computer Method for Constructing School Time Tables. Comm. ACM 7 (1964) 160—163.

DANTZIG, G. B.: [1] Discrete Variable Extremum Problems. Operations Res. 5 (1957) 266—277.

—: [2] On the Significance of Solving Linear Programming Problems with Some Integer Variables. Econometrica 28 (1960) 30—44.

—: [3] Linear Programming and Extensions. Princeton/NJ: Princeton Univ. Press 1963.

DANTZIG, G. B., D. R. FULKERSON and S. M. JOHNSON: [1] Solution of a Large-Scale Travelling-Salesman Problem. Operations Res. 2 (1954) 393—410.

—: [2] On a Linear Programming Combinatorial Approach to the Travelling-Salesman Problem. Operations Res. 7 (1959) 58—66.

DANTZIG, G. B., and A. J. HOFFMAN: [1] Dilworth's Theorem on Partially Ordered Sets. Linear Inequalities and Related Systems. Ann. Math. Studies, vol. 38. Princeton/NJ: Princeton Univ. Press 1956.

DELEANU, A., and P. L. HAMMER (IVĂNESCU:) [1] Optimal Assignement of Numbers to Vertices by Pseudo-Boolean Programming. Bull. Math. Soc. Sci. Math. R. P. Roumaine 7 (55) (1963) 149—150.

Denis-Papin, M., and Y. Malgrange: [1] Exercices de calcul booléien avec leurs solutions. Paris: Ed. Eyrolles 1966.

Desbazeille, G.: [1] Exercices et problèmes de recherche opérationnelle. Paris: Dunod 1964.

Dilworth, R. P.: [1] A Decomposition Theorem for Partially Ordered Sets. Ann. of. Math. 51 (1950) 161—166.

Dinkelbach, W.: [1] Die Maximierung eines Quotienten zweier linearer Funktionen unter linearen Nebenbedingungen. Z. Wahrscheinlichkeitstheorie und Verw. Gebiete 1 (1962) 141—145.

Drăguşin, C.: [1] Transformations for the Simplification of Multi-Poles (in Romanian). Diploma Thesis, University of Bucharest 1964.

Egerváry, J.: [1] On Combinatorial Properties of Matrices (in Hungarian). Math.-Fiz. Lapok 38 (1931) 16—28.

—: [2] Combinatorial Methods for Solving Transportation Problems. Mag. Tud. Akad. Mat. Kut. Intéz. Közleményei 4 (1959) 15—28.

Faure, R., and Y. Malgrange: [1] Une méthode booléienne pour la résolution des programmes linéaires en nombres entiers. Gestion, No. spécial, avril 1963.

—: [2] Nouvelles recherches sur la résolution des programmes linéaires en nombres entiers. Gestion, No. spécial, juin 1965, pp. 371—375.

Flament, C.: [1] L'analyse booléenne de questionnaires. Math. Sci. Hum., No. 12 (1966) 3—10.

Flood, M. M.: [1] The Travelling-Salesman Problem. Operations Res. 4 (1956) 61—75.

Ford Jr., L. R., and D. R. Fulkerson: [1] Maximal Flow through a Network. Canad. J. Math. 8 (1956) 399—404.

—: [2] A Simple Algorithm for Finding Maximal Network Flows and an Application to the Hitchcock Problem. Canad. J. Math. 9 (1957) 210—218.

—: [3] Network Flow and Systems of Representatives. Canad. J. Math. 10 (1958) 78—85.

—: [4] Flows in Networks. Princeton/NJ: Princeton Univ. Press 1962.

Fortet, R.: [1] L'algébre de Boole et ses applications en recherche opérationnelle. Cahiers Centre Etudes Recherche Opér. 1, No. 4 (1959) 5—36.

—: [2] Applications de l'algèbre de Boole en recherche opérationnelle. Rev. Française Recherche Opér. 4 (1960) 17—26.

—: [3] Résolution booléenne d'opérations arithmétiques sur les entiers non négatifs et applications aux programmes linéaires en nombres entiers. Rapport SMA. Paris, mars 1960.

Fulkerson, D. R.: [1] Note on Dilworth's Theorem for Partially Ordered Sets. Proc. Amer. Math. Soc. 7 (1956) 701—702.

—: [2] A Network Flow Feasibility Theorem and Combinatorial Applications. Canad. J. Math. 11 (1959) 440—451.

—: [3] Flows in Networks. Recent Advances in Mathematical Programming. Edited by R. L. Graves and P. Wolfe. New York: McGraw Hill 1963, pp. 319 to 332.

—: [4] Scheduling in Project Networks. Rand Corp. Mem. RM-4137-PR. June 1964.

Fulkerson, D. R., and H. J. Ryser: [1] Widths and Heights of (0, 1)-Matrices. Canad. J. Math. 13 (1961) 239—255.

Gale, D.: [1] A Theorem on Flows in Networks. Pacific J. Math. 7 (1957) 1073 to 1082.

GASPAR, T.: [1] Programming the Algorithm for the Minimization of Pseudo-Boolean Functions for a MECIPT-1 Computer (in Romanian). Stud. Cerc. Mat. 19 (1967) 1135—1148.

—: [2] An ALGOL 60 Program for the Minimization of Pseudo-Boolean Functions. Rev. Roumaine Math. Pures Appl. (in press).

GILL, A.: [1] Introduction to the Theory of Finite-State Machines. New York: McGraw Hill 1962.

GOMORY, R. E.: [1] An Algorithm for Integer Solutions to Linear Programs. Princeton IBM Math. Research Project, Techn. Rep. No. 1, 1958. Republished in Recent Advances in Mathematical Programming, edited by R. L. Graves and P. Wolfe. New York: McGraw Hill 1963.

—: [2] Essentials of an Algorithm for Integer Solutions to Linear Programs. Bull. Amer. Math. Soc. 64 (1958) 275—278.

GOTLIEB, C. C.: [1] The Construction of Class-Teacher Time-Tables. IFIP Congress (1963) 73—77.

GRENIEWSKI, H.: [1] Boole-Kolmogoroff Algebra and Input-Output Analysis. Communication at the First European Conference of TIMS and ES. Warsaw, September 1966.

HAKIMI, S. L.: [1] Optimum Distribution of Switching Centers in a Communication Network and Some Related Graph Theoretic Problems. Operations Res. 13 (1965) 462—475.

HALL, P.: On Representatives of Subsets. J. London Math. Soc. 10 (1935) 26—30.

HALL JR., M.: [1] Distinct Representatives of Subsets. Bull. Amer. Math. Soc. 54 (1948) 922—926.

HALMOS, P. R., and H. E. VAUGHAN: [1] The Marriage Problem. Amer. J. Math. 72 (1950) 214—215.

HAMMER (IVĂNESCU), P. L.: [1] Programmation polynomiale en nombres entiers. C. R. Acad. Sci. Paris 258 (1964) 424—427.

—: [2] On the Minimal Decomposition of Finite Partially Ordered Sets in Chains. Rev. Roumaine Math. Pures Appl. 9 (1964) 897—903.

—: [3] Systems of Pseudo-Boolean Equations and Inequalities. Bull. Acad. Polon. Sci. Sér. Sci. Math. Astronom. Phys. 12 (1964) 673—680.

—: [4] The Method of Successive Eliminations for Pseudo-Boolean Equations. Bull. Acad. Polon. Sci. Sér. Sci. Math. Astronom. Phys. 12 (1964) 681—683.

—: [5] Outline of the Pseudo-Boolean Method for Integer Polynomial Programming. Bull. Acad. Polon. Sci. Sér. Sci. Math. Astronom. Phys. 12 (1964) 685—686.

—: [6] Some Network-Flow Problems Solved with Pseudo-Boolean Programming. Operations Res. 13 (1965) 388—399.

—: [7] Pseudo-Boolean Programming with Special Constraints. Elektron. Inform. Kybernetik 1 (1965) 165—183.

—: [8] Pseudo-Boolean Programming and Applications. Lecture Notes in Mathematics No. 9. Berlin/Heidelberg/New York: Springer 1965.

—: [9] Dynamic Programming With Bivalent Coefficients. Proc. Conf. on Math. Methods in Economic Research, Smolenice (Czechoslovakia). Praha 1966, pp. 191—208; Mat. Vesnik (Beograd) 3 (18) (1966) 87—100.

—: [10] Optimal Plant Location — A Pseudo-Boolean Approach, Communication at the British-Israeli Conference on Operations Research, Haifa, 1967; Submitted for publication to Management Sci.

HAMMER (IVĂNESCU), P. L., and I. ROSENBERG: [1] Application of Pseudo-Boolean Programming to the Theory of Graphs. Z. Wahrscheinlichkeitstheorie und Verw. Gebiete 3 (1964) 167—176.

HAMMER (IVĂNESCU), P. L., I. ROSENBERG and S. RUDEANU: [1] On the Determination of the Minima of Pseudo-Boolean Functions (in Romanian). Stud. Cerc. Mat. 14 (1963) 359—364.

—: [2] Application of Discrete Linear Programming to the Minimization of Boolean Functions (in Russian). Rev. Math. Pures Appl. 8 (1963) 459—475.

HAMMER (IVĂNESCU), P. L., and S. RUDEANU: [1] On Solving the Transportation Problem by the Egerváry method. I. (in Romanian). Com. Acad. R. P. Romîne 11 (1961) 773—778.

—: [2] On Solving the Transportation Problem by the Egerváry Method. II. (in Romanian). Stud. Cerc. Mat. 14 (1963) 59—67.

—: [3] Minimization of Switching Circuits in Actual Operation. IFAC Symposium on Hazard and Race Phenomena in Switching Circuits, Circular Letter No. 14. Bucharest 1965; Rev. Roumaine Math. Pures Appl. 12 (1967) 407—444.

—: [4] Programmation pseudo-booléenne. I. Le cas linéaire. C. R. Acad. Sci. Paris Sér. A—B, 263 (1966), 164—167.

—: [5] Programmation pseudo-booléenne. II. Le cas non linéaire. C. R. Acad. Sci. Paris Sér. A—B, 263 (1966) 217—219.

—: [6] Pseudo-Boolean Methods for Bivalent Programming. Lecture Notes in Mathematics No. 23. Berlin/Heidelberg/New York: Springer 1966.

—: [7] A Pseudo-Boolean Approach to Matching Problems in Graphs, With Applications to Assignment and Transportation Problems. Théorie des Graphes, Journées Internationales d'Etude, Rome, juillet 1966. Paris/New York: Dunod, and Gordon & Breach 1967, pp. 161—175.

—: [8] Extensions of Pseudo-Boolean Programming. Mat. Vesnik (Beograd) 4 (19) (1967) 113—118.

—: [9] A Pseudo-Boolean Viewpoint on Systems of Representatives. Applicationes Math. (Wroclaw) (in press).

—: [10] Pseudo-Boolean Programming. Communication to the Symp. on Appl. of Math. to Economics, Bucharest 1967.

—: [11] Pseudo-Boolean Programming and an Induced Equivalence of Combinatorial Problems. Communication at the Proc. Intern. Symp. on Math. Programming, Princeton 1967 (to appear).

HARARY, F., R. Z. NORMAN and D. CARTWRIGHT: [1] Structural Models. An Introduction to the Theory of Directed Graphs. New York/London/Sydney: Wiley 1965.

HARPER, L. H.: [1] Optimal Assignments of Numbers to Vertices. J. Soc. Indust. Appl. Math. 12 (1964) 131—135.

HEALY JR., W. C.: [1] Multiple Choice Programming. Operations Res. 12 (1964) 122—138.

HOFFMAN, A. J.: [1] Some Recent Applications of the Theory of Linear Inequalities to Extremal Combinatorial Analysis. Proc. Symp. Appl. Math. 10 (1960).

HOFFMAN, A. J., and H. W. KUHN: [1] On Systems of Distinct Representatives. Linear Inequalities and Related Systems. Ann. Math. Studies, vol. 38. Princeton/NJ: Princeton Univ. Press 1956.

—: [2] Systems of Distinct Representatives and Linear Programming. Amer. Math. Monthly 63 (1956) 455—460.

IHDE, G.-B.: [1] Aussagenlogik und Boolesche Algebra in der Unternehmensforschung. Dissertation. Göttingen 1966.

INAGAKI, Y., and K. SUGINO: [1] An Application of Pseudo-Boolean Programming to the State Assignment of Sequential Circuits with High Reliability. Mem. Fac. Eng. Nagoya Univ. 17 (1965) 173—184.

IVĂNESCU, P. L. see HAMMER (IVĂNESCU), P. L.

KARP, R. M.: [1] A Note on the Application of Graph Theory to Digital Computer Programming. Information and Control 3 (1960) 179—190.

KAUFMANN, A.: [1] Méthodes et modèles de la recherche opérationnelle. Tome 2. Paris: Dunod 1964.

KAUFMANN, A., and Y. MALGRANGE: [1] Recherche des chemins et circuits hamiltoniens d'un graphe. Rev. Française Recherche Opér. 7 (1963) 61—73.

KIRCHGÄSSNER, K.: [1] Über Zuteilungsprobleme einer Universität. Referat auf der Jahrestagung der DGU. Braunschweig 1963.

KÖNIG, D.: [1] Theorie der endlichen und unendlichen Graphen. Leipzig: Akad. Verlagsges. mbH 1936. English translation: New York: Chelsea 1950.

KRÁL, J.: [1] The Formulation of the Problem of Program Segmentation in Terms of Pseudo-Boolean Programming. Sbornik Výzkumnych Prací Ústavu výpočtové techniky ČSAV-ČVUT, pp. 36M—44M Praha 1966.

KUHN, H. W.: [1] The Hungarian Method for the Assignment Problem. Naval Res. Logist. Quart. 2 (1955) 83—97.

—: [2] Variants of the Hungarian Method for Assignment Problems. Naval Res. Logist. Quart. 3 (1956) 253—258.

KÜNZI, H. P., and W. OETTLI: [1] Integer Quadratic Programming. Recent Advances in Mathematical Programming. Editors R. L. Graves and P. Wolfe; New York: McGraw-Hill 1963, 303—308.

LAND, A. H., and A. G. DOIG: [1] An Automatic Method of Solving Discrete Programming Problems. Econometrica 28 (1960) 497—520.

LINDSEY II, J. H.: [1] Assignment of Numbers to Vertices. Amer. Math. Monthly 71 (1964) 508—516.

LITTLE, J. D. C., K. G. MURTY, D. W. SWEENEY and K. CAROLINE: [1] An Algorithm for the Travelling Salesman Problem. Operations Res. 11 (1963) 972—989.

LUNC, A. G.: [1] The Application of Boolean Matrix Algebra to the Analysis and Synthesis of Relay-Contact Networks (in Russian). Dokl. Akad. Nauk. SSSR 70 (1950) 421—423.

—: [2] Algebraic Methods of Analysis and Synthesis of Relay-Contact Networks (in Russian). Izv. Akad. Nauk SSSR, Ser. Mat. 16 (1952) 405—426.

MACIŠČAK, K., and D. A. POSPELOV: [1] The Problem of the Optimal Distribution of Programs to Parallel-Working Computers (in Russian). Kievskiĭ Dom Naučno-Tehničeskoĭ Propagandy. Seminar. Teoria Avtomatov. Kiev 1964, 26—37.

MAGHOUT, K.: [1] Détermination des nombres de stabilité et du nombre chromatique d'un graphe. C. R. Acad. Sci. Paris 248 (1959) 3522—3523.

—: [2] Une méthode pour la résolution des programmes linéaires. C. R. Acad. Sci. Paris 250 (1960) 2510—2512.

—: [3] Une méthode pour la résolution des programmes linéaires. Programmes paramétriques. C. R. Acad. Sci. Paris 250 (1960) 2837—2839.

—: [4] Algorithme pour la résolution des programmes linéaires. C. R. Acad. Sci. Paris 252 (1961) 3186—3188.

—: [5] Une nouvelle méthode de résolution des programmes quadratiques. C. R. Acad. Sci. Paris 252 (1961) 3381—3383.

—: [6] Applications de l'algèbre de Boole à la théorie des graphes et aux programmes linéaires et quadratiques. Cahiers Centre Etudes Recherche Opér. 5 (1963) 21—99.

MAĬSTROVA, T. L.: [1] Linear Programming and the Problem of Minimizing Normal Forms of Boolean Functions (in Russian). Problemy Peredači Informacii No. 12 (1962) 5—15.

MALGRANGE, Y.: [1] Recherche des sous-matrices premières d'une matrice à coefficients binaires. Application à certains problèmes de graphe. Recherche des formes normales minimales d'une fonction booléienne. Deuxième Congrès de l'AFCALTI. Octobre 1961, pp. 231—242.

MANN, H. B., and H. J. RYSER: [1] Systems of Distinct Representatives. Amer. Math. Monthly 60 (1953) 397—401.

MANNE, A. S.: [1] Plant Location Under Economies-of-Scale; Decentralization and Computation. Management Sci. 11 (1964) 213—235.

MARCUS, S., and EM. VASILIU: [1] Mathématiques et phonologie. La théorie des graphes et le consonantisme de la langue roumaine. Rev. Math. Pures Appl. 5 (1960) 319—340 and 681—704.

MARTOS, B.: [1] Hyperbolic Programming (in Hungarian). Mag. Tud. Akad. Mat. Kut. Intéz. Közleményei 5 (series B) No. 4 (1960) 383—406.

MCCLUSKEY JR., E. J.: [1] Error-Correcting Codes — A Linear Programming Approach. Bell System Tech. J. 38 (1959) 1485—1512.

—: [2] Minimum-State Sequential Circuits for a Restricted Class of Incompletely Specified Flow Tables. Bell System Tech. J. 41, No. 6 (1962) 1759—1768.

MEDIAN, S.: [1] On the Minimization of Relay Switching Circuits in Actual Operation (in Romanian). Instit. Politehnic Braşov, Lucrări ştiinţifice No. 8, 1965 (in press).

MENDELSOHN, N. S., and A. L. DULMAGE: [1] Some Generalizations of the Problem of Distinct Representatives. Canad. J. Math. 10 (1958) 230—241.

MIHOC, G., and E. BALAS: The Problem of Optimal Time-Tables. Rev. Roumaine Math. Pures Appl. 10 (1965) 381—388.

MITITELU, S.: [1] On the Synthesis of Repetition-Free (p, q) Switching Multi-Terminals (in Romanian). Stud. Cerc. Mat. (in press).

MOISIL, GR. C.: [1] Axiomatization of the Theory of Simplification of Combination Automata. Proceedings of the 2nd IFAC Congress, Basle, August 1963.

—: [2] An Algebraic Theory of the Actual Operation of Relay Switching Circuits. IFAC Symposium on Hazard and Race Phenomena in Switching Circuits. Circular Letter No. 5, Bucharest 1964.

—: [3] An Extension of the Quine Simplification Theory for the Two-Terminals with Break-Before-Make Relay Contacts. IFAC Symposium on Hazard and Race Phenomena in Switching Circiuts. Circular Letter No. 10, Bucharest 1964.

—: [4] An Extension of the Quine Simplification Theory for the Two-Terminals with Make-Before-Break Relay Contacts. IFAC Symposium on Hazard and Race Phenomena in Switching Circuits. Circular Letter No. 13, Bucharest 1964.

—: [5] Théorie structurelle des automates finis. Paris: Gauthier-Villars 1967.

—: [6] On Certain Representations of the Graphs Occurring in Transportation Problems (in Romanian). Com. Acad. R. P. Romîne 10 (1960) 647—652.

NÉMETI, L.: Das Reihenfolgeproblem in der Fertigungsprogrammierung und 87—99. Linearplanung mit logischen Bedingungen. Mathematica (Cluj) 6 (29) (1964)

NÉMETI, L., and F. RADÓ: [1] Ein Wartezeitproblem in der Programmierung der Produktion. Mathematica (Cluj) 5 (28) (1963) 65—95.

ORE, O.: [1] Theory of Graphs. Amer. Math. Soc. Coll. Publ. vol. 38, Providence 1962.

POSPELOV, D. A.: [1] On a Method for Constructing Computing Systems (in Russian). Kievskiĭ Dom Naučno-Tehničeskoi Propagandy. Seminar. Teoria Avtomatov. Kiev 1964, pp. 3—24.

PYNE, I. B., and E. J. MCCLUSKEY JR.: [1] An Essay on Prime Implicant Tables. J. Soc. Indust. Appl. Math. 9 (1961) 604—631.

RADÓ, F.: [1] Linear Programming with Logical Conditions (in Romanian). Com. Acad. R. P. Romîne 13 (1963) 1039—1042.

RADÓ, F.: [2] Un algorithme pour résoudre certains problèmes de programmation mathématique. Mathematica (Cluj) 6 (29) (1964) 105—116.

ROSENBERG, I.: [1] Das Minimum der realen Funktionen auf dem Kartesischen Produkt von endlichen Mengen. Rev. Roumaine Math. Pures Appl. 10 (1965) 1377—1383.

ROSENBLATT, D.: [1] On the Graphs and Asymptotic Forms of Finite Boolean Relation Matrices and Stochastic Matrices. Naval Res. Logist. Quart. 4 (1957) 151—167.

—: [2] On Linear Models and the Graphs of Minkowski-Leontief Matrices. Econometrica 25, April 1957.

—: [3] On the Graphs of Finite Idempotent Boolean Relation Matrices. J. Research of Nat. Bureau Standards-B. Math. and Math.-Phys. 67 B (1963) 249—256.

ROY, B.: [1] Cheminement et connexité dans les graphes; application aux problème d'ordonnancement. Metra, Série spéciale No. 1, 1962.

ROY, B., PH. T. NGHIEM and P. BERTIER: [1] Programmes linéaires en nombres entiers et procédure S. E. P. Metra 14 (1965) 441—460.

ROY, B., and B. SUSSMANN: [1] Les problèmes d'ordonnancement avec contraintes disjonctives. SEMA, Rapport de Recherche No. 9 bis, octobre 1964.

RUDEANU, S.: [1] Boolean Equations and their Applications to the Study of Bridge Circuits. I. Bull. Math. Soc. Sci. Math. Phys. R. P. Roumaine 3 (51) (1959) 445—473.

—: [2] On the Definition of Boolean Algebras by Means of Binary Operations (in Russian). Rev. Math. Pures Appl. 6 (1961) 171—183.

—: [3] Boolean Equations and their Applications to the Study of Bridge Circuits. II. (in Romanian). Com. Acad. R. P. Romîne 11 (1961) 611—618.

—: [4] Boolean Functions and Sheffer Functions (in Russian). Rev. Math. Pures Appl. 6 (1961) 747—759.

—: [5] On Solving Boolean Equations by the Löwenheim Method (in Romanian). Stud. Cerc. Mat. 13 (1962) 295—308.

—: [6] Determination of the Hamiltonian Circuits of a Graph by R. Fortet's Method (in Romanian). Com. Acad. R. P. Romîne 12 (1962) 661—666.

—: [7] Remarks on Motinori Gotō's Papers on Boolean Equations. Rev. Roumaine Math. Pures Appl. 10 (1965) 311—317.

—: [8] Solutions non redondantes des équations booléennes. Bull. Math. Soc. Sci. Math. Phys. R. P. Roumaine 7 (55) (1963) 45—49.

—: [9] On Solving Boolean Equations in the Theory of Graphs. Rev. Roumaine Math. Pures Appl. 11 (1966) 653—664.

—: [10] Axiomatization of Certain Problems of Minimization. Studia Logica 20 (1967) 37—61.

—: [11] Notes sur l'existence et l'unicité des noyaux d'un graphe. II. Application des équations booléennes. Rev. Française Recherche Opér. 10 (1966) 301—310.

—: [12] Irredundant Solutions of Boolean and Pseudo-Boolean Equations. Rev. Roumaine Math. Pures Appl. 11 (1966) 183—188.

RYSER, H. J.: [1] Combinatorial Mathematics. The Carus Math. Monographs No. 14. Published by the Math. Assoc. America 1963.

SHANNON, C. E.: [1] The Zero-Error Capacity of a Noisy Channel. Composium on Information Theory. IRE-Transactions 3, 3, 1956.

SHERMANN, G. R.: [1] A Combinatorial Problem from Scheduling University Classes. J. Tennessee Acad. Sci. (1963) 116—117.

SIMONNARD, M.: [1] Programmation linéaire. Paris: Dunod 1962.

SLAGLE, J. R.: [1] An Efficient Algorithm for Finding Certain Minimum Cost Procedures for Making Binary Decisions. J. Assoc. Comput. Mach. 11 (1964) 253—264.

STAHLKNECHT, P.: [1] Zur elektronischen Berechnung von Stundenplänen. Elektron. Datenverarbeitung H. 2, 1964.

TOMESCU, I.: [1] A Method for the Determination of the Total Conductibilities of a Multi-Terminal (in Romanian). Stud. Cerc. Mat. 17 (1965) 1109—1115.

—: [2] On Certain Theorems for Simplifying Switching Circuits (in Romanian) An. Univ. Bucureşti Ser. Şti. Natur. Mat.-Fiz. 15 (1966) 155—160.

—: [3] A Method for the Determination of the Total Conductibilities of a Switching Circuit (in Russian). Rev. Roumaine Math. Pures Appl. (in press).

—: [4] Sur les méthodes matricielles dans la théorie des réseaux. C. R. Acad. Sci. Paris Sér. A 263 (1966) 826—829.

TUCKER, A. W.: On Directed Graphs and Integer Programs. IBM Math. Res. Project. Technical Rep. Princeton/NJ: Princeton Univ. Press 1960.

UDAGAWA, K., and Y. INAGAKI: [1] State Assignment Method of Reliable Sequential Circuits by Means of Dynamic Programming (in Japanese). J. Inst. Elec. Comm. Japan 47, No. 6 (1964) p. 40.

UDAGAWA, K., Y. INAGAKI and M. GOTŌ: [1] A State Assignment of Reliable Sequential Circuits Using Quadratic Programming (in Japanese). J. Inst. Elec. Comm. 47, No. 11 (1964) p. 1731.

VITAVER, L. M.: [1] Finding the Minimal Colouring of the Vertices of a Graph Using Boolean Powers of Incidence Matrices (in Russian). Dokl. Akad. Nauk SSSR 147 (1962) 758—759.

WEISSMAN, J.: [1] Boolean Algebra, Map Coloring and Interconnections. Amer. Math. Monthly 69 (1962) 608—613.

WITZGALL, C.: [1] An All-Integer Programming Algorithm with Parabolic Constraints. J. Soc. Indust. Appl. Math. 11 (1963) 855—871.

Supplementary Bibliographies

A. Boolean Equations and Generalizations

AKERS, S. B.: [A 1] On a Theory of Boolean Functions. J. Soc. Indust. Appl. Math. 7 (1959) 487—498.

ANDREOLI, G.: [A 1] Un problema di partizioni di insiemi e certi sistemi simmetrici di equazioni booleane. Giorn. Mat. Battaglini 89 (1961) 1—13.

—: [A 2] Una proprietà caratteristica per le soluzione dei sistemi di equazioni booleane e loro discussione. Ricerca (Napoli) 13, Maggio-Agosto (1962) 1—9.

ANGSTL, H.: [A 1] Über Gleichungen in der Aussagenlogik. Kontrolliertes Denken, Untersuchungen zum Logikkalkul und zur Logik der Einzelwissenschaften. Ed. Albert Menne, Alex. Wilhelmy, H. Angstl (rotaprint). München: Kommissions-Verlag Karl Alber 1951.

BALAKRAN: [A 1] The General Equation in the Algebra of Logic. J. Indian Math. Soc. 3 (1911) 213—218.

BARR, M.: [A 1] Consistency and Solvability of a System of Boolean Equations. Amer. Math. Soc. Notices 7 (1965) 1003.

BAZILEVSKIĬ, JU. JA.: [A 1] Transformation and Solution of Logical Equations (in Russian). Voprosy Teorii Mat. Mašin 2 (1962) 107—212.

BERNSTEIN, B. A.: [A 1] Note on the Condition that a Boolean Equation Have a Unique Solution. Amer. J. Math. 54 (1932) 417—418.

BIRKHOFF, G.: [A 1] Lattice Theory. Amer. Math. Soc. Coll. Publ. vol. 25, New York 1948; reprint 1961.

BEAUFAYS, O.: [A 1] Sur la résolution de l'équation booléenne à deux inconnues. Revue A, 6 (1964) 146—148.

BOOLE, G.: [A 1] The Mathematical Analysis of Logic. Cambridge 1847.

—: [A 2] An Investigation into the Laws of Thought. London 1854.

CARRUCCIO, E.: [A 1] Equazioni logiche nel calcolo delle proposizzioni. Bol. Un. Mat. Ital. 18 (3) (1963) 44—56.

CARVALLO, M.: [A 1] Détail sur la résolution des équations de Boole en vue du codage. Publ. Instit. Statist. Univ. Paris 13 (1964) 21—44.

—: [A 2] Principes et applications de l'analyse booléenne. Paris: Gauthier-Villars 1965.

ĆETKOVIĆ, S.: [A 1] Solution of an Infinite System of Set Equations (in Serbian). Bull. Soc. Math. Phys. Serbie 4 (1952) 51—59.

COUTURAT, L.: [A 1] L'algèbre de la logique. 2ème éd. Paris 1914.

ÈĬNGORIN, M. JA.: [A 1] On Systems of Equations of Algebraic Logic and the Synthesis of Discrete Control Systems with Feed-Back (in Russian). Izv. Vysš. Učebn. Zaved. Radiofizika 1 (1958) 169—184.

—: [A 2] On the Synthesis of Certain Control Systems with Feed-Back, Based on Symmetric Systems of Equations of Algebraic Logic (in Russian). Izv. Vysš. Učebn. Zaved. Radiofizika 5 (1962) 588—601.

—: [A 3] On the Synthesis of Discrete Control Systems, Based on Systems of Delay-Equations of Algebraic Logic (in Russian). Izv. Vysš. Učebn. Zaved. Radiofizika 6 (1963) 810—832.

—: [A 4] On Two Geometrical Interpretations of the Solutions of Systems of Equations of Algebraic Logic and of the Positions of their Physical Models (in Russian). Izv. Vysš. Učebn. Zaved. Radiofizika 6 (1963) 1071—1075.

GOODMAN, A. W.: [A 1] Set equations. Amer. Math. Monthly 72 (1965) 607—613.

GOODSTEIN, R. L.: [A 1] The Solution of Equations in a Lattice. Proc. Roy. Soc. Edinburgh A 67 (1966/67) 231—242.

GOTŌ, M.: [A 1] On the General Solution of a Logical Equations with Many Unknowns (in Japanese). Bull. Electr. Lab. 20 (1956) 81—87, 152.

— [A 2] Various Types of the General Solution of a Multi-Valued Logical Algebraic Equation with Many Unknowns (in Japanese). Bull. Electr. Lab. 20 (1956) 671—682, 707—709.

—: [A 3] General Solution of the Logical Algebraic Equation with Many Unknowns. XII-th Gen. Assembly URSI, Commiss. 6, Boulder Colo, No. 51 (1957).

GOTŌ, M., and Y. KOMAMIYA: [A 1] Theory of Synthesis of Sequential Networks by Means of the Solution of Logical Functional Equations and Logical Algebraic Equations. Information Processing, Paris 1959, p. 426.

GREBENŠČIKOV, V. N.: [A 1] Coalitions of Systems of Equations of the Boolean Algebra and their Solutions (in Russian). Dokl. Akad. Nauk SSSR 141 (1961) 1317—1319.

GRIGORIAN, JU. I.: [A 1] Algorithm for the Solution of Logical Equations (in Russian). Jurn. vyčislit. mat. i mat. fiziki. 2 (1962) 186—189.

HAMMER (IVĂNESCU), P. L.: [A 1] Systems of Pseudo-Boolean Equations and Inequalities. Bull. Acad. Polon. Sci. Sér. Sci. Math. Astronom. Phys. 12 (1964) 673—679.

—: [A 2] The Method of Successive Eliminations for Pseudo-Boolean Equations. Bull. Acad. Polon. Sci. Sér. Sci. Math. Astronom. Phys. 12 (1964) 681—683.

HAMMER (IVĂNESCU), P. L., and S. RUDEANU: [A 1] The Theory of Pseudo-Boolean Programming. Lecture Notes in Mathematics vol. 23. Berlin/Heidelberg/New York: Springer 1966.

ITOH, M.: [A 1] On the Lattice of n-Valued Functions (in Japanese). Techn. Reports Kyushu Univ 28 (1955) 96—101.

—: [A 2] On the General Solution of the n-Valued Function-Lattice (Logical) Equation in One Variable (in Japanese). Techn. Reports Kyushu Univ. 28 (1956) 239—243.

ITOH, M.: [A 3] On the General Solution of the n-Valued Function-Lattice (Logical) Equations in Several Variables (in Japanese). Techn. Reports Kyushu Univ. 28 (1956) 243—246.

—: [A 4] On the General Solution of the Boolean (Two-Valued Logical) Equation in Several Variables (in Japanese). Techn. Reports Kyushu Univ. 28 (1956) 246—248.

—: [A 5] On the General Solution of the Three-Valued Logical Equation (in Japanese). Techn. Reports Kyushu Univ. 28 (1956) 248—256.

—: [A 6] General Solution of the General n-Valued Logical Equation. Rev. Un. Mat. Argentina Az. fiz. argent. 18 (1958) 181.

—: [A 7] On Boolean Equations with Many Unknown Elements and Generalized Poretsky's Formula (in Japanese). Techn. Reports Kyushu Univ. 30 (1957) 211—217. English translation in Rev. Univ. Nac. Tucumán 12 (1959) 107—112.

IVĂNESCU, P. L. see HAMMER (IVĂNESCU), P. L.

JEVONS, W. S.: [A 1] Pure Logic, or the Logic of Quality apart from Quantity. Stanford, London 1864.

JOHNSON, W. E.: [A 1] Sur la théorie des équations logiques. Bibl. Congrès Intern. Philosophie 3 (1901) 185—199.

KLÍR, J.: [A 1] Solution of Systems of Boolean Equations (in Tcheque). Apl. Mat. 7 (1962) 265—272.

LALAN, V.: [A 1] Equations fonctionnelles dans un anneau booléen. C.R. Acad. Sci. Paris 230 (1950) 603—605.

LEDLEY, R. S.: [A 1] Digital Computational Methods in Symbolic Logic, with Examples in Biochemistry. Proc. Nat. Acad. Sci. USA 41 (1955) 498—511.

—: [A 2] Boolean Matrix Equations in Digital Circuit Design. IRE Trans. Electronic Computers 8 (1959) 131—139.

—: [A 3] Digital Computers and Control Engineering. New York: McGraw Hill 1960.

LÖWENHEIM, L.: [A 1] Über das Auflösungsproblem im logischen Klassenkalkul. Sitzungsber. Berl. Math. Gesellschaft 7 (1908) 89—94.

—: [A 2] Über die Auflösung von Gleichungen im logischen Gebietkalkul. Math. Ann. 68 (1910) 169—207.

—: [A 3] Über Transformationen im Gebietkalkul. Math. Ann. 73 (1913) 245 bis 272.

—: [A 4] Gebietdeterminanten. Math. Ann. 79 (1919) 223—236.

MAITRA, K. K.: [A 1] A Map Approach to the Solution of a Class of Boolean Functional Equations. Comm. Electronics No. 59 (1962) 34—36.

MARTELLOTTA, R.: [A 1] Funzioni booleane implicite. Ricerca (Napoli) 12, Settembre-Dicembre 1961, 11—25.

NADLER, M., and B. ELSPAS: [A 1] The Solution of Simultaneous Boolean Equations. IRE Trans. Communication Theory 7, No. 3 (1960).

NORDIO, S.: [A 1] Contributo allo studio dei sistemi di equazioni logiche. Tecnica italiana 27 (1962) 621—626.

PARKER, W. L., and B. A. BERNSTEIN: [A 1] On Uniquely Solvable Boolean Equations. Univ. Calif. Publ. Math. (NS) 3, No. 1 (1955) 1—30.

Poretski, P. S.: [A 1] On the Methods for Solving Logical Equalities and on the Converse Method of Mathematical Logic (in Russian). Bull. Soc. Math. Phys. Kasan 2 (1884) 161—330.

—: [A 2] Concerning the Booklet of G. Volkonski "Logical Calculus" (in Russian). Bull. Soc. Math. Phys. Kasan 7 (1889) 260—268.

—: [A 3] La loi des racines en logique. Revue Math. (Turin) 6 (1896) 5—8.

—: [A 4] Sept lois fondamentales de la théorie des égalités logiques. Bull. Soc. Math. Phys. Kasan 8 (1898) 33—103, 129—216.

—: [A 5] Exposé élémentaire de la théorie des égalités logiques à deux termes a et b. Rev. Métaphysique et de Morale (1900) 169—188.

—: [A 6] Théorie des égalités logiques à trois termes a, b et c. Bibl. Congrès Intern. Philosophie 3 (1901) 201—233.

—: [A 7] Quelques lois ultérieures de la théorie des égalités logiques. Bull. Soc. Math. Phys. Kasan 10 (1902) 50—84, 191—230.

—: [A 8] Théorie des non-égalités logiques. Bull. Soc. Math. Phys. Kasan 13 (1903) 80—119, 127—184.

—: [A 9] Appendice sur mon nouveau travail «Théorie des non-égalités logiques». Bull. Soc. Math. Phys. Kasan 14 (1905) 118—131.

—: [A 10] Théorie conjointe des égalités et non-égalités logiques. Bull. Soc. Math. Phys. Kasan 1910.

Postley, J.: [A 1] A Method for the Evaluation of a System of Boolean Algebraic Equations. Math. Tables and Other Aids to Computation 9 (1955) 5—8.

del Re, A.: [A 1] Lezioni sulla Algebra della Logica. Napoli 1907.

Rouche, N.: [A 1] Some Properties of Boolean Equations. IRE Trans. Electronic Computers 7 (1958) 291—298.

Rudeanu, S.: [A 1] Boolean Equations and their Applications to the Study of Bridge Circuits. I. Bull. Math. Soc. Sci. Math. Phys. R. P. Roumaine 3 (51) (1959) 445—473.

—: [A 2] Boolean Equations and their Applications to the Study of Bridge Circuits. II (in Romanian). Com. Acad. R. P. Romîne 11 (1961) 611—618.

—: [A 3] On the Determination of Boolean Algebras by Means of Binary Operations (in Russian). Rev. Math. Pures Appl. 6 (1961) 171—183.

—: [A 4] On the Solution of Boolean Equations by the Löwenheim Method (in Romanian). Stud. Cerc. Mat. 13 (1962) 295—308.

—: [A 5] The Determination of the Hamiltonian Circuits of a Graph by the Method of R. Fortet (in Romanian). Com. Acad. R. P. Romîne 12 (1962) 661—666.

—: [A 6] Remarks on Motinori Gotō's Papers on Boolean Equations. Rev. Roumaine Math. Pures Appl. 10 (1965) 311—317.

—: [A 7] Solutions non redondantes des équations booléennes. Bull. Math. Soc. Sci. Math. Phys. R. P. Roumaine 7 (55) (1963) 45—49.

—: [A 8] On Solving Boolean Equations in the Theory of Graphs. Rev. Roumaine Math. Pures Appl. 11 (1966) 633—664.

—: [A 9] Sur les équations booléennes de S. Ćetković. Publ. Inst. Math. Belgrade 6 (20) (1966) 95—97.

—: [A 10] On Tohma's Decompositions of Logical Functions. IEEE Trans. Electronic Computers EC-14 (1965) 929—931.

—: [A 11] Irredundant Solutions of Boolean and Pseudo-Boolean Equations. Rev. Roumaine Math. Pures Appl. 11 (1966) 183—188.

—: [A 12] On Functions and Equations in Distributive Lattices. Proc. Roy. Soc. Edinburgh (in press).

Schröder, E.: [A 1] Operationskreis des Logikkalkuls. Leipzig 1877.

—: [A 2] Vorlesungen über die Algebra der Logik. Bd. I, § 4, Leipzig 1890.

SCHUBERT, E. L.: Simultaneous Logical Equations. Comm. and Electronics No. 46 (1960) 1080—1083.

SEDMAK, V.: [A 1] Sur un système d'équations ensemblistes. Bull. Soc. Math. Phys. Serbie 7 (1955) 217—218.

SEMON, W.: [A 1] A Class of Boolean Equations. Sperry Rand Research Corp. SRRC-RR-62-17, August 1962.

SKOLEM, T.: [A 1] Über die Symmetrisch Allgemeinen Lösungen im identischen Kalkul. Skr. norske Vid.-Akad. Oslo, No. 6, 1—32. Fund. Math. 18 (1932) 61—76.

SPRINKLE, H. D.: [A 1] Some Equations over Complete Boolean Algebras. Amer. Math. Monthly 8 (1956) 607.

STAMM, E.: [A 1] Beitrag zur Algebra der Logik. Monatsh. Math. 22 (1911) 137—149.

—: [A 2] Über Relativfunktionen und Relativgleichungen. Monatsh. Math. 38 (1931) 147—166.

STANOJEVIĆ, C.: [A 1] On a System of Set-Theoretical Equations (in Serbian). Bull. Soc. Math. Phys. Serbie 4 (1952) 39—41.

SVOBODA, A.: [A 1] An Algorithm for Solving Boolean Equations. Information Processing Machines No. 9 (1963) 271—281.

SVOBODA, A., and K. ČULIK: [A 1] An Algorithm for Solving Boolean Equations (in Russian). Avtomat. i Telemeh. 25 (1964) 374—381.

TAUTS, A.: [A 1] Universal Logic (in Russian). Trudy Fiz. Astron. Tartu No. 19 (1962).

—: [A 2] Solutions of Logical Equations of Propositional Calculus (in Russian). Trudy Fiz. Astron. Tartu No. 20 (1962).

—: [A 3] Solution of Logical Equations of the First Order Predicate Calculus (in Russian). Trudy Fiz. Astron. Tartu No. 24 (1964).

TOHMA, Y.: [A 1] Decomposition of Logical Functions Using Majority Decision Elements. IEEE Trans. Electronic Computers 13 (1965) 698—705.

TOMS, R. M.: [A 1] Systems of Boolean equations. Amer. Math. Monthly 73 (1966) 29—35.

WANG, H.: [A 1] Circuit Synthesis by Solving Sequential Boolean Equations. Z. Math. Logik Grundlagen Math. 5 (1955) 291—322.

WENDELIN, H.: [A 1] Untersuchungen zur Mengenalgebra. J. Reine Angew. Math. 188 (1950) 78—99.

—: [A 2] Ein Vergleichskriterium für Ausdrücke in Booleschen Verbänden und einige Anwendungen. J. Reine Angew. Math. 188 (1950) 147—149.

WHITEHEAD, A. N.: [A 1] A Treatise on Universal Algebra, with Applications. Cambridge 1898.

—: [A 2] Memoir on the Algebra of Logic. Amer. J. Math. 23 (1901) 139—165, 297—316.

ŽELEZNIKAR, A.: [A 1] Solvability Problems of Propositional Equations. Glasnik Mat.-Fiz. Astronom. 15 (1960) 237—244.

—: [A 2] Behandlung logistischer Probleme mit Ziffernrechner. Glasnik Mat.-Fiz. Astronom. 17 (1962) 171—179.

ZEMANEK, H.: [A 1] Die Lösung von Gleichungen in der Schaltalgebra. Archiv Elektr. Übertragung 12 (1958) 35—44.

B. Boolean Algebra and Switching Theory*

ADELFIO JR., S. A., and C. F. NOLAN: [B 1] Principles and Applications of Boolean Algebra for Electronic Engineers. London: Iliffe Books 1966.

ARNOLD, B. H.: [B 1] Logic and Boolean Algebra. Englewood Cliffs/NJ: Prentice-Hall 1962.

* For a more complete bibliography see the English version of GR. C. MOISIL [B 1].

BARTEE, T. C., I. L. LEBOW and I. S. REED: [B 1] Theory and Design of Digital Machines. New York: McGraw-Hill 1962.

BEAUFAYS, O.: [B 1] Leçons d'algèbre logique. Bruxelles: Presses Universitaires de Bruxelles 1964.

BERKELEY, E. C.: [B 1] Symbolic Logic and Intelligent Machines. New York: Reinhold 1959.

CALDWELL, S. H.: [B 1] Switching Circuits and Logical Design. New York: Wiley 1958. Russian translation. Moskva: Izdatel'stvo po innostrannoi literatury 1962.

CARVALLO, M.: [B 1] Monographie des treillis et algèbre de Boole. Paris: Gauthier-Villars 1962.

—: [B 2] Principes et applications de l'analyse booléenne. Paris: Gauthier-Villars 1965.

CURTIS, H. A.: [B 1] A New Approach to the Design of Switching Circuits. Princeton/NJ: Van Nostrand 1962.

DENIS-PAPIN, M., R. FAURE and A. KAUFMANN: [B 1] Cours de calcul booléien appliqué. Paris: Albin Michel 1963.

DENIS-PAPIN, M., and Y. MALGRANGE: [B 1] Exercices de calcul booléien avec leurs solutions. Paris: Ed. Eyrolles 1966.

DUBISCH, R.: [B 1] Lattices to logic. New York/Toronto/London: Blaisdell 1964.

FLEGG, H. G.: [B 1] Boolean Algebra and its Application. London/Glasgow: Blackie 1964.

GINSBURG, S.: [B 1] An Introduction to Mathematical Machine Theory. Reading/Mass.: Addison-Wesley 1962.

GAVRILOV, M. A.: [B 1] Theory of Switching Circuits (in Russian). Moskva: Izdatel'stvo Akademii Nauk SSSR 1950. Czech translation. Praha: SNTL 1952. German translation. Berlin: VEB Verlag Technik 1953.

GLUŠKOV, V. M.: [B 1] Synthesis of Digital Automata (in Russian). Moskva: Fizmatgiz 1962.

GREA, R., and R. HIGONNET: [B 1] Etude logique des circuits électriques. Paris: Berger Levrault 1955. Revised English edition: New York: McGraw-Hill 1959.

HOERNES, G. E., and M. F. HEILWEIL: [B 1] Introduction to Boolean Algebra and Logic Design. New York/San Francisco/Toronto/London: McGraw Hill 1964. French translation. Paris: Dunod 1966.

HOHN, F. S.: [B 1] Applied Boolean Algebra. An Elementary Introduction. New York: The MacMillan Co. 1960.

HUMPHREY, W. S.: [B 1] Switching Circuits with Computer Applications. New York: McGraw-Hill 1958.

HURLEY, R. B.: [B 1] Transistor Logic Circuits. New York: Wiley 1961.

KEISTER, W., A. RITCHIE and S. H. WASHBURN: [B 1] The Design of Switching Circuits. Princeton/Toronto/New York/London: Van Nostrand 1951; further editions 1952, 1953, 1955, 1956.

KOBRINSKIĬ, N. E., and V. A. TRAHTENBROT: [B 1] Introduction to the Theory of Finite Automata (in Russian). Moskva: Fizmatgiz 1962.

KUNTZMANN, J.: [B 1] Algèbre de Boole. Paris: Dunod 1965.

LEDLEY, R. S.: [B 1] Digital Computers and Control Engineering. New York: McGraw-Hill 1960.

LIVOVSKI, L.: [B 1] Finite Automata with Hydraulic and Pneumatic Logical Elements (in Romanian). Bucureşti: Editura Academiei RPR 1963.

MALEY, G. A., and J. EARLE: [B 1] The Logic Design of Transistor Digital Computers. Englewood Cliffs/NJ: Prentice-Hall 1963.

MARCUS, M. P.: [B 1] Switching Circuits for Engineers. Englewood Cliffs/NJ: Prentice-Hall 1962.

MCCLUSKEY JR., E. J.: [B 1] Introduction to the Theory of Switching Circuits. New York: McGraw-Hill 1965.

MCCLUSKEY JR., E. J., and T. C. BARTEE: [B 1] A Survey of Switching Circuit Theory. New York: McGraw-Hill 1962.

MILLER, R. E.: [B 1] Switching Theory (2 vol.). New York/London/Sydney: Wiley 1965—1966.

MOISIL, GR. C.: [B 1] The Algebraic Theory of Switching Circuits (in Romanian). Bucureşti: Editura Tehnică 1959. Russian translation: Moskva: Izdatel'stvo po innostrannoĭ literatury 1962. Czech translation Praha: Nakladatelství Československé Akademie Věd 1965. English translation: New York: Pergamon Press 1967.

—: [B 2] Relay Switching Circuits under Direct Control (in Romanian). Bucureşti: Editura Academiei RPR 1959.

—: [B 3] Sequential Operation of Ideal Relay Switching Circuits (in Romanian). Bucureşti: Editura Academiei RPR 1960.

—: [B 4] Transistor Circuits (2 vol.) (in Romanian). Bucureşti: Editura Academie RPR 1961—1962.

—: [B 5] An Algebraic Theory of the Actual Operation of Relay Switching Circuits (in Romanian). Bucureşti: Editura Academiei RPR 1965 (see also: IFAC Symposium on Hazard and Race Phenomena in Switching Circuits Circular Letter No. 5, Bucharest 1964).

—: [B 6] The Algebraic Theory of Relay Switching Circuits (in Romanian). Bucureşti: Editura Tehnică 1965.

—: [B 7] Théorie structurelle des automates finis. Paris: Gauthier-Villars 1967.

MULLER, D. E.: [B 1] Asynchronous Switching Theory. Univ. of Michigan 1958.

NASLIN, P.: [B 1] Circuits à relais et automatismes à sèquences. Paris: Dunod 1958.

PERRIN, J.-P., M. DENOUETTE and E. DACLIN: [B 1] Méthodes modernes d'étude des systèmes logiques. Ecole Nationale Supérieure de l'Aéronautique. Centre d'Etudes et de Recherches en Automatismes 1966.

PHISTER JR., M.: [B 1] Logical Design of Digital Computers. New York: Wiley 1958.

POSPELOV, D. A.: [B 1] Logical Methods for the Analysis and Synthesis of Circuits (in Russian). Moskva/Leningrad: Izdatel'stvo Energija 1964.

RIGHI, R.: [B 1] Algebre booleane con applicazzioni alla teoria degli automatismi a contatto. Lezioni tenute all'Instituto superiore delle Poste e Telecomunicazioni, Roma 1961.

ROGINSKIĬ, V. N.: [B 1] Elements of Structural Synthesis of Relay Switching Circuits (in Russian). Moskva: Izdatel'stvo Akademii Nauk SSSR 1959.

RUDEANU, S.: [B 1] Axioms for Lattices and Boolean Algebras (in Romanian). Bucureşti: Editura Academiei RPR 1964.

RUTHERFORD, D. E.: [B 1] Introduction to Lattice Theory. New York: Hafner 1965.

SCOTT, N. R.: [B 1] Analogue and Digital Computer Technology. New York: McGraw-Hill 1960.

SOUBIES-CAMY, H.: [B 1] Les techniques binaires et le traitement de l'information. Paris: Dunod 1961.

WHITESITT, J. E.: [B 1] Boolean Algebra and its Applications. Reading/Mass.: Addison-Wesley 1961.

YOUNG, F. H.: [B 1] Digital Computers and Related Mathematics. NewYork: Ginn 1961.

Author Index

Subject Index

721/49/67 — III/18/203